THE ENLIGHTENMENT CYBORG:
A HISTORY OF COMMUNICATIONS AND CONTROL
IN THE HUMAN MACHINE, 1660–1830

ALLISON MURI

The Enlightenment Cyborg

A History of Communications and
Control in the Human Machine,
1660–1830

UNIVERSITY OF TORONTO PRESS
Toronto Buffalo London

© University of Toronto Press Incorporated 2007
Toronto Buffalo London
Printed in Canada

ISBN 10: 0-8020-8850-3
ISBN 13: 978-0-8020-8850-5

∞

Printed on acid-free paper

Library and Archives Canada Cataloguing in Publication

Muri, Allison, 1965–
    The Enlightenment cyborg : a history of communications and control
    in the human machine, 1660–1830 / Allison Muri.

    Includes bibliographical references and index.
    ISBN 13: 978-0-8020-8850-5 (bound)
    ISBN 10: 0-8020-8850-3 (bound)

    1. Cyborgs – History – Textbooks.   2. Human-machine systems –
    History – Textbooks.   I. Title.

    TA167.M86 2006   303.48'3   C2006-906162-9

University of Toronto Press acknowledges the financial assistance
to its publishing program of the Canada Council for the Arts and the
Ontario Arts Council.

This book has been published with the help of a grant from the Canadian
Federation for the Humanities and Social Sciences, through the Aid to
Scholarly Publications Programme, using funds provided by the Social
Sciences and Humanities Research Council of Canada.

University of Toronto Press acknowledges the financial support for
its publishing activities of the Government of Canada through the
Book Publishing Industry Development Program (BPIDP).

for Patricia and Philip,
Neaera and Marya,
for Ray
and for Club La Mettrie

Imagination is the soul, since it plays all the roles of the soul ... It is imagination ... which adds the piquant charm of voluptuousness to the tenderness of an amorous heart; which makes tenderness bud in the study of the philosopher and of the dusty pedant, which, in a word, creates scholars as well as orators and poets. Foolishly decried by some, vainly praised by others, and misunderstood by all; it follows not only in the train of the graces and of the fine arts, it not only describes but can also measure nature.

L'homme machine, 1747

Contents

Illustrations follow pages 150 and 182

Acknowledgments

First, thanks to Raymond Stephanson and Larry Stewart for that remarkable graduate seminar so many years ago that set this course of research in motion. Many thanks to Robert Markley, Peter Stoicheff, Andrew Taylor and to Peter Walmsley for their encouraging and perceptive comments on various chapters. Thanks, too, to the two anonymous readers at University of Toronto Press whose constructive remarks resulted in a much improved manuscript. Thank you especially to Ray, who read these chapters many times over, and whose encouragement and support were unfailing.

Thanks also to Richard A. Banvard, Paul Bidwell, Hilary Clark, Paul Coddington, Ron Cooley, Luba Frastacky and Anne Dondertman, Len Findlay, Gordon Fulton, Danielle Garnier, Stephen Greenberg, Peter J. Knapp, Terry Matheson, Siobhan McMenemy, Miles Muri, Taran Schindler, and Susan Walker.

I am grateful to the Social Sciences and Humanities Research Council of Canada and the Arts Research Board of McMaster University for financial assistance. The English Department at McMaster University generously provided resources for the postdoctoral research that led to the completion of this manuscript.

THE ENLIGHTENMENT CYBORG

1 Introduction

This book is best begun with a caveat: there is no such thing as the Enlightenment cyborg.

To imagine the twentieth-century cybernetic organism in terms of a period encompassing the years 1660 to 1830 could seem anachronistic to say the least. From any perspective the man-machine of the Enlightenment, a relatively short-lived figure in early modern philosophy and medicine, bears no resemblance to the physical reconstruction of the human form possible today, or to the imagined monsters of flesh, metal, and electronic circuitry featured in science fiction and film. The cyborg as a living organism modified by technology coincides with the emergence of mechanized war, mass production, electronic communications, and reproductive technologies in the past century. Its moment of origin could be situated in 1948 with the publication of Norbert Wiener's *Cybernetics*, a schema for a new science of cybernetics, or perhaps in 1960 when Manfred Clynes and Nathan Kline coined the term *cyborg* to describe a self-regulating human-machine system in outer space. Alternatively, the cyborg as icon of techno-humanity such as Darth Vader in the *Star Wars* series, the Borg and Geordi La Forge in *Star Trek: The Next Generation*, the entire human populace in *The Matrix*, or the Terminator in the films by that name could be said to have its beginnings when the human-machine construct was popularized in such works as Anne McCaffrey's *The Ship Who Sang* (1969), the film *Cyborg 2087* (1966), or Martin Caidin's novel *Cyborg* (1972), which inspired the television series *The Six Million Dollar Man* (1974–8). Japanese film and fiction also featured cyborgs in the 1960s: *Kaitei daisenso* (1966), also known as *Water Cyborgs*, was a film based on the book by Masami Fukushima, and *Saibogu 009* (1966), in English *Cyborg 009*, featured early Japanese anime. What

do these futuristic texts have to do with the history of the Enlightenment? On the surface, cyborgs are creatures born of the unique meld of
technology and organism in the twentieth century. Cyborgs look towards
what is yet to come.

D.S. Halacy's *Cyborg: Evolution of the Superman* was published in 1965.
Manfred Clynes wrote in the Introduction: 'A new frontier is opening
which allows us renewed hope.' He suggested that it was 'not merely
space, but more profoundly the relationship between "inner space" and
"outer space" – a bridge ... between mind and matter, beginning in our
time and extending into the future.'[1] In his book *As Man Becomes Machine* (1970), also aimed at a popular audience, David Rorvik celebrated
the near future when the brain would be implanted 'in a mechanical
"body" ... or "drained" into a computer and then discarded ... the
ultimate melding of man and machine, constituting a victory for both
entities: man-mechanized, machine-humanized.' Together, Rorvik wrote,
man and machine could become 'something more than they were as
single organizations of matter, peacefully coexisting in an evolutionary
symbiosis.'[2] In the 1980s a darker vision of the imminent and radical
transformation of human spirit, mind, or consciousness in relationship
to matter, embodiment, and technology emerged. The human-machine
hybrid as an icon of a troubled, dark, corrupt, or postapocalyptic future
made its appearance in the cyberpunk fiction made famous by William
Gibson's *Neuromancer* (1984) and in films such as *Terminator* (1984),
RoboCop (1987) (see figure 1), and *Cyborg* (1989), as well as *Tetsuo* (1988)
and others in Japan. The television series *Star Trek: The Next Generation*
(1987–94) offered both terrifying and promising versions of humans
enmeshed with computerized machines in the twenty-fourth century.
From its inception, the cyborg's significance and meaning had little
to do with history: for scientists and creative artists alike the cyborg
heralded the future.

The cyborg became a frequently cited symbol of radical change in
academic discourse, especially following the publication of Donna
Haraway's 'Manifesto for Cyborgs' in 1985. She postulated the potential
'techno-digestion' of the 'organic, hierarchical dualisms ordering discourse in "the West" since Aristotle still ruled,' arguing that the 'dichoto-
mies between mind and body, animal and human, organism and machine,
public and private, nature and culture, men and women, primitive and
civilized are all in question ideologically.'[3] As David Tomas has suggested, the cyborg 'represents a radical vision of what it means to be
human in the western world in the late 20th century.'[4] Theorists, both

radical and conservative, and representing a wide spectrum of political affiliations, have agreed that we are about to be, or have already been, utterly and irrevocably changed by the technologies merging humans and machines. 'We cannot go back ideologically or materially,' Haraway wrote in her influential manifesto.[5] A decade later Sherry Turkle suggested, 'as we stand on the boundary between the real and the virtual, our experience recalls what the anthropologist Victor Turner termed a liminal moment, a moment of passage when new cultural symbols and meanings can emerge,' such as the 'multiple viewpoints [that] call forth a new moral discourse.'[6] Sven Birkerts lamented rather more apprehensively that electronic technology would result in physical metamorphosis from what it means to be human: 'Could it be that *we* are changing, evolving, and beckoning that future toward us? ... maybe the meaning and purpose of being human is itself undergoing metamorphosis ... But evolution is evolution, and no amount of nostalgia can temper its inexorability.'[7] Bruce Mazlish similarly (although optimistically) wrote, 'Humans are on the threshold of decisively breaking past the discontinuity between themselves and machines,' a process that is related to 'evolving human nature ... an evolving identity, secured in the process of adaptation to "nature."'[8] In the arena of academic theory, too, the cyborg has predicted our supposedly post-human future.

Even if one were to reflect on the cyborg's position in the context of larger historical movements, the logical conclusion might be that the nineteenth century – age of modernity, of manufacture and medicine, of the motor and the engine, of electric machines and electric communications networks – has a more obvious relationship to the postmodern cyborg figure than does the pre-industrial Enlightenment. So why write of a non-existent Enlightenment cyborg? The term itself suggests a historical continuity falsely devoid of ruptures, transformation, and change through the long period separating the mechanistic philosophy espoused during the seventeenth and eighteenth centuries and the biotechnological viewpoint that dominates various discourses in industrialized countries today. Of course, no unbroken lineage exists to be traced from an ancestor man-machine to its offspring cyborgs; what demands our attention, however, are the shared assumptions concerning the perceived relationships of human to mechanism, material embodiment to human spirit, and mind to matter in both the early modern and postmodern conceptions of the human machine. These shared assumptions form the primary basis of this cyborg history. Cartesian subjectivity and the Enlightenment project have been almost obligatory in defining the cyborg.

As Claudia Springer has argued, 'the Cartesian mind/body duality is ultimately eclipsed by the concept of the cyborg. Rather than accomplish an ideal Enlightenment universe where human reason takes center stage, the cyborg undermines the very concept of "human."'[9] But Descartes' physiological explanations for mind-body mechanisms were argued against, modified, and ultimately eclipsed quite early in the Enlightenment: indeed, many English physiologists forthrightly rejected Cartesian philosophy. And to suggest that human reason – still valued, revered, and celebrated now in all its limitations as it was then – took centre stage in all Enlightenment versions of an ideal universe is to disregard much of the rich and varied literature of that period.

If the complexity of early modern thought is reduced to Cartesian dualism, or a blind agenda of 'progress,' or binary dualistic hierarchies, the metaphor of the cyborg as postmodern harbinger of impending change can imply a future coloured by any given pair of political glasses. A feminist writer critiquing patriarchal domination proclaims the cyborg as a welcome sign for future emancipation. A cultural critic attempting to demonstrate the immorality of science and technology claims that computer networks have stolen our consciousness or that the human has been rendered obsolete. Another claims the cyborg as a sign of our commercialism, materialism, and technological hubris, while another argues that the redesigned bodies of cyborgs will have no place at Judgment Day when the dead are to be resurrected. Yet another predicts a transcendent immortality when we upload our minds to computers. The surprising blindness to the complexity of the cyborg's history has been a product of politicized visions of radical change, and of the always marketable rhetoric of impending apocalypse or utopia. My proposal is that to understand how radical the cyborg figure is – or if it is, indeed, a radically postmodern reconfiguration of human identity – we should begin to examine what the man-machine meant to early modern conceptions of human identity.

A cyborg history situated in the Enlightenment is important if only because the 'mechanization of the world picture,' to use E.J. Dijksterhuis's term,[10] led to medical and biological interpretations of the human as machine. It is also important, however, because the cyborg in a curiously conventional formula has represented for so many writers a postmodern, 'post-Enlightenment,' or post-Cartesian figure. This study seeks to address two problematic areas in postmodern theory and cyberstudies of the past decades: that is, the surprising absence of an adequate history of the cyborg figure in the figure of the human machine of the seventeenth

and eighteenth centuries and the, perhaps less surprising, misappropria-
tion of the Enlightenment in postmodernist readings of the cyborg.

The Problem of 'Modernity' and Moralizing in
Postmodern Cyborg Discourse

After the Second World War, and particularly in such works as Theodor
Adorno's and Max Horkheimer's *Dialectic of Enlightenment*, both the
Enlightenment and the omnipresent technology that was an eventual
product of a long history of scientific and technological discovery came
to represent totalitarianism, control, and domination in both cultural
studies and literary theory. More recently Michel Foucault's influential
works have interpreted early modernity as an imposition by practitioners
of science and medicine of their professional authority over popular
knowledge. Jean Baudrillard has argued in *Simulations* and in *Simulacra
and Simulation* that, as a result of mass media and other forms of mass
cultural production, reality has been superseded by simulacra, thus
shifting human experience from reality to a more sinister 'hyperreality.'
Such readings have influenced circles of theory to the point where the
terms *Enlightenment* and *reason*, and sometimes *technology* can be used as
tokens signifying white imperialist, hegemonic, or patriarchal tyranny
and subjugation. For many writers the cyborg has accordingly represented either an overcoming or a rearticulation of the arrogances or ills
of Western imperialism and 'progress' inscribed throughout the early
modern period. Since its inception the cyborg has been a sign of evolution beyond the perceived boundaries and limitations not only of being
human but also of being modern. 'There is a startling temporal and
geographical correlation,' according to the editors of *The Cyborg Handbook*, 'between cyborgism and postmodernism.'[11] Indeed, the appearance of the cyborg in contemporary culture and literature has been
described overwhelmingly as a 'post'-(insert critical noun here) phenomenon, and as a politicized insignia of an often ill-defined postmodernism. The cyborg consciousness has been defined not so much by
evidence as by the necessity of a given theoretical stance with which a
complicated and nuanced history prior to the twentieth century is in-
compatible.

I am proposing that such a history is necessary to combat the oversimplified and frequently incendiary rhetorics of utopia and despair that
have tended to characterize theories of human identity in a technologized
environment. The changes that have occurred as a result of our material

technology are evident in the evolving conformations of our machines and our landscapes; they are also obvious in the means by which we move from place to place, organize our communities, modify and augment our bodies, and communicate with one another. But have the 'extensions' of man indeed induced a new consciousness that was for the last decades of the twentieth century increasingly labelled not only 'postmodern' but also 'post-human'? What does the human-machine relationship mean in terms of the expression of our humanity? If we have evolved or are evolving a new postmodern and post-human consciousness, our cyborg literature should reflect not only new material technologies but also newly wrought themes specific to post-human nature. The literature of cyborgs, however, reflects not so much a new ontology as a new material technology propping the latest politicized iterations of familiar and still compelling questions about human nature. When writers question what it means to be human by defining perceived differences between 'modern' and 'postmodern,' they forget perhaps that during Jonathan Swift's day the same question was defined by perceived differences between 'ancient' and 'modern.' Moreover, many of the questions then being asked of the scientific and medical categorizations of the human body-mind as machine are similar to those in our own day: Who is to be the authority on knowledge? What will happen to morality and decency when publishing and communications are not policed or vetted by accepted moral authority? In the material terms of morality, if our bodies are material and mechanical – action and reaction – then how can we inscribe systems of morality? Does God have a place in a mechanical universe? And perhaps foremost, is our 'spirit' solely the product of the machinery of our bodies? If cyborg consciousness is uniquely post-human, it should also be characterized by acknowledging these continuities with past expressions of our quest to understand our communications, our bodies, and our humanity.

The human machine has been an enduring figure in our philosophical, medical, and literary imaginations for centuries. The history of the man-machine is well known: why, then, the apparent lack of historical consciousness in our postmodern theory? The answer lies, regrettably, in our all-too-human responses to the lure of the 'new.' I say regrettably because the presentist bias of such scholarship has been prone to mistaking predictions of the future for original contributions to academic discourse. I say regrettably also because the obsession to be at the vanguard of intellectual pursuit all too often has resulted in examining only the here and now, without investigating how the metaphors and allegories of the past define our own imaginings of present-day realities. That there has been a far more sophisticated and diverse conversation about the human-machine since Descartes' speculations than generally has been acknowledged would complicate the claims that have been made for the postmodern condition. To some it will appear that I am oversimplifying current trends in criticism, but it would not, perhaps, be an exaggeration to say that the cyborg represents a postmodern shift from so-called modern ideals because it presents an opportunity to imagine a new and better world in the twenty-first century where boundaries of race, gender, or geography become permeable, or because it presents an opportunity to predict a troubled and fragmented future where stability, spirituality, and identity are threatened. In either case the cyborg may be used to critique a social order created by modern agendas of progress.

The cyborg's identification with fragmentation of identity and loss of subjectivity or with postmodern identity frequently follows Fredric Jameson's portrayals of the postmodern condition as characterized by superficiality, 'the transformation of reality into images,' schizophrenia, and fragmentation.[12] Cyborg theory as a category of the postmodern has also been influenced by Baudrillard's ominous projection of contemporary society – distinct from the 'counterfeit' period from the Renaissance to the Industrial Revolution and the 'production' era of industrialism – as a 'hallucinatory' world of simulacra destroying 'every ideal distinction between true and false, good and evil' – an era in which capital 'was the first to feed ... on the destruction of every referential, of every human goal, which shattered every ideal distinction between true and false, good and evil.'[13] Presented as a sign of secular culture governed by technological and corporate structures, the postmodern cyborg is also mapped as the pitting of 'scientific' against what we might call humanist values, as in Jean-François Lyotard's designation of *modern* as 'any science that legitimates itself with reference to a metadiscourse ... making an explicit appeal to some grand narrative' and *postmodernism* as 'incredulity toward metanarratives.'[14] A form of this version of the postmodern is evident in Haraway's 'theory in a postmodernist ... mode,' which deploys the cyborg as a means of detachment from and challenge to historical dualistic 'traditions of "Western" science and politics – the tradition of racist, male-dominant capitalism.'[15] The politicized and multiple meanings of postmodernism have been transferred to the cyborg as emblem of a writer's political stance.

Anne Balsamo, writing in 1988, called the cyborg '*the* postmodern icon.'[16] Claudia Springer suggested that 'transgressed boundaries, in fact, define the cyborg, making it the consummate postmodern concept.'[17] She also claimed that 'at the same time that the cyborg represents the triumph of the intellect, it also signifies obsolescence for human beings and the dawn of a posthuman, post-Enlightenment age.'[18] Sherry Turkle in a similar vein suggested that computer-mediated environments, characterized by terms such as '"decentered," "fluid," "non-linear," and "opaque,"' directly contrast with 'modernism, the classical world-view that has dominated Western thinking since the Enlightenment ... characterized by such terms as "linear," "logical," "hierarchical," and by having "depths" that can be plumbed and understood.'[19] Joseba Gabilondo, in *The Cyborg Handbook*, however, wrote that 'the cyborg is not the general, postmodern form of subjectivity created by multinational capitalism but rather the hegemonic subject position that its ideology privileges. The old subject position of "Man," although displaced in its hegemony by the cyborg, still operates in a very productive way. When Foucault proclaimed the death of "Man" in 1966, he did not realize that capitalism does not get rid of its old technologies and apparatuses; instead it exports them to the Third World.'[20] Chela Sandoval reasoned that 'theorists of globalization engage with the introduction of an oppositional "cyborg" politics as if these politics have emerged with the advent of electronic technology alone, and not as a requirement of consciousness in opposition developed under previous forms of domination.'[21] In contrast, Robbie Davis-Floyd and Joseph Dumit described their 1998 book *Cyborg Babies* as 'postmodern anthropology.'[22] For Bruce Grenville, however, the cyborg is 'an uncanny image that reflects our shared fascination and dread of the machine and its presence in modern culture.'[23] In *Cyborg Citizen*, a text proposing a new system of government for humans as cyborgs, Chris Hables Gray wrote about the crisis of 'our specific postmodern condition' in contrast to the period of modernity: 'We do not live in the seemingly stable modern world our grandparents did ... At the root of all this change is that great creation of the modern era: *technoscience* ... The main argument for the validity of postmodernism is modernism. The label 'modernism' is quite established, generally applying to grand narratives that are either irrational (racism, nationalism, high art) or hyperrational (technoscientific progress).'[24] The label 'modernism,' however, cannot constitute an irrefutable argument for the validity of the stereotype. Meanwhile, Steve Mann – author of *Cyborg: Digital Destiny and Human Possibility in the Age of the Wearable Computer: The Fascinating and True Story of an Inventor, a Visionary and the World's First*

Cyborg (2001) – declared in 2003 that the age following the destruction of the World Trade Center was 'Post-Cyborg.' According to Mann, 'the cyborg-age of yesterday is connected with ideas of post-modernism, deconstructionism, and posthumanism (itself, somewhat related to the ideas of cyborgism). But these ideas, along with culture jamming, as well as my own sur/sousveillance situationist street theatre, have become ineffective in the contemporary age of Terror.'[25] Is the cyborg modern or postmodern, or have we already evolved, in what Mann and his colleagues have archly called the post-postcorporeal era, from '"pomo" (cyborgism) to "popo" (postcyborgism)'?[26] Does the cyborg reflect and reinscribe earlier identities and assumptions, or does it reform and revolutionize human identity? The fact is, no one knows for sure, notwithstanding the many pronouncements that have been made upon the future cyborg identity over the past few decades. Nevertheless, as these examples have shown, the cyborg as delegate of postmodernity has been invoked in all manner of attack against a mythologized modernity.

In this sense, one 'origin' of the cyborg can be traced to its birth as a politicized emblem in critical theory. Two significant voices of theory from the mid-1980s demarcate the polar extremes of cyborg studies as postmodern morality theory. Haraway is probably best known for her proposal that the cyborg is a theoretical model for empowerment of the postmodern subject, representing the breakdown of binaries such as mind/body, spirit/matter, male/female, nature/culture, or self/other supposedly inscribed throughout the history of 'the West's escalating dominations of abstract individuation.'[27] Baudrillard, at approximately the same time, was focusing on technology as a sign of humanity's descent into social chaos and perversion. In his essay, 'The Ecstasy of Communication,' Baudrillard claimed that the supposed redundancy of the body due to technologies of communication had resulted in a loss of 'real' connections of community. What was once 'a public scene or true public space,' he wrote, is now 'gigantic spaces of circulation, ventilation and ephemeral connections.'[28] Baudrillard's play on the word *obscene* as the movement away from the scene of the three-dimensional material body and landscape to a hyperreal simulation relied upon the denotation of obscene as indecent and morally offensive. Further, while claiming that 'this is not necessarily a negative value judgment,' he emphasized the disappearance of passion in the obscene world of information – of 'hazard,' 'chance,' and 'vertigo' as opposed to the 'passion,' 'investment,' 'desire,' and 'expression' of a previous era.

Following Haraway's initiative, cyberfeminists such as Springer described the 'thrill of escape from the confines of the body,' where the cyborg transforms 'the self into something entirely new.'[29] Sadie Plant argued that multidimensional cyberspace with all its multiplicities could be an egalitarian alternative to the unitary outlook of Western man, presenting new opportunities for communication and exchange.[30] Allison Fraiberg speculated that 'the cyborg notion of transgressed boundaries and leaky distinctions finds its immunological referent in the discourses of AIDS' and suggested that 'a large scale recognition of this resituated interconnectedness, and the subsequent resurfacing of the body – of some – might begin to shift ... sites of authority [thus] ending the scientific community's holding of people for ransom.'[31] Dale Spender suggested that electronic media could break down traditional divisions: 'We can see how national boundaries have been undermined, how even subject divisions are being dislodged. It could be that rigid divisions of gender are also being weakened by this tendency to get rid of the old certainties.'[32]

In the more pessimistic vein of Baudrillard, Michael Heim has argued that 'the darker side [of computers] hides a sinister melding of human and machine. The cyborg, or cybernetic organism, implies that the conscious mind steers ... our organic life. Organic life energy ceases to initiate our mental gestures. Can we ever be fully present when we live through a surrogate body standing in for us? The stand-in self lacks the vulnerability and fragility of our primary identity ... The more we mistake the cyberbodies for ourselves, the more the machine twists ourselves into the prostheses we are wearing.' In cyberspace, claimed Heim, human ethics 'languish,' and cyborg existence 'may amplify an amoral indifference to human relationships.'[33]

Sven Birkerts described human identity in a digital culture as 'fractured and fragmented, dissolved and distracted' since the human-machine interface has disturbingly obliterated the supposedly stable identity or 'coherent inwardness' associated with the 'once-grand growth' of humanism.[34] Arthur Kroker and Michael Weinstein more hysterically contended that the 'hardwiring' of our bodies with the machine signals a cultural disintegration of apocalyptic proportions: 'If we cannot escape the hard-wiring of (our) bodies into wireless culture, then how can we inscribe primary ethical concerns onto the will to virtuality? How can we turn the virtual horizon in the direction of substantive human values: aesthetic creativity, social solidarity, democratic discourse, and economic justice?'[35] Vivian Sobchack also at one time suggested an extreme version of the moral hazards of 'merging' with the machine: 'The two-dimensional, binary superficiality of electronic space at once distorts and liberates the activity of consciousness from the gravitational pull and orientation of its hitherto embodied and grounded existence ... It denies or prosthetically transforms the spectator's physical body so that subjectivity and affect free-float or free-fall or free-flow across a horizontal/vertical grid ... Devaluing the physically lived body and the concrete materiality of the world, electronic presence suggests that we are all in imminent danger of becoming merely ghosts in the machine.'[36]

In other words, for the conservative cybertheorist, if the human mind or spirit were to merge with the machine, our souls would separate from the cathedral of humanistic flesh – lost to technology rather than given over to human morality or ethics. Our very humanity is under threat of dissolution. For the more hopeful cyberfeminist or technophile, merging with the machine has evoked the possibility of freedom from our very faulty humanity and the limitations of our bodies. And for a small few the cyborg has represented more of the same old hegemony of the Western world. In almost any case, however, there has been a notable lack of historical context for the discourse surrounding this subject.

**The Problem of Descartes, Dualism, and 'Enlightenment':
Subjectivities in Cyborg Discourse**

Older human-machines can help us grasp the subjectivities of new ones, but they require closer attention than the cursory invocation of representative images that they have so far received. The problematic dualistic Cartesian subjectivity that stands for modern or Enlightenment identity is itself a literary construction that varies in meaning depending upon the critical stance for which the cyborg acts as an emblem. René Descartes' distinction between mind and body – the unitary subject separate from the material world of three-dimensional bodies – informs the literature of the cyborg, even though Descartes problematized his own metaphysics by proposing that the soul is united to all parts of the body in his *Traité des passions* (1650), and even though subsequent medical discoveries linking the material processes of both body and brain to the mind (or soul, as Descartes called it), have been far more influential in the medical-technical understanding and reformatting of organic bodies than Descartes' mind-body dualism ever was. Indeed, Thomas Willis's physiological theories about the nervous system and comparative anatomy dominated medical thought for nearly a century, while Descartes' philosophy – influential as it was – came to be recognized as more theoretical than anatomical and was dismissed fairly early by many English natural philosophers. Does the significance of Descartes in cyborg theory have to do simply with the fact that his philosophical texts are studied in the humanities far more often than are the medical texts of his contemporaries and successors? That is, are notions of cyborg identity primarily constructed through textual tradition and therefore subject to the whims of theoretical vogue rather than the relationships of actual machines and real bodies?

Indeed, the Cartesian implications for the cyborg identity have no stable meaning, and their symbolic significance fluctuates depending upon whether one uses the cyborg image to symbolize loss of community and coherent identity, radical change, or egalitarianism and individual empowerment. For Jay David Bolter, for example, the nexus of human mind and electronic space indicated radical change: 'A philosophy of mind for the coming age of writing will have to recognize the mind as a network of signs and will see that network spreading out beyond the individual mind ... Such a philosophy may be nothing less than the end of the ego, the end of the Cartesian self as the defining quality of humanity.'[37] Ten years later, however, Bolter and Richard Grusin concluded that 'technologies such as virtual reality do not simply repeat but in fact remediate the Cartesian self.'[38] David Brande also argued that the cyborg is a 'trope for the dissolution of the Cartesian subject.' Further, 'the denaturing of the Cartesian subject and the articulation of cyborg or fragmented consciousness ought to be seen as an effect of changing modes and relations of production and of changes in the division of labor.'[39]

For William Macauley and Angel Gordo-López, however, the cyborg reaffirmed the Cartesian subject as a step towards the 'next evolutionary stage' where 'the apparent abandonment of the body is akin to the technological refurnishing of the cave.'[40] Quoting Heim's conjecture that 'at the computer interface, the spirit migrates from the body to a world of total representation ... without a grounding in bodily experience. You can lose your humanity at the throw of the dice,'[41] Catherine Waldby mistakenly argued that 'the impulse towards the supernatural which haunts the cybernatural is clearly locatable within the same Cartesian dualism which informs medicine's ideas of life and death.'[42] Springer, conversely, suggested that 'the idea of the cyborg is simultaneously a culmination of Descartes's separation of reason from emotion and a supersession of that opposition. At the same time that the cyborg represents the triumph of the intellect, it also signifies obsolescence for human beings ... the cyborg appears to rest on a dichotomy between mind and body, but it actually supersedes the dichotomy and makes it anachronistic in a new vision of fusion and symbiosis with electronic technology.'[43]

The 'Cartesian subject' has been used to uphold various political claims; however, Cartesian dualism has had little if any influence on the medical understanding of physiology and technologies that have created living cyborgs. Cartesian rational philosophy was popular in late-seventeenth- and early-eighteenth-century Europe, but empirical studies in natural philosophy through the eighteenth century disproved Descartes' hydraulic system of animal spirits in the hollow nerves. Isaac Newton, in turn, provided an important insight into nervous function, which was superseded by more accurate knowledge of the mechanisms of nervous and muscular physiology. Regrettably, much of postmodern cyborg theory seems to require a caricature of early modernity where Enlightenment reason or Enlightenment dualisms or the Cartesian self can act as some kind of cultural-theory shorthand for a past identity which we have transcended, or by which we are haunted and imperilled. In some cases, Enlightenment, and a Cartesian sovereign subjectivity are used as if they were equivalent and interchangeable values.

Consider the following comment by David Cox, which is by no means exceptional in analyses of cyborg consciousness: 'The Enlightenment and its giddy claims to the sole "take" on the human condition are reinforced with every computer generated urban planning layout, every blockbuster movie – particularly those with elaborate computer graphics and most other representations which seek to privilege the individual as a sovereign, isolated subject.'[44] Such reductionist gestures are possible primarily because of an entrenched tradition in contemporary theory that assumes an established history of cyberculture as outcome of a poorly defined Enlightenment/Cartesian subjectivity. David Brande, for example, has conflated Cartesian with Enlightenment to describe the new cyborg subjectivity: 'The denaturing processes of science and technology, as well as those of recent theory and cultural postmodernism, vacate the space carved out by and for Cartesian and Enlightenment versions of subjectivity, while they create multiple spaces for different forms of being – or rather, becoming – in the world, the (necessarily) variable forms of posthumanity, including the cyborg.'[45]

Another example is Sandy Stone's claim that virtual systems had created citizens with a textually mediated physicality in 'a discursive space that has quasi-Cartesian concomitants.' Certainly, we have a social system that relies heavily upon digitized documentation, but why raise the spectre of Descartes when writing about the collection of data used to express specific personal information in the twentieth century? (It is unclear whether this reference is to spatial coordinates – the Cartesian plane – or to Descartes' dualistic definitions of body and spirit.) Stone also argued that the visual comprehension of how things work and the rational comprehension of their function receded as the working structures of machines disappeared under 'smooth and shiny' decorative surfaces from the 1930s on, resulting in 'a phantasmatic interiority in tension with the perspicuous surface' of machines. At this point, Stone concluded, 'we are already far from the Enlightenment ontologization of the relationship between knowledge and perception.'[46] Once again, it is unclear how the Enlightenment functions in this argument.

Of the many and diverse theories of knowledge and perception espoused during the period, one of the most influential was that of John Locke, in *An Essay Concerning Human Understanding*. Locke, however, said that in addition to the sensation of external objects providing the mind with ideas of sensible qualities, a second source of ideas contributes to human knowledge: the mind's own operations provide ideas that cannot be derived from objects outside of the body (i.e., perception, thinking, doubting, believing, reasoning, knowing, willing). If we cannot see the workings of a machine, but can imagine them, just how far are we from an Enlightenment 'ontologization'? Moreover, what is the relationship of the 'chthonic interior space' of twentieth-century toasters and vacuum cleaners to the equally mysterious interior spaces of ancient or early modern machines that also were built with great care to conceal their inner workings? Why refer to the Enlightenment at all? Too much in the invocation of Descartes and the Enlightenment is assumed as universally understood, and too much left unexplained. Besides being highly speculative, such theories have tended to hypostatize the 'cyber' as the demarcation from or the outcome of a presumably universal modernity: 'It is no accident that the modern has become postmodern as human changes to cyborg,' reads the introduction to *The Cyborg Handbook*.[47] Sadie Plant characterized cyborg identity as a necessary breaking away from modernity, remarking rather cryptically that 'the cyborg has no history, but that of the human is rewritten as its past.' Plant contended – quoting Foucault – that modernity is marked by 'an explosion of numerous and diverse techniques for achieving the subjugation of bodies and the control of populations, marking the beginning of an era of "bio-power"' and suggested instead a process of 'forgetting the past, which is also the abandonment of truth and the dismemberment of authority.'[48] These meagre abridgments of modern history do little to illuminate the cyborg construct, and much to obfuscate its past.

My object is not to denounce these arguments, which have made valuable contributions to examinations of contemporary society; but I do want to question the impoverished notions of a rich early history that have been propagated through presentist analyses of cyberculture and cyborgs. Statements like these inspired my initial research into the 'Enlightenment cyborg,' not because they were inherently wrong but because such claims were historically inadequate, they were inevitably assumed to constitute some irrefutable and accepted proof of past identities, and they were for the most part utterly unsupported by evidence. The only conclusion at which one could arrive from such comments was that Enlightenment or Cartesian subjectivities are little more than a scholarly habit.

A New Schema for Cyborg Theory

This book is founded on the pugnacious premise that cybertheory as a politicized postmodern practice has done an injustice to history. Even if we were to accept the rather doubtful claim especially prominent through the latter decades of the twentieth century that cyborg subjectivities are a radical transformation of human identity, we should note that the figure of the cyborg is a metaphoric, academic, and political construction as much as it is a physical one. The historical relationships of the man-machine and cyborg complicate the prevalent and frequently simplistic equations of then versus now; however, any satisfactory history has been ignored, especially if the critical motive is to attack the Enlightenment or modern 'agenda' that has become the scapegoat for the problems of contemporary society. Finger-wagging structuralist and postmodernist readings of the Enlightenment have caricatured the rise of mathematics, science, and technology as instruments of exclusion, hegemony, and disciplinary power. The surprising blindness to the complexity of the cyborg figure's relationship to the early modern man-machine must be understood within the context of publishing trends of academic theory, of postmodernism, of radical change, and the especially marketable rhetoric of impending apocalypse or utopia as the twentieth century drew to a close, all of which have exacerbated the tendency to ignore the cyborg's affinity with and debt to a long literary tradition. The uncritical assumptions that the moderns universally accepted rigid categories of stable human identity and binary categories, or of an unacknowledged conspiracy of the Enlightenment to control and dominate society through reason, science, and mathematical logic should not form an unchallenged basis for cyborg theory. That there has been, from the late seventeenth century through to the present, a far more sophisticated and diverse conversation about the human machine than is generally acknowledged would have complicated the claims being made for the postmodern condition of cyborgs.

Studies of the cyborg over the past few decades, while offering constructive examinations of present issues, have not examined its important historical antecedents in the eighteenth-century figure of the man-machine. Beyond summary references to Descartes, Hobbes, or occasionally to the infamous French *philosophe* Julien Offray de la Mettrie – who made the ironically provocative and profoundly serious claim that man *is* a machine – the cyborg would seem to have little history of note prior to its 'being spat out of the womb-brain of its war-besotted parents in the middle of the last century of the Second Christian Millennium.'[49] Some treatments of the cyborg posit origins that imply a complex history;[50] others are problematic because of the tendency to summarize the cyborg's 'origin story' in less than two pages (some going as far back as prehistoric man the toolmaker), or they are phrased solely in terms of the Cartesian mind-body dichotomy and humanlike automata, or they comprise broad overviews of the eighteenth-century man-machine within a few paragraphs. Such summary histories provide glimpses of a fascinating dialogue about humans as machines, but they cannot illuminate the complex story of what was in fact a variety of representations of the human-machine, all of which were very much subject to dispute and support, approval, and condemnation.

Donna Haraway's argument that the cyborg in a sense has no origin story or mythical narrative of origins occasionally has been interpreted to mean it has no history at all prior to the mid-twentieth century, and only recently has her early essay inspired counterclaims and examinations of the cyborg's early modern origins. Jonathan Sawday, for example, reads Haraway as claiming 'cyborgs *have* no forerunners,' and he invokes Shakespeare's *Coriolanus* as insight into an 'alternative manifesto' in terms of 'the impossible history of the cyborg which Haraway has sketched for us.'[51] Gill Kirkup and her colleagues, in *The Gendered Cyborg: A Reader,* submit that 'it is questionable whether Haraway's cyborg is an ahistorical construct,'[52] and accordingly include in the reader

Londa Schiebinger's 'Taxonomy for Human Beings' on the system of nomenclature devised by Carolus Linnaeus in the eighteenth century. Haraway herself has posited that the historical connections of contemporary subjectivities and technoscience are traceable to the production of gender in Robert Boyle's seventeenth-century air-pump experiments.[53] These relatively recent re-examinations of cyborg history provide provocative analyses of crucial aspects of the early modern cyborg: the 'long history of [metaphorical] intersection' of bodies and engines,[54] the historically gendered interpretations of our position in a system of natural zoology, and the situated knowledges of such organizations as the Royal Society, which excluded women and labouring men while making claims to objectivity.[55] We are, nevertheless, still missing a detailed and substantial examination of a significant aspect of the cyborg construct. The cyborg is not *like* a machine; nor can it be defined through automata or unprogrammed prosthetic devices or gender boundaries. The cyborg *is* an organic machine that is steered or governed by a homeostatic mechanism. The conception of the cyborg is the moment of the human being *as* a machine, defined, powered, governed, steered, and motivated by the same forces as machines.

The Problem of Definition

The narrow specificity of my proposed definition – returning to the original meaning of cyborg as steered organism – is necessitated by the multitude of definitions for cyborg that include any characteristic from having a wooden leg to writing a cheque with a pen, having been vaccinated, wearing contact lenses, or reading a text from a computer screen. Perhaps the one predominant characteristic of the cyborg has been the uncertainty and ambiguity of defining it. The difficulty of identifying what the cyborg is and what it means to contemporary subjectivity may best be illustrated by Haraway's continued refinements of her first influential definition. In her 1985 manifesto she identified the cyborg as 'a cybernetic mechanism, a hybrid of machine and organism, a creature of social reality as well as a creature of fiction.' Cyborgs are 'creatures simultaneously animal and machine, who populate worlds ambiguously natural and crafted,' she wrote. Haraway itemized three crucial boundary breakdowns that made her manifesto's analysis possible: 'the boundary between human and animal is thoroughly breached ... Late twentieth-century machines have made thoroughly ambiguous the difference between natural and artificial, mind and body, self-developing and externally-designed, and many other distinctions that used to apply to organisms and machines, [and] the boundary between physical and non-physical is very imprecise for us.'[56] In 1989 Haraway reiterated two of the three 'leaky distinctions' that she had noted in her manifesto. A cyborg exists, she suggested, when 'two kinds of boundaries are simultaneously problematic: 1) that between animals (or other organisms) and humans, and 2) that between self-controlled, self-governing machines (automatons) and organisms, especially humans (models of autonomy). The cyborg is the figure born of the interface of automaton and autonomy.'[57]

In her foreword to *The Cyborg Handbook* in 1995, Haraway argued that Gaia, the earth seen through the eyes of global atmospheric research as a dynamic, self-regulating, and homeostatic system, is a cyborg – as is Schwarzenegger's Terminator, the rat with its osmotic pump featured in the seminal article by Clynes and Kline on the cyborg, and *Mixotricha paradoxa*, a microorganism found in the hindgut of a South Australian termite. Here, Haraway claimed that 'cyborgs are about particular sorts of breached boundaries 'that confuse a specific historical people's stories about what counts as distinct categories crucial to that culture's natural-technical evolutionary narratives.' In this piece she seemed to resist definition, commenting, 'whatever else it is, the cyborg point of view is always about communication, infection, gender, genre, species, intercourse, information, and semiology.'[58] The cyborg also exists now, Haraway pointed out, in genomic and electronic databases and in cyberspace. It would seem that the cyborg is everything, and perhaps there is truth in this: the cyborg in this sense is an interpretational framework for critique. Haraway's 1997 definition was almost directly opposed to the first: 'The cyborg is a cybernetic organism, a fusion of the organic and the technical forged in particular, historical, cultural practices. Cyborgs are not about the Machine and the Human, as if such Things and Subjects universally existed. Instead, cyborgs are about specific historical machines and people in interaction that often turns out to be painfully counterintuitive for the analyst of technoscience.'[59] This variety of explanations illustrates the difficulty of articulating the cultural significance of the cyborg as a representative icon. The cyborg mutates; its features are formed by a political genome that subtly transforms itself over time. The cyborg defies definition particularly when we recognize that categories of human and machine, or animal and human, or matter and non-matter historically have not been so clearly demarcated as Haraway's original essay and numerous others have implied.

The cyborg figure resists easy cultural categories, especially if it is meant to demonstrate an overgeneralized radical change in human identity or subjectivity.

What *is* the nature of humanity, of identity, and of the self? What *was* it? And why should we accept that it has radically changed? Is it human selfhood that has changed, or only some depictions of it in fiction, film, and academic theory? Obviously, our understanding of nature and our ability to manipulate it is constantly reforming our *explication* of human identity; it would seem that the theoretical changes to human subjectivity are not only a technological reformatting of human social systems but also a textual scholarly production. The question that this book asks in response to these diverse generalizations of early cyborg theory is the following: What does the historical record tell us? What if we were to take, for example, the second of Haraway's definitions above and apply it to La Mettrie's *L'homme machine*, published in 1747? La Mettrie said, foremost, that there is no boundary between physical and non-physical: We can see, he wrote, in arguing against a distinction between matter and immaterial spirit, that 'there is only one substance in the universe and that man ... is to the ape and the cleverest animals what Huygens's planetary clock is to one of Julien Leroy's watches' – that is, man is merely a better machine. However much philosophers 'may wish to exalt themselves,' he argued, 'they are basically only animals and vertically crawling machines.' As for autonomy, 'we think, and we are even honest citizens, only in the same way as we are lively or brave; it all depends on the way our machine is constructed.'[60] How do we situate those claims against the theories that characterize the late twentieth-century human-machine identity as not only postmodern but also post-industrial, post-Enlightenment, post-nature, post-gender, post-national, post-identity, post-cultural, post-biological, post-evolutionary, or post-human? One might argue with La Mettrean hyperbole that the postmodern cyborg existed in 1747! Obviously, it did not – and this is why Haraway's last distinction that cyborgs are about specific historical machines and people in interaction seems to me a more satisfying approach than to suggest that the cyborg is created by body modifications, or that boundaries between human and machine existed in modernism and evaporated with the arrival of postmodernity. Although the cyborg figure as an expression of transgressed boundaries and the fragmented, decentred, or multiple self in a corporate and capitalistic 'wired' world is certainly a plausible and meaningful category, the presupposition that it is a specifically postmodern entity is worth exploring, if only because the assumption is so pervasive and yet largely unexplained. How accurate, or even helpful in the understanding of our own humanity, is the proposition 'postmodernity is to cyborg as modern is to human'?[61]

The cyborg of theory, fiction, and film is largely a make-believe beast: a politicized icon, a pretender for the future, a metaphor of hope or despair, a literary figure, analogical and symbolical. This is not to say that real cyborgs do not exist. Cybernetic organisms (in the sense of having exogenous regulatory feedback mechanisms) are a part of everyday life in many nations of the world. The cyborg, then, is a discursive entity with two interrelated histories: the history of how ordinary living bodies and artificial mechanisms have been so defined as to enable the interior 'grafting' of pacemakers or myoelectric limbs or pharmaceutical implants, and also the history of the imagined human-machine hybrid upon which fiction and theory have imposed a radically altered or evolved identity, spirit, morality, or consciousness. This book problematizes the claim for impending evolution that has been a defining characteristic of cyborg speculation, not because human identity is fixed, but because it is impossible to predict the universality or the revolutionary import of a so-called cyborg identity for the future of human morality, spirituality, or identity. Moreover, we may be able to understand more about the cyborg as a trope by examining its precedence in the period of the man-machine. I look to the seventeenth and eighteenth centuries and suggest that the boundaries between what is human and what is machine have not been obscured so much as the language used to define both human and machine has converged: common terms in informatics and cybernetics or physiology and technology enable us to define and to predict the workings of bodies and machines using the same language. Bionanotechnology is premised on the assumption that organic and inorganic assemblages, man-made machines and natural machines, are built of the same material components defined and constrained by the same laws and described by the same language.

The cyborg figure exists, as did the man-machine, not because boundaries between human and machine have dissolved but because of the assumption extending back to ancient Greek philosophy of an essential unity of matter, whether machine, nature, or organism. Both man-machine and cyborg exist because of the important assumption, established in the Enlightenment, that humans can be defined in the same terms and by the same physics as machines – that is, the assumption that the relationships of matter, energy, and force are common to both natural and artificial organisms. The question of electronic or digital representation and the imagined disembodiment or virtuality of cyborgs is secondary in this study to the primary focus on the physiological mechanisms governing the body. As we will see, however, the two are closely related and interdependent in the image of the cyborg. For the purposes of the following history, I have largely limited my inquiry to two crucial and related characteristics of the present-day cyborg figure. First is the metaphor of steering or government of the body, without which the term *cyborg* would not exist. Second is the closely related mechanistic assumption that delineates both the man-machine and the cyborg: that in its most basic form, the matter of living beings and of inert materials is essentially the same – that identity, consciousness, or soul, like the body, is a product of physical mechanism. These assumptions produced the nascent definitions for living bodies that we use today in science and medicine. They also instantiated nascent cyborg narratives, both of optimism for future progress due to mechanization or technology, and of resistance to the idea of the human spirit being reduced to physical mechanism.

It is unnecessarily limiting to exaggerate the novelty of the issues of identity and social values raised by the cyborg figure, particularly if we examine our own literature for traces indicating a revolutionary change in human consciousness. Both cyborg fiction and cyborg theory tend to incorporate and reinscribe certain western European narratives about human identity, while cyborg technology and biomedicine reconstruct the interaction of humans and machines with unprecedented sophistication. Is there evidence that these new interactions have radically transformed postmodern consciousness? Although the cyborg figure should be understood by examining what is unique about our contemporary circumstances, which unquestionably derive from technological change, the postmodern cyborg is also (under other names) a creature of history. If we are to interpret cyborgs politically, we need to resituate our studies to examine the conditions of their origins. It is of course tempting to suggest that the cyborg represents something new and radically different, 'a new map, a new way to conceive of power and identity,'[62] but it is equally important to attempt to understand the extent to which our present constructs of identity are dependent upon, and indeed reconfigurations of, those of the past. The human-machine has been an enduring figure in our literary and medical imaginations for hundreds of years and, while the technology of the past three hundred years has changed dramatically, our literature has continued to express common themes centring on the meaning and value of the human spirit in a mechanized world.

The Enlightenment Cyborg

The seventeenth and eighteenth centuries witnessed transitions to a capitalistic and commercialized print culture; from an early machine culture to an early factory culture; from spiritualized theories of aether as the quintessential rarefied air of the gods and of human spirit to physical and physiological theories of aether as a subtle but material fluid subject to mechanical properties; from medical representations of the female body as an inverted and less perfect version of the male to the female body as a different mechanism from the male, specifically built for reproduction; from the human body as microcosm to the body as clockwork mechanism guided by the soul; and finally to body and mind alike as material mechanism. The telescope and microscope helped to change the perception of bodies and their relationships to one another. The planets, once described as perfect spheres moving within perfect concentric celestial spheres, were now seen to be imperfect bodies travelling in mathematically described ellipses. The microscope revealed a plethora of minute bodies, never before imagined, in a drop of water. Proto-cyborg-theorist Robert Hooke began his *Micrographia: or Some Physiological Descriptions of Minute Bodies Made by Magnifying Glasses* (1665) with a description of the point and the plane. What had been the smallest indivisible unit in geometry – the ideal and perfect point – or the perfect geometric element that has length but not width or thickness – the line – now could be seen, represented by the point of a needle and a printed period, and by the edge of a razor, as distorted and scarred bodies with dimension. What had been invisible could no longer be imagined as ideal and perfect, and literally new worlds became visible. Hooke's premise was that God's mechanisms at this level of magnification were perfect, while human artifice was dramatically flawed. Both, nevertheless, were material made visible to reveal previously unimagined new worlds at an infinitesimal scale. The new technologies of seeing, along with new knowledge of chemistry and mechanics, were to usher in a sense of the body as a myriad of infinitesimal mechanical bodies. Hooke suggested that the limitations of human perception had restricted human knowledge but these 'infirmities' could be corrected 'with *Instruments,* and, as it were, the adding of *artificial Organs* to the natural.' This had already been accomplished in recent years, he concluded, by the invention of optical glasses: 'By this the Earth it self, which lyes so neer us, under our feet, shews quite a new thing to us, and in every *little particle* of its matter, we now behold almost as great a variety of Creatures, as we were able before to reckon up in the whole *Universe* it self.' With this new 'mechanick knowledge,' he wrote, 'we may perhaps be inabled to discern all the secret workings of Nature, almost in the same manner as we do those, that are the productions of Art, and are manag'd by Wheels, and Engines, and Springs, that were devised by humane Wit.'[63] The extension of limited human senses in order to perceive material bodies was one key to understanding the mechanisms of the universe – and to perfecting those created by humans. The other was through experimental and mathematical demonstrations of what could not be directly perceived. The microscope had proven there was another material dimension to the universe. Now through chemistry, optics, and later in the eighteenth century, discoveries in magnetism and electricity, philosophers attempted to demonstrate experimentally the smallest particles of matter giving life and motion to the universe.

In the 1660s Robert Boyle (1627–91) set out to develop his mechanical corpuscular theory based on matter and motion. The materialist interpretations of human consciousness in the works of René Descartes (1596–1650), Thomas Willis (1582–1666), Thomas Hobbes (1588–1679), and John Locke (1632–1704) together with the more notorious controversy of freethinker Anthony Collins (1676–1729) and Samuel Clarke (1626–1701), would have Jonathan Swift and his fellow Scriblerians, principally John Arbuthnot and Alexander Pope in *The Memoirs of Martinus Scriblerus,* responding decades later with mockery and ridicule. Descartes' dualism of the thinking being (*res cogitans*) and the extended being (*res extensa*) was an attempt to accommodate immaterial spirit with known anatomy and the new philosophy of matter, but his mechanistic *bête machine,* based on the premise that animals had no 'rational soul,' invited the inevitable realization of La Mettrie's infamous *homme machine*: the human body and consciousness both as wholly material entities. Crucial bridges between these two images were the comparative neuroanatomy initiated by Willis, an Oxford doctor, and the mathematical demonstrations of nature's mechanical properties instigated by Sir Isaac Newton. Willis established that forms of knowledge and advanced mental capacity were the functions of corporeal mechanism. Only a few years after Descartes' death, Willis argued in *Two Discourses Concerning the Soul of Brutes* (1683) that the corporeal soul extended from the brain throughout the body. His explication of the government of the human body was eminently distinct from that of Descartes, notably in that he postulated that there were two souls in the human 'Engine': the immortal and 'wholly Etherial' rational soul and the sensual and material corporeal soul extended from the brain throughout the body – constantly at war with one another. An enduring significance of Willis's work was that he studied the mind/soul as material *bodily* processes and mechanisms. He described the animal spirits as subtle particles moving like waves or like rays of light, which 'transmitted, as through Perspective glasses, the impressions of sensible things and the instincts of motions to be performed.'[64] But it was Newton's enquiry on matter in the 1706 *Opticks*, Query 24 (later renumbered as 31) that reinforced the notion of a hypothetical and quasi-material aether of infinitesimal particles. Light, magnetism, electricity, and nervous function could all be understood as related quasi-material mechanisms from this point. Willis's and Newton's conceptions of aether contributed significantly to our understanding of human-machine bodies – far more significantly than did Cartesian dualism.

By the mid-eighteenth century the material properties that enabled the processes of human will and action were regularly hypothesized, investigated, and empirically tested. Both Bryan Robinson in his *A Dissertation on the Aether of Sir Isaac Newton* (1743) and Browne Langrish in his 'Croonian Lectures on Muscular Motion' (1747) conjectured that the nervous fluid might be understood as the quasi-material aether that Newton had described in Query 31. George Cheyne wrote in his best-selling book *The English Malady* (1733) that the body may be explained as a machine controlled and enlivened by a substance that seemed to him most likely to be that same Newtonian aether. In 1749 David Hartley's influential mechanical physiology of the mind and morality, *Observations on Man* was published; it, too, was influenced by Newton's Query 31. La Mettrie's bold claim in *L'homme machine,* published anonymously in Holland in 1747, became the most fully articulated argument of the period for animal motion, morality, learning, speech, and rationality as solely material mechanisms. Denis Diderot also suggested a monistic interpretation of mind and body in depicting the human nervous system, in his unpublished play *Le Rêve de d'Alembert* (1769), as a web extending beyond the body to the contiguous objects in the world and universe sensed by the individual.

Later the discovery of electricity and its effects on organic bodies compelled the conversation about body mechanics and animation to change its vocabulary. Samuel Taylor Coleridge, who embraced and then rejected Hartley's mechanical physiology, could stand as an emblem of a groundswell against a mechanistic world-view, while physiologists continued to examine and explain the human body in terms of its material mechanisms. Mary Godwin Shelley and William Blake in their unique ways imagined the horror of a new mechanistic form of human creativity usurping the natural order of God and threatening human spirituality. An instructive account of changing attitudes towards and investigations into the mechanisms of muscular motion is provided by the annual Croonean lectures of the Royal Society, established through the bequest of Dr Croune upon his death in 1684. The long eighteenth century also saw the invention of James Watt's feedback device that would become known as the governor. All of these changes contributed in various ways to a discourse of human-machines characterized by both suspicion and hope, and they contributed to a growing body of medical, philosophical, and technical literature that has increasingly defined the creative processes of God or nature and of humans in the same terms.

This study focuses in part on some of the work of the French materialist *philosophes*, but primarily upon the works of English writers of the Restoration and Regency periods who were important contributors to the history of contemporary medicine and biotechnology, as well as to literary and cultural studies. The (often distinctly anti-Cartesian) English materialist representations of human consciousness and volition that I will examine here did not have the same anti-clerical and combative goals as the radical *philosophes* did in their version of the man-machine. The members of the Royal Society, whose works represent an important part of the English tradition of the human-machine, were more cautious than was La Mettrie in any claims they made about the soul; nevertheless, a dialogue about the mechanistic approach to human physiology can be observed throughout the period in a variety of genres from the philosophical and empirical studies of Isaac Newton and Thomas Willis towards the end of the seventeenth century, to the eighteenth-century philosophies of Locke and Hartley, and to the fiction and poetry of Swift, Pope, Arbuthnot, Sterne, Coleridge, and Mary Shelley. Accordingly, metaphysics plays a lesser role in this history than do empirical mechanics, iatrochemistry, or anatomy, and the expressive responses to such studies in poetry and fiction.

All of these texts tell a story different from the overarching assumption that, since the late seventeenth century, inhabitants of the Western world have all been carting around a singular version of the 'Cartesian ego,' a mind separate from the body's material, a ghost in the machine.

What becomes readily apparent through reading these and other inquiries and experimental demonstrations by the Royal Society throughout the eighteenth century is that the image of physical and mental processes as mechanical never waned, but the question of the soul became increasingly peripheral to medical discourses that were intent upon understanding the physical phenomena of bodies. The immaterial and immortal soul – which long preceded Descartes – remained a tradition in metaphysical or literary discourses, but in terms of physiology and medical sciences it receded in relevance. The medical and technological intervention in the workings of human bodies that creates cyborgs is not a legacy of Cartesian dualism, but an emphatically monistic one where human identity is profoundly physical and embodied. Willis's positing of an embodied form of knowledge and an embodied soul/mind in both the animal and human engine marks the beginning of a pronounced rejection in the physiological sciences of the Cartesian dualism of body and mind. Descartes' man-machine with an immaterial soul at the centre of the brain in the pineal gland (see figure 2) is only one part of a long history, even if it dominates contemporary fictive representations of the cyborg. Mamoru Oshii's images of the cyborg in the anime movie *Ghost in the Shell* (modelled on the manga series by Masamune Shirow) offer an example: the cyborg Motoko Husanagi's consciousness and identity are embedded, along with the fragment of brain that remains from her original body, in a wholly artificial cybernetic body (see figure 3). Husanagi and fellow operatives are charged with investigating an electronic sentient agent capable of infiltrating and taking over the minds of its human victims through the computer network. She later abandons her body completely and continues to exist in the computer network. References to the relationships between early modern philosophy and contemporary theory are more explicit in the anime *Ghost in the Shell 2: Innocence* (2004), which features Coroner Haraway, obviously modelled after cyborg theorist Donna Haraway: 'Descartes didn't differentiate man from machine, animate from inanimate,' comments the cyborg detective Batou to Coroner Haraway and police officer Togusa. 'He lost his beloved five-year-old daughter and then named a doll after her, Francine. He doted on her. At least that's what they say.' Later the renegade hacker Kim explains to Batou and Togusa, 'Dolls ... make us face the terror of being reduced to simple mechanisms and matter. In other words, the fear that, fundamentally, all humans belong to the void ... Further, science, seeking to unlock the secret of life, brought about this terror. The notion that nature is calculable inevitably leads to the conclusion that humans too, are reducible to basic, mechanical parts.' To this Batou replies, quoting La Mettrie's *L'homme machine*: 'The human body is a machine which winds its own springs. It is the living image of perpetual motion.' Kim's response is: 'In this age, the twin technologies of robotics and electronic neurology resurrected the 18th century theory of man as machine. And now that computers have enabled external memory, humans have pursued self-mechanization aggressively, to expand the limits of their own functions. ... The mirage of life equipped with perfect hardware engendered this nightmare.' The anime reiterates the common conflation of Cartesian dualism with eighteenth-century philosophy, and reaffirms that to a great extent the controversies surrounding the cyborg in fiction, film, and theory continue to be a process of coming to terms with and reincorporating the debates over the various definitions of human consciousness that were born in the seventeenth and eighteenth centuries.

The terms of the debates over the man-machine were not of course – situated as they were within specific historical moments, cultural environments, and political relationships – equivalent to today's debates over the cyborg. Significant aspects do, however, recur in the metaphors and analogies of both our science and our literature in complex and intriguing ways. Robert Markley, one of the few scholars who have examined cyberculture's history in eighteenth-century thought, has shown the importance of the work of Gottfried Wilhelm Leibniz (1646–1716) as a historical predecessor to cyberspace or virtual reality. In his essay 'Boundaries: Mathematics, Alienation, and the Metaphysics of Cyberspace,' Markley argues that cyber-technologies do not 'offer a breakthrough in human, or cyborgian, evolution, but merely (though admittedly) a seductive means to reinscribe fundamental tensions within Western concepts of identity and reality.'[65] The polyvocal discourse of philosophy and physiology in early modern literature and science is part of cyborg history. Rather than a totalitarian enterprise to impose systems of control and domination upon human bodies, these discourses tended towards monism *and* dualism, vitalism *and* mechanism, acceptance of *and* resistance to mechanistic perspectives of nature, culture, and humankind.

Philosophies of the man-machine and the responses to these writings in poetry and fiction throughout the Enlightenment engendered a multifaceted rhetoric of selfhood and embodiment that has helped to shape our own medical and fictional discourses of the cybernetic-organic human being. The early definitions of the human body-mind as a mechanism steered, controlled, or governed by material and physical processes of communication have yet to be examined in contemporary cybertheory. Nor have the mechanistic philosophies and mechanical technology of the early modern period been considered together as constituting a history of the human-machine hybrid. The early intellectual premises that suggested human bodies, like machines, could be analysed in terms of predictable physical properties contributed to the contemporary techno-medical understanding of humans as mechanisms of pumps, valves, levers, and hinges, and as homeostatic systems governed by feedback mechanisms.

What has to a great extent inspired this investigation into what we might call the 'Enlightenment cyborg' is the confidence of so many theorists through the past few decades in the premise that networked communications mean inevitable and profound changes to individual and social consciousness. The image of a dominant mechanism – whether steersman, pilot, sovereign, or governor – controlling the individual body, the applicability of machine mechanisms to the workings of the human machine, and the supposed reconstitution of body and body politic due to ungoverned or networked communications systems are dominant metaphors in theories of postmodern transformations. Yet these are all constructs fashioned by or derived from early modern tropes for the human-machine. If we position the history of cyborgs as part of the history of physiology as a system of organic mechanisms, we can see the cyborg not as Cartesian automaton with a soul but more accurately as a human-machine moved by the energy of atoms or particles, and controlled by a circuit of communications.

I use the term *Enlightenment*, then, with some caution, in order to emphasize the tremendous variety of ideas and debates throughout the long eighteenth century that contributed to our fictional, theoretical, and real cyborgs. To imagine the cyborg's history we must be mindful that the English Enlightenment, which influenced English literary traditions as well as industrial and physiological ones, did not operate on a solely Cartesian framework, and involved the creation of other machines to mimic human thought and labour than the admittedly evocative automata admired by Descartes. Such a history cannot ignore the important contributions of philosophers such as Descartes, La Mettrie, and Leibniz, but the focus here is upon a particular tradition of influence in English medical history where the language describing the mechanisms of bodies or minds and the language describing the mechanisms of machines merged. This study deals extensively with British natural philosophy and literature, if only to plot the development of a non-

Cartesian strain of philosophy that forever changed the understandings and relationships of bodies and machines. It also focuses on the English tradition in order to follow the course of human-machine analogies in the context of English literature and literary studies that have contributed so heavily to the image of the cyborg in American popular culture. Another extremely important aspect of cyborg history that cannot be addressed here, regrettably, is the history of Japanese cyborgs in film, anime, and other multimedia arts. What I hope to demonstrate in this study of the cyborg's role in the Western European literary tradition is that the images and interpretations of the cyborg – at least in the film, fiction, and cultural studies of the Western English-speaking world – are part of the tradition of literature that began in the age of sensibility as a reaction to mechanical interpretations of human knowledge and understanding. Accordingly, they echo some of the ideas that characterized the literature of sensibility and the Romantics. The long eighteenth century's science of mechanics promoted by the Royal Society; the experimental and mathematical philosophy established through the influences of Boyle and Newton; the predominance of feedback or self-regulation in the history of English literature, medical philosophy, and engineering; and the responses to these images in fiction and poetry together gave rise to a tradition of assumptions about humans as machines that continue to inform English Literature and criticism.

2 Matter, Mechanism, and the Soul

A living body is not a meer aggregate of flesh, bones, & c. but an exquisitely contrived, and very sensible engine, whose parts are easily set a work by proper, though very minute, agents, and may, by their action upon one another, perform far greater things, than could be expected from the bare energy of the agents, that first put them into motion.

Robert Boyle, 'Of the Reconcileableness of
Specific Medicines to the Corpuscular Philosophy' (1744)

What makes humans something more than material lumps put in motion by mere energy? How does our embodiment mesh with that aethereal something that makes us more than the sum of our parts? If humans are the product of purely material processes, where does God fit in, and free will, and morality? What of the immortal soul? These were the inconvenient problems with the mechanistic construct of the man-machine through the late seventeenth and eighteenth centuries. They were unanswerable, debated at length, and never ultimately resolved. But even if the man-machine seemed a terribly limited description of humans endowed with vitality, life, and morality, it was nevertheless a significantly threatening presence in the philosophy of early modern Europe. At St Mary-le-Bow on 4 April 1692, for the sake of 'proving the Christian Religion against notorious Infidels,' Richard Bentley (1662–1742) read the second of eight Boyle Lectures that he had been charged to deliver that year on the folly and unreasonableness of atheism. In this particular sermon, entitled 'Matter and Motion Cannot Think: Or, a Confutation of Atheism from the Faculties of the Soul,' Bentley argued against the dangerous materialist speculation that thought, will, and perception are caused by the motions of atoms. His lecture would, he said, prove that these functions were enabled by an immaterial substance, the soul, separate from the gross matter of the body. This, in turn, would clearly demonstrate the existence of a 'Supreme and Spiritual Being.' If such faculties as thought and perception were inherent in matter, Bentley protested, what 'monstrous absurdities would follow?' Either stocks and stones would be percipient and rational; or men, composed of too many unique thinking and sensitive atoms to comprise a unified sense of mind and body, would have no feeling or perception at all. Bentley reduced the irreligious notion of sentient matter to ludicrousness, postulating that if 'Motion in general or any degree of its velocity can beget Cogitation; surely a Ship under sail must be a most intelligent Creature; though while she lies at Anchor, those Faculties be asleep.' To imagine that 'meer Body may produce Cogitation and Sense,' or that the comprehension exhibited by animals can 'be performed by the pure mechanism of their Bodies,' is to suggest that humans themselves are merely 'Engines of a finer make and Contexture.'[1] That he would fix his attention for an entire sermon on the motions of atoms or particles suggests the significant threat the man-machine posed to the system of faith and government that provided Bentley, Chaplain to the Lord Bishop of Worcester, both salary and spiritual certainty.

Bentley's rhetorical hyperbole about sentient ships was surely both humorous and more than adequately persuasive for his audience, but by the twentieth century the image of an intelligent ship would become both a realized mechanism and a much fantasized image of potential futures.[2] We are now capable of building ships that steer themselves via electromechanical systems in a disturbing semblance of human thought. We are also capable of steering and changing the human body and consciousness with chemical compounds or prosthetic devices, all made possible by the knowledge of molecular and electronic motions and reactions. Bentley's sermon demonstrates the historical precedents for the problematic representation of human spirit, will, and reason as material mechanisms at the molecular scale. Today, our dystopian film, fiction, and critical theory alike reiterate his image of the monstrous results when mere mechanical matter is supposed to influence judgment and will: the cyborg or sentient machine is persistently represented as highly rational, but lacking certain degrees of feeling, perception, or morality. The uncertainty now, as it did then, concerns what governs or directs sentient will and reason. What will differentiate future humans from future monstrosities in the eerie form of human-like automata?

What will separate humankind from the very machines we are creating in our own image? These questions are at the heart of cyborg discourse, and they are affiliated with the discourse of the man-machine: the natural philosophy of the man-machine and the biomedical physiology of cyborgs both were based on the premise that the body and mind are products of the organization, actions, and reactions of material particles. In theorizing and fictionalizing both the cyborg and the man-machine, however, competing definitions of mind and body, or even matter, have made these figures morally, spiritually, and politically emblematic. The cyborg and the man-machine are powerful signifiers of the potential diminution (or liberation) of human spirit, soul, consciousness, or will if choices, rational thought, and morality can be reduced to the flux of atoms or particles.

The complex relationship of the language and history of the man-machine to those of the cyborg necessitates a series of definitions: What is the man-machine? What is a cyborg? What is matter? What is mechanical? What is soul or spirit ... or body for that matter? The sections in this chapter attempt to synthesize these disparate and interrelated concepts by taking three separate approaches to defining several key categories. I begin with a brief summary of how cyborg identity may be understood to be the result of the specificity of our language for the material particles of our bodies. The following section examines the transition from the vague balance of spirits in the humoral body to the new approaches of understanding the body as comprised of particles whose motions are governed by mechanics. The section after that traces the downward path of aether conceived as the diviner air of the gods to aether as a mechanical medium pervading the atmosphere, earth, and organic bodies alike, and finally aether as the electrical fluxes and forces that confer life, vitality, and consciousness to material bodies.

At the most reductive level, the notion of the cyborg figure as a logical progression from the rebirth of atomistic theories in the seventeenth and eighteenth centuries might seem to invoke a problematic version of the inevitable evolution of medical and technical progress. However, it is crucial to this study to acknowledge that the erratic development of – and resistance to – the philosophies of atoms and particles in the early modern man-machine helped us eventually to construct, mythologize – and resist – the electrochemical and material spirit of today's technologized body. The new philosophy serves as a beginning for this cyborg history: the early philosophies of mechanics, matter, and motion, and the more controversial interpretation of humans as engines enlivened by mechanical forces of matter in aethereal particles, transformed the body from a spiritualized microcosmos governed and directed by the stars to a mechanism governed by the motions of material particles.

This history is of both the body as machine and body as a material particle with mechanical properties. The mathematical abstraction of the body, which has been celebrated and feared as 'virtual flesh,' paradoxically required materialization of its vital, sensory, and motor functions. Once conceptualized as a microcosm containing both matter and vaguely defined immaterial spirit, the increasingly material body and psyche would be defined using enhanced technologies of vision for examining ever smaller bodily mechanisms, improved technologies of chemical experiment and analysis for examining bodily reactions, and mathematical abstractions for measuring, calculating, and predicting biochemical probabilities.

But what was the body? And what is the body now? Both a concept opposed to mind or soul, and an object such as a planet, a microorganism, or a particle, *body* in the seventeenth or eighteenth centuries (or the twenty-first century) presents no small task to define. In 1724 Nathan Bailey's *Universal Etymological Dictionary* defined body as a 'compound of Matter and Form: A Party of Soldiers; a Company of People,' as a simpleton, as a vessel for distillation, as a mass of metal in gunnery, and finally as a geometrical magnitude with dimensions of length, breadth, and thickness. Bailey's definitions reproduce those provided in Edward Phillips's *The New World of Words: or, Universal English Dictionary* (1706) and John Kersey's *Dictionary Anglo-Britannicum* (1708), except for the new addition of 'simpleton' and an important clarification of geometrical body, which 'is opposed to Body as it is a Term in Natural Philosophy, inasmuch as the First is penetrable, and the Second impenetrable.' In the 1730 edition, Bailey delineated body primarily as a theoretical impenetrable material in nature rather than a notional penetrable form in geometry: 'as defin'd by *Naturalists* [a body is] a solid, extended, palpable substance, compos'd of *matter, form*, and *privation*, according to the *Peripateticks*: 2. of an Assemblage of hooked heavy atoms, according to the *Corpuscularians* and *Epicureans*; of a certain quantity of extension according to *des Cartes*; of a system or association of solid, massy, hard, impenetrable, moveable particles, ranged or disposed in this or that manner according to Sir *Isaac Newton*; ... others define body to be that which has extension, resistance, and is capable of motion.' Following this series of definitions, and before the meanings having to do with geometry, Bailey introduced the distinction between body and soul: 'with regard to animals, [body] is used in opposition to the soul ... in which sense body makes the subject of Anatomy.' Phillips in 1706 had seen fit to distinguish between body and soul only in the definition for *difference*: 'Difference (in *Logick*) signifies an Essential Attribute ... and is the Universal Idea of that *Species*. Thus *Body* and *Spirit* or *Soul*, are two Species of Substance ... in a Body we find Impenetrability and Extension, in a Spirit the power of thinking and reasoning; so that the difference of a Body is impenetrable Extension, and the difference of a Spirit is Cogitation or Thought.' It would seem that, as body became more commonly defined as a matter of impenetrable solidity, there was a corresponding imperative to incorporate spirit or soul into the definition.

This theory is not perfect, as the twenty-fourth edition of Bailey's dictionary, 'carefully enlarged and corrected by Edward Harwood,' omits reference to spirit, but consider Samuel Johnson's influential *Dictionary of the English Language* (1755). Johnson tells us that the word body is of Saxon derivation and that it 'originally signified the height and stature of a man.' In his first four definitions of body, Johnson emphasized matter versus non-matter: body could mean 'the material substance of an animal, opposed to the immaterial soul,' or 'matter, opposed to spirit.' A body could be 'a person; a human being; whence *somebody*, and *nobody*' or, fourth, 'reality; opposed to representation.' The symbolic or abstract body of a joint power, an army, or a corporation follows, and finally, Johnson's eleventh definition, where in geometry body meant 'any solid figure.' Although one cannot reach unqualified conclusions by comparing these works in isolation, these sets of definitions several decades apart nevertheless provide a telling instance of how the old assumed boundaries between body and spirit were being negotiated. If for Johnson the distinction between material body and immaterial soul was paramount in defining a body, for Bailey's dictionary in 1724 body was foremost a compound of matter and form. The introduction of impenetrability seems to have necessitated the acknowledgment and confirmation of the soul. As a mathematical construct in the new philosophy, a body's materiality and extension had to be assumed. Incorporeal spirit was not part of the equation, and understandably so; but as we will see, this tension between the permeating and unknowable spirit and the impenetrable and finite body continued to be problematic throughout the history of the man-machine and the cyborg.

To understand the body as a machine was not merely to study it as a number of related mechanisms – as say, an engine – in terms of gross anatomical body components but also as a system of minute particles. The chemical and mathematical descriptions of bodies both dematerialized the body as mappable, calculable, mathematical abstraction, and materialized the subtle spirit animating the world. The human soul in the seventeenth and eighteenth centuries became part of a physical, material mechanism. In the early 1600s physiology was still dominated by Galen (129–ca. 199), who had combined the humoral doctrine with his own observations of the brain and nerves, and initiated a longstanding theory of animal spirits stored in the ventricles and flowing through the hollow nerves of the body to effect sense and motion. The animal spirits were a very subtle, fine substance that mediated between the material body and the immaterial soul: they linked the material and mental realms and thus actualized life, motion, and sensitivity. Robert Burton in *The Anatomy of Melancholy* (1621) described *spirit* as a 'most subtle vapour, which is expressed from the blood, and the instrument of the soul, to perform all his actions; a common tie or medium between the body and the soul.' For Burton, following the systems of Hippocrates and Galen, the body's physiological and mental state resulted from the balance of the four humours and it was animated by the three souls: the natural or nutritive, the vital, and the animal or sensitive. The animal spirits, separate from the corruptible flesh, flowed from the hollow ventricles in the brain through the nerves to 'the subordinate members, giv[ing] sense and motion to them all.'[3] The Galenic version of ventricles as storage tanks for the animal spirits would prompt debate well into the eighteenth century over whether the nerves were hollow or contained fluid,[4] at least in part because Galenic physiology, faithful to the idea of an immaterial soul, attempted to establish that these spirits functioned as a link between gross physical mechanism and the mental and vital activities of the soul.[5] As physicians and anatomists began to examine the body more closely, however, the doctrine of the hollow nerve gradually lost favour, as did the belief that an immaterial soul could act on and effect material mechanisms. By 1800 animal spirits were beginning to lose their sacrosanct position, to be incorporated into the base material of the body. Thomas Willis's theory of the 'soul of brutes' common to man and beast (see chapters 3 and 4) established that forms of knowledge and advanced mental capacity were functions of an 'energetical' but nevertheless corporeal mechanism.

No longer did immaterial spirits, governed by the heavens and God's will, animate the blood and enliven the heart (see figure 4). With William Harvey's treatise on the circulation of the blood, the heart was seen to be a mechanism, a pump (see figure 5). The new anatomy examined the body as a composite of mechanistic systems, displacing the older humoral-astrological versions of physical and mental functions determined by celestial bodies under God's direction (see figures 6 and 7). We can observe a general trend in anatomy from depictions of the body as a holistic entity inhabited by living spirit, even while dissected and presumably dead (see figures 7 and 8; see also figures 31 and 39), to the body as a system of material components given life by the material organization of the system itself (see figures 9 through 11; see also figures 32 and 33). An engine can be defined as 'any mechanical complication, in which various movements and parts concur to one effect,' according to Samuel Johnson. Isolated, the components of the body-as-engine were merely dead material, but once assembled, they composed a living machine. The inclusion of the body's systems along with other machines governed by geometry and physics in John Harris's *Lexicon Technicum* (1708) is evidence of this new mechanical understanding of the body in terms of its material organization: 'As to *Physick* and *Natural Philosophy* and those admirable helps to the understanding of Nature, which *Geometry*, applied to *Physical Enquiries* hath of late afforded us, and to which indeed we are chiefly indebted to that Prodigious Mathematician Sir. *Isaac Newton*: I have endeavoured to give you every-where the Marrow and Substance of it.' What were once described vaguely as 'qualities' Harris collected under the terms '*Electricity, Solidity, Elasticity, Effluviums, Magnetism, Light and Colours, &c.*'[6] New understanding of the minutia and mathematics of material bodies was inexorably merging what had once been clearly separate as matter and non-matter. The eye, that 'mirror of the soul,' was now a mechanism that refracted corpuscles or beams of light to create an impression like a camera obscura or magic lantern casting images on the wall of a darkened room. Isaac Newton's *Opticks*, first published in 1704, explained this phenomenon as a mathematical formula equivalent to the mechanism of glass lenses (see figure 12). Not only were the lens and humours of the eye like a glass device, but Newton postulated that the pictures were propagated in the same way for either organic or inorganic mechanism, by the motions of the minute particles that make up matter. Newton's expanded and revised *Queries* of later editions posited the possibility that the medium called *aether* might be the medium not only for universal gravitation, rays of light, electric and magnetic effects, and vision but also for human will, sensitivity, and motion. To understand the body as a machine was to come to an understanding of all matter as composed of innumerable, infinitesimal material particles. Whether the particles moved or were moved, and whether they were infinitely divisible or ultimately indivisible were central problems in the philosophy that would form the basis of the man-machine ideology. In any case, the material atom was recognized as likely to be the smallest component of both the man-machine and inert matter; further, in both cases it might be responsible for energetical motions such as fermentation or digestion, chemical combustion, magnetism and electricity, or involuntary and voluntary muscular motions. Robert Boyle's comparison of the mechanism of human motion to that of an engine such as a musket rested upon this assumption: 'I consider the body of a living man, not as a rude heap of limbs and liquors, but as an engine consisting of several parts so set together, that there is a strange and conspiring communication betwixt them, by virtue whereof, a very weak and inconsiderable impression of adventitious matter upon some one part may be able to work on some other distant part, or perhaps on the whole engine, a change far exceeding what the same adventitious body could do upon a body not so contrived.' The 'faint motion of a man's little finger on a piece of iron,' explained Boyle, will have no effect if the iron is not part of an engine, but the communication of that faint impression upon trigger, spring, cock, and flint against steel to open the pan and strike fire upon the gunpowder will throw out a heavy leaden bullet 'with violence enough to kill a man at seven or eight hundred foot distance.' This was the power of inert matter assembled so that larger mechanisms received 'impressions' and communicated them to smaller and invisible atomic mechanisms: 'the engine of an human body is so framed,' Boyle concluded, that it is similarly 'capable of receiving great alterations from seemingly slight impressions of outward objects.'[7] This conceptualization of particles or atoms was a quarrelsome philosophy, since it attempted to explain not only how the body worked on a visible level but also on an invisible level – at the level that had previously been explained as immaterial or spiritual: the level of the soul, of knowledge, of the will, and of identity.

Here arises a problem of terminology: throughout this book I investigate early modern usage of *soul* as a precursor to various terms related to the mind. Associated as it is with religious concepts of immortality and spirituality, soul generally is not used to describe human consciousness today, whereas mind and soul were often used interchangeably in earlier periods. Willis coined the term 'Psycheology' for his 'Discourse of the Soul'[8] and attributed to the 'Rational Soul' the various qualities that we attribute to the mind or consciousness such as knowledge, imagination, memory, appetite, sense, motion, passions, and instincts. Descartes in his *Meditations on Philosophy*, subtitled 'In which Are Demonstrated the Existence of God and the Immortality of the Soul,' discusses faculties that we would associate with the mind such as judgment, reason, or intellect. Similarly in his *Treatise on Man* the soul perceives movement, size, distance, colours, sounds, and smells, and it feels pleasure, pain, hunger, thirst, joy, sadness, and other passions. Descartes notes here that the pineal gland, identified as the seat of the soul in his *The Passions of the Soul*, is the seat of the imagination and of common sense.[9] Bentley, too, characterizes the soul as 'an immaterial Substance in us ... essentially distinct from our Bodies' that – much as we characterize the brain and central nervous system today – 'thinks and apprehends, and reflects and deliberates; that determines and doubts, consents and denies; that wills, and demurrs, and resolves, and chooses, and rejects; that receives various sensations and impressions from external objects, and produces voluntary motions of several parts of our Bodies.' Bentley's object was to demonstrate that this 'Soul or Spirit ... doth necessarily evince the existence of a Supreme and Spiritual Being.'[10] Nevertheless, the material character of what was then characterized as the soul would slowly become an accepted model of human consciousness (although given other names in medicine and physiology). John Locke called understanding 'the most elevated Faculty of the Soul'[11] and attributed to the soul various other faculties that we today associate with consciousness and identity.

In 1737 the Scottish metaphysician Andrew Baxter (1686?–1750) anonymously published a lengthy treatise to prove the immateriality of the soul through the 'principles of reason and philosophy'; he defined *soul* as 'that which is active and percipient in us, be it what it will ... its activity appears in moving the body, and in the command it hath over its own thoughts; its perceptivity, in being sensible of the action of matter upon it, and of its own internal actions of thinking.'[12] Almost two decades later La Mettrie wrote *L'homme machine* (published in Holland) to prove precisely the opposite, that the soul was material: 'All the parts of the soul can be properly reduced to imagination alone, which forms them all,' he argued, 'and thus ... judgement, reason and memory are only parts of the soul.' These faculties are solely, he said, 'the wonderful and incomprehensible result of the organisation of the brain' – and then he promptly fled the country fearing for his life. *Soul*, argued La Mettrie, is only a word that we cannot define, and so 'a good mind should use [it] only to refer to that part of us which thinks.'[13] The faculties of the soul were not generally in dispute; whether it was material, mortal, or motive were the crucial questions concerning the human machine.

I use *soul* to refer to the historical version of what we today call consciousness, identity, or mind. The term carries with it intimations of immortality, morality, and spirituality that also subtly define the cyborg of contemporary fiction and theory even though soul or spirit is sometimes only implied. I occasionally refer to *spirit*, which also carries with it a religious connotation of immortality and human transcendence along with more secular connotations of individual character, will, or courage. Spirit is often used by twentieth-century writers to suggest a specifically human characteristic that transcends animal or machine or cyborg, and it is in the context of an examination of the historical connections of mind and soul as superior to mechanism that I use the term here.

Defining the Cyborg: Molecules, Electrons, and Spirit

Although Bentley's derisive dismissal of such a concept as sentient matter would have little credibility as a line of reasoning today, when we understand memories and thought as fluxes of ions associated with calcium, magnesium, potassium, glutamate, and so on, it is nevertheless true that the religious connotations of immaterial spirit and soul in the human machine remained a common theme in twentieth-century musings upon cyborg consciousness. As I have discussed elsewhere, this theme is explicit in recent films such as *Johnny Mnemonic* or *The Matrix*, and in academic theory such as the work of Jean Baudrillard.[14] From the moment *cyborg* became a word, it signalled a complicated relationship between bodies and souls. Manfred E. Clynes and Nathan S. Kline, in their 1960 article 'Cyborgs and Space,' were concerned with eliminating the internal mechanisms of the body as much as possible in order to create new possibilities for the mind and spirit. These two scientists speculated that exogenous regulation of body temperature, 'human "fuel" consumption,' cardiovascular control, pressure, fluid balance, feeding, and other problems of existing for prolonged periods within a spaceship would help to 'provide an organizational system in which such robotlike problems [as the body's natural mechanisms] are taken care of automatically and unconsciously, leaving man free to explore, to create, to think, and to feel.' Solving these problems, they concluded, 'will not only mark a significant step forward in man's scientific progress, but may well provide a new and larger dimension for man's spirit as well.' The old trope of material body separate from higher-order mind or spirit pro-

vided a rationale for the extreme interference and disturbing manipulation of the body's systems as envisaged by Clynes and Kline. The rhetoric
of elevated spirit, however, is destabilized by their matter-of-fact materialist view of human minds, describing options to correct 'psychophysiological problems' such as using 'psychic energizers' to keep the astronaut
continuously awake and fully alert, and 'an emergency osmotic pump
containing one of the high-potency phenothiazines together with reser-
pine' in case of a psychotic episode.[15]

The curious meld of older notions of spirit with new versions of
technologically constructed consciousness was also apparent in Marshall
McLuhan's claims during that time that the receding of the mechanical
age meant a new trajectory for human spirit, in what he called the global
'extension of human consciousness.' 'In this electric age,' claimed
McLuhan, 'we see ourselves being translated more and more into the
form of information, moving toward the technological extension of
consciousness. That is what is meant when we say that we daily know
more and more about man. We mean that we can translate more and
more of ourselves into other forms of expression that exceed ourselves.'
This technological extension of our nervous system, McLuhan predicted,
would change not only human consciousness but also human spirituality: 'Electricity points the way to an extension of the process of consciousness itself ... that could confer a perpetuity of collective harmony and
peace.'[16] What McLuhan was conflating, either consciously or unconsciously, was a religious version of human spirit with a mechanistic
association of the mind as extended, electrified matter. Speculations
upon the spirits of so-called post-humans extending beyond the body in
electric or electronic media have been reiterated regularly in fiction,
film, and theory for several decades: the relationship between material
embodiment and immaterial identity, consciousness, or soul is a fundamental trope of cyborg literature. But rarely do we examine the cultural
history of these relationships of extension, matter, and the immaterial
forces now called electrical energy and consciousness in the figure of the
cyborg.

Cyborg and man-machine are both politicized by the intensity of
debate over the dualism of mind and matter in philosophy and theory.
Paradoxically, however, in biomedicine and biotechnology the cyborg
exists precisely because these terms have been rendered largely irrelevant. The notion of spirit as distinct from body continues to appear in
discourses ranging from biotechnology to cyberpunk fiction; however,
the chemical and physiological language used to define both mind and

matter, and to describe and predict their behaviour, has no need of the terminology of an older language that employed such dualisms simply because it lacked specific terms and crucial knowledge of atomic or molecular structures. One does not need to debate the difference between mind and matter to create functional nanotechnology, however important the distinction may be in ethical terms; one instead needs to understand, for example, how RNA molecules can become motors – that is, made to do mechanical work by binding to the energy-bearing molecule ATP.[17] In this sense, the cyborg is the convergence of the language used to describe humans and machines as much as it is the convergence of the material body and machine. The relationship between cyborg hardware (the body itself) and cyborg software (the steering or programming of body-mind functions) is both a metaphoric and a material outcome of the gradual refinement of those early definitions of matter, mechanic, mind, and body.

The definitions for the material of our bodies and the mechanisms of their functions have resulted in an entity called cyborg because now, for the first time in human history, organisms and artificial mechanisms can merge at the molecular level. Now that technicians can perceive, manufacture, and control the previously invisible working parts of the human machine, nature and artifice merge: the organic body has become a mechanism that humans can encode as text, manufacture, and, debatably, discover and copyright (e.g., U.S. Patent No. 5,397,696 on a cell line from a Hagahai person from Papua, New Guinea). Furthermore, 'beginning in summer 1991 and through the next several months, the National Institutes of Health (NIH) applied for patents on thousands of human DNA sequences believed to represent approximately 5 percent of all human genes.'[18] In general terms the codes used to define the human body in medical-scientific discourse have changed over several hundred years from being primarily metaphor and simile (the nerves, arteries, and veins are like rivers; the heart is like a pump) to an increasing dominance by alphanumeric symbol (e.g., the precise mathematical determination of the constituents and contaminants of blood or urine, or the precise sequence of cellular DNA as sequences of letters). In *Anatomy of Melancholy* (1621), Burton, following Galen's system, described the 'vegetal' aspect of the soul – the part that enables growth – as acting like a lodestone, a lamp, the flame of a torch, or a boiling pot. These analogies compared invisible forms of energy to the vitality of the living body, and comprised a figurative language that was necessary because a more precise language and analytical tools were lacking. Like the *Ma'aseh*

Toviyah (1708) of Tobias Cohen, which depicts the human body as a house warmed by the simmering cauldron of the stomach (see figure 13), this language illustrated a general idea about the workings of the human body as analogous to other natural processes that exhibited motion and heat. These analogies were based on an understanding of natural physical and chemical processes that developed in parallel to describe both human and machine systems.

The contemporary post-human cyborg is a result of the remarkable refinement of these centuries-old mechanistic views of human bodies. These images are obviously comparative: they do not suggest that humans and ordinary tools or machines are equivalent. Throughout the seventeenth and eighteenth centuries there was a growing acceptance of the premise that natural and artificial machines were made of the same stuff – that the basic building blocks of all material, organic or inorganic, were extremely fine particles or atoms acting in predictable ways according to laws of motion. Although today we retain the mechanical analogies of filters in our bowels, a pump with valves in our circulatory system, or hinges in our limbs, our bodily processes and machine processes both might also be rewritten as numerical formulae or programming code that would describe, for example, organic molecules, myoelectric control of a prosthetic limb, or the sequence of proteins in nucleic acid (i.e., symbols such as $H_2NCONH_{2(s)}$; $0/1$; $\pm$; or AGCTCTTCTC). The distinction between natural and organic body components and artificial machine components at this level of symbolic definition is slight or even non-existent. NH_2CONH_2 is urea, a natural product of the body. Differing by only a molecule of oxygen and the ratio of hydrogen and nitrogen in its molecular structure benzoguanamine, $C_9H_9N_5$ is used in creating resins for laminating agents and adhesives. Ohm's law applies to both the electric circuits of nerve membranes in the body and to electric circuits in machines: $V = I \times R$, that is, the voltage difference V between two points is equal to the current I multiplied by the constant R, the resistance of a given pathway. We describe our nervous system, as we describe our electronic computer systems, as a complex interaction of on/off switches. Our bodies themselves are encoded in an on/off language, a four-letter alphabet of binary pairs, A-T and C-G, which makes it possible not only to imagine coding the body's programs as we do computer programs but to also construct computers from DNA.[19] The artificial myoelectric control of a heartbeat, or the sequence of proteins naturally manufactured by DNA, can be calculated and predetermined by formulae that blur any distinction between biological and artificial processes. We have a symbolic language that does not differentiate between the calcium ion involved in an electrochemical reaction to make a person's arm move and the calcium ion that reacts to produce an inert mass of calcium carbonate. At the molecular scale, nanobiotechnology unites engineering and biological sciences to design in situ medical devices and drug delivery. The combined universality and specificity of the language we use to identify and control the most basic elements of bodily structures and machine structures alike is what makes humans cyborgs, and this universality and specificity of language is what formed the earliest conceptions of the man-machine – a conceptualization of the processes of mind and body alike as the result of material structures and workings.

Defining the Man-Machine I: Mechanicks and Matter

Until the late seventeenth century Galenic medicine and older traditions going back to Socrates treated the human body as a microcosm created from the four elements of the universe: earth, water, air, and fire composed our bones, blood and other fluids, breath, and bodily heat. The elements corresponded to the four humours in our bodies, namely, blood, black bile, yellow bile, and phlegm, which in turn were governed by the planets in the celestial sphere (see figures 6 and 7). This ancient tradition dominated medical and physiological viewpoints throughout the Middle Ages; it represented health, life, conception, motion, and vitality as ultimately inexplicable spiritual forces. The new mechanical philosophy that dominated physiological theories from the 1660s onward saw these instead as potentially explicable material forces and motions. It would be impossible to define precisely what the mechanical philosophy was, or what the man-machine was – as impossible as defining what matter itself was. Competing philosophers quibbled over the Cartesian billiard-ball model of matter as impenetrable atoms that were infinitely divisible, governed by laws of geometry, and set in motion by God's fiat; or Newton's indivisible particles, also governed by laws of geometry and calculation, and cohered by attractive and repulsive forces. Regardless of the differences and disputes associated with the new mechanical philosophy, however, mechanism was gradually and surely eroding the occult qualities of zodiac-man governed by celestial bodies.

More and more, the human body came to be understood as a system determined by the laws of mechanics. For example, Walter Charleton

(1620–1707), physician-in-ordinary to Charles I and subsequently to Charles II in exile, wrote in *Natural History of Nutrition, Life, and Voluntary Motion* (1659) that the principle of life or soul is a 'Naturall Heat,' or 'Vital Flame,' or 'a certain Fire,' while the muscles of the body act purely as mechanism. In the section 'Of Voluntary Motion, or the Use of the Muscles,' Charleton explains that the Soul seated upon her throne in the brain is the source of motion but, being immaterial, cannot produce any effect. Motion is 'impressed' upon the muscles by the 'transmission of the Animal Spirits from the Brain.' The animal spirits, material but nevertheless so subtle that they approach the nature of the soul, are 'proportionate to the immediate energy of the Soul, than either the grossness of the member, or muscles ordained to move it.' Motion is the result of the intermediate agent of animal spirits, which have 'transmitted' a 'swift and speedy Impulse from the ... incorporeal Agent, the soul, [to] those corporeal instruments, the Muscles.'[20] The human-machine is thus a mechanical system enlivened by energy. Muscular motion, like that of other machines, is purely mathematically describable forces and geometry (figures 14, 15, and 16 demonstrate aspects of the man-machine as a machine comparable to other labouring machines). What was perhaps more contested and problematic, as we will see in later chapters, was the man-machine as 'rational machine.'

Understandably, even into the eighteenth century physicians attempted to reconcile an older version of the humoral-zodiacal body with new explanations for anatomical and psychic functions. In 1724 the physician John Maubray (d. 1732) advised his readers of the power and influence of the planets over the body's health: Mars excites the yellow bile, he observed, as Saturn 'exasperates' melancholy, Luna increases phlegm, and Sol and Jupiter 'govern' the blood. On the formation of the fetus, Maubray noted that while the vegetative soul (responsible for growth) was generated from the commixture of male and female seed in the womb, the sensitive (feeling) soul proceeded from the organized matter of the fetus, and the rational (thinking and immortal) soul was infused into the body by God. On this infusion, Maubray refused to speculate, since it had been for so long such a matter of controversy, but he did hope that he might use the principles of both philosophy and religion to reinforce one another and prove a satisfactory explanation for all. To that end, Maubray commented that both astrologers and philosophers would agree that the three 'occult Qualities, and secret Influences of celestial Bodies' that infused and influenced human bodies – namely, heat, light and motion – act alongside other natural causes and extend into the womb where they foment, govern, promote, preserve, and determine the processes of conception. The celestial bodies do not touch the fetus, he explained, but act remotely through these secret influences (God, Maubray is careful to emphasize, is the primary agent that causes the action).[21] The notion that occult qualities of stars and planets could influence human bodies from afar, however, was losing favour, as the new experimental philosophy sought to find demonstrable and material explanations for natural processes. Heat, light, and motion were key problems in the new mechanical philosophy that conceptualized the human body as an engine of hydraulics and pumps, levers and wedges, boilers giving off energizing heat, lenses and optic machines, and fine material called animal spirits. Most philosophers were leery of any suggestion of occult qualities.

Although the image of the man-machine is not entirely circumscribed by the characteristics of classical mechanical philosophy, it was born and grew alongside these early attempts to describe the forces and functions of nature in mathematical terms of matter and motion. And, although there is not a complete connection between the man-machine and materialism, the materialist premise that whatever exists is matter or is determined by matter, is also a crucial part of the man-machine's history. Defining the man-machine concept thus becomes a negotiation of terms – not only are we concerned with the question of mind versus matter, but with that of mechanism versus materialism. Robert E. Schofield has attempted to differentiate between mechanist and materialist schools of thought,[22] but as John Yolton has explained, this differentiation is not only confusing but also misleading. Yolton distinguishes rather between 'the traditional immaterialists, who defended a passive concept of matter' and the 'new materialists, who made matter active.'[23] Whether matter had an active principle inherent in itself would become a central issue for the man-machine as a defining paradigm of human action and vitality. *Mechanism* and *materialism* were extremely problematic terms in describing human function, even though they were considered by the seventeenth-century physiologists in terms of active forces not necessarily in exclusion of will, morality, or God. But to consider a human being as mere matter in motion was – and remains today – a troubling premise. The image of the man-machine would be augmented and challenged by later theories of sensibility, sensitivity, and vitality. Nevertheless, while the old debate that is represented by the apparently opposing terms mechanism and vitalism has been a controversy that occasionally flares, strictly adhering to the opposition is not advantageous to defining the characteristics of either the man-machine or the cyborg.

If we take mechanics to mean simply 'the geometry of motion,' as Samuel Johnson defined it, then the man-machine as a mechanical system did indeed lack an adequate explanation for the origins of motion. However, the early modern examination of the human as purely mechanical was also an assumption that chemical reactions were mechanical motions. There were many quibbles within the new philosophy: some thought atoms were flexible and even divisible, while others insisted they were not. For this reason some philosophers used the Epicurean term *atom* whereas others preferred *corpuscle*; but such differences of opinion presumed a prevailing acceptance of the new theories of nature – including human bodies – as mechanisms of material particles with active properties. Even if the original cause of the motions of particles could not be determined, as Marie Boas noted in her still useful exposition of the mechanical philosophy, by 1800 or so 'there were no real scientists who failed to accept one of the available particulate theories of matter.'[24] The evolution of mechanical philosophy was a matter of building upon, correcting, expanding, and delving further into rather than comprehensively rejecting earlier premises. The eighteenth-century chemist William Cullen concluded, for example, that 'upon the whole, chemistry not only considers a set of properties different from these which the mechanical philosophy considers, but a set of properties also that may depend on different circumstances.'[25] Cullen stayed within the realm of Newtonian natural philosophy, however, noting also the 'two great principles, the attractive & repulsive,' which are 'the source of motion & change.'[26] With the failure of mechanical models to explain electromagnetism, mechanical philosophy ultimately fell into some disrepute, but mechanism by no means disappeared from scientific investigation. The early modern atomists conceived a theory of those particles, but could not come to a consensus as to whether the forces pulling them together were intrinsic or external. If today we take for granted the attractive forces between molecules in a protein chain in muscle or of ions in a neural reaction in the brain, it is due to the experimental examinations, proofs, and refutations of early mechanistic theories of particles or atoms. Present-day theories explain how the smallest constituents of matter – whether molecules, electrons, atoms, or quarks – move; nevertheless, molecules and even electrons are considered as material things, with weight and properties that are determined math-

ematically and experimentally, and despite the arrival of quantum physics, chemists continue to speak of the mechanisms of chemical reactions,
both organic and inorganic. The cyborg identity retains the influence of
mechanical theory in the simple and yet crucially important understanding that our bodies are composed of infinitesimal and dynamic material
particles that are both attracted to and repelled by one another by some
kind of force at a distance.

As mechanics itself moved beyond definition solely as geometry of
motion, so too did the man-machine. The human engine was undeniably living and moving, thinking and sensing, and undeniably vital. The
Cartesian system, where matter and motion were the only acceptable
explanations, would ultimately be unsatisfactory for human physiology,
partly because it seemed to diminish the presence of God and partly
because it simply was inadequate to explain so many phenomena of the
natural world. Indeed, mechanism was rarely so pure and geometrical as
Descartes imagined it. The complex interrelationship of Newton's philosophical and religious formulations of the attractive and active principles
enmeshed with mechanical ones – the secret universal animating or
vegetative spirit that was the means of gravity, growth, generation, fermentation, and vitality – would have a far more influential and profound
authority than that of the Cartesian framework. On this, as Betty Jo
Teeter Dobbs has commented: 'a modern thinker may be inclined to
assume that Descartes chose the better part [with his purely mathematical models of a mechanical universe], but in fact the natural philosophy
that he claimed to have established with mathematical certainty was soon
overthrown by Newton's. Descartes's mathematico-deductive method
was not adequately balanced by experiment, observation, and deduc-
tion; Newton's was.' Moreover, to contrast mechanism with vitalism or
sensibilities is perhaps overly simplistic from the viewpoint of a history of
science. As Dobbs has argued, the mechanical philosophy that has so
often been representative of the Newtonian revolution was probably not
the dominant system in Newton's own time but rather only one system
that was 'complementary rather than competitive.'[27] To Newton – and to
many of his followers – God *acted* in the world and that belief did not
negate or transcend mechanism.

A century later the surgeon and anatomist John Hunter (1728-93)
would hypothesize: 'an organ is a peculiar conformation of matter, (let
that matter be what it may) to answer some purpose, the operation of
which is mechanical; but, mere organization can do nothing, even in
mechanics, it must still have something corresponding to a living prin-

ciple; namely, some power.' The complete living body, Hunter argues, is made up of 'body, blood, and motion ... out of which arises a principle of self-motion; a motion totally spent upon the machine,' for the body requires the motion of the blood and the blood requires the motion of the body in order to live. The heart 'may be considered as a truly mechanical engine; for although muscles are the powers in an animal, yet these powers are themselves often converted into a machine, of which the heart is a strong instance; for from the disposition of its muscular fibres, tendons, ligaments, and valves, it is adapted to mechanical purposes; which make it a complete organ or machine in itself.' Hunter's elaborations upon the 'materia vitae diffusa' that permeates the solids and fluids of the whole body to provide it with powers of preservation, susceptibility of impression in the brain, and reciprocal action in the body, reinforce the concept of the body as a machine to which power has been added. Indeed, Hunter's specification that 'nothing material is conveyed' when the nerves communicate 'the impression or action' from brain to body and vice versa prefaces his delineation of the brain as 'a mass of ... matter ... constituting an organ in itself'[28] for the purpose of receiving impressions and producing action. For Hunter, the body fluids and solids, including the brain and nerves, are material and mechanical systems: whatever powers may be the living principle or vital material, Hunter repeatedly and explicitly terms the living body a *machine*. Presumptions about the terms *mechanical, mechanism,* and *machine,* however, have inevitably and mistakenly diminished the image of the man-machine to a philosophy that seems to oppose thought, reason, free will, sensitivity, orthodox Christianity, or human spirituality.

Today we understand mechanics as having a number of categories: in general the term denotes the theoretical and practical knowledge concerned with inventing and building machines, and calculating their efficiency. Mechanics also refers to the branch of applied mathematics known as classical or Newtonian mechanics, concerned with the motion and equilibrium of bodies and the action of forces. Classical mechanics includes kinematics (the study of objects in motion), dynamics (the study of the action of force in producing or varying motion), and statics (the study of rest or equilibrium under the action of forces). Paradoxes or inconsistencies in certain instances led to the development of quantum mechanics, a theory of the motion and interaction of particles (particularly subatomic particles) that incorporates the concept of wave-particle duality. Statistical mechanics is the application of statistics to the field of mechanics, in terms of modelling and predicting the behaviour of large numbers of particles, atoms, molecules, and so on, when subjected to a force.[29]

'There cannot be fewer principles than the two grand ones of mechanical philosophy, matter and motion,' Boyle exclaimed in 'The Excellency and Grounds of the Mechanical Hypothesis' appended to *The Excellency of Theology, Compared with Natural Philosophy* (written in 1665 and published 1674). Understanding 'mechanicks' entailed a process of studying and explaining the geometrical or mechanical properties of matter and forces mathematically, not only in terms of the large and visible structures of levers and pulleys, but also in terms of the smallest particles of which matter was made. As Boyle explained, in his revised and expanded *Some Considerations Touching the Usefulness of Experimental Naturall Philosophy* (1671), the new philosophy examined the laws of motion for invisible as well as visible bodies in the material world: 'I do not here take the term mechanicks in that stricter and more proper sense, wherein it is wont to be taken, when it is used onely to signify the doctrine about the moving powers (as the beam, the leaver, the screws, and the wedge,) and of framing engines to multiply force: but I here understand the word mechanicks in a larger sense, for those disciplines, that consist of the applications of pure mathematicks to produce or modify motion in inferior bodies: so that in this sense they comprise not only the vulgar staticks, but divers other disciplines, such as the centrobaricks, hydraulicks, pneumaticks, hydrostaticks, balisticks, &c.' The new philosophy of matter in motion was known by a number of names, and according to Boyle's definition, mechanics was a term that on certain levels corresponded to a nascent form of modern physics.

In *The Christian Virtuoso* (1690) Boyle characterizes the new philosophy as 'the philosophy, which is most in request among the modern virtuosi, and which by some is called the new, by others the corpuscularian, by others the real, by others (though not so properly) the atomical, and by others again the Cartesian, or the mechanical philosophy.' This aspect of mechanics delved into the mysteries of various forms of motion that may be responsible for growth and vitality, and Boyle emphatically stated that faulty reason – not the philosophy – was to blame if atheists or libertines used the philosophy to demonstrate their claims. Experimental philosophy, he argued, revealed such divine excellency in the fabric of the universe that nothing 'so incompetent and pitiful a cause as blind chance, or the tumultuous justlings of atomical portions of senseless matter'[30] seemed a probable origin or first cause. Newton (around 1674) had already made a clear distinction between vegetation, from the Latin *vegetare*, to animate or enliven (reactions between exceedingly fine material particles in vital processes that lead to generation) and mechanism (reactions of the 'grosser' particles of matter that unite to give texture to such objects as bones, flesh, wood, fruit, etc.).[31] In this 1674 manuscript Newton described the operations that take place between the 'grosser' particles as 'vulgar chemistry,' simply the mechanical coalition or separation of particles without any 'vegetation.' The subtler particles, he postulated, gave life and activity to the earth in active processes between particles, which he called ferments. These particles were material, but the forces between them were variously described by Newton as force, spirit, or medium, and were ambiguously corporeal or incorporeal. The 'aetherial breath' of the earth, which he characterized as 'a great animal,' was a 'subtle spirit' and 'nature's universal agent, her secret fire'; it was also, importantly, the 'material soul of all matter.' Newton realized that this substance, or 'active spirit,' which in chemical reactions generated heat and light, might be 'the body of light.' He considered this in the *Opticks*: 'Are not gross Bodies and Light convertible into one another; and may not Bodies receive much of their Activity from the Particles of Light which enter their Composition?' Newton began to speculate that the subtle and active spirit that permeated matter and held it together in a living form might be 'electrical attraction unexcited.' In his manuscript 'De vita & morte vegetabili' he stated explicitly that 'the particles of bodies coalesce and cohere in diverse ways by means of electric force ... but with the ceasing of life that vital attraction ceases and its absence immediately begins the action of dying which we call corruption and putrefaction.'[32] In his works on the nervous system, Willis had years before characterized animal vitality as subtle aethereal spirits that comprised quasi-material particles moving through the body in waves or like light; he had also characterized the human nervous system as a vital ferment like a chemical reaction in a chemist's vessel. Boyle, Willis, and Newton all explicitly tried to work God into the material system of living, reproducing, and thinking animal engines that their experiments implied. The definition of *mechanical* in terms of human functions shifted with cyborgian indeterminacy to accommodate both political and religious beliefs.

One might argue that the mechanical man-machine disappeared with new formulations of vitality and sensibility; but the human-machine disappeared to the same extent that the machine did. The mechanisms for both developed into something more active and vital as the knowledge of chemistry and electricity became more specific, predictable, and manageable. While electrical and chemical machines were being developed, so were our bodies coming to be understood as active electrical and chemical machines. As we understand our bodies we understand our machines, and as we understand our machines so we understand our bodies.

To return to the mechanical man-machine, then: what was mechanics in the eighteenth century, and how did it apply to machines and to human bodies? Boyles' definition was well entrenched by the early 1700s. Harris's lexicon of technical terms defined 'Mechanicks' as 'the Geometry of Motion, a mathematical science, which shews the effects of powers, or moving forces, so far as they are applied to engines, and demonstrates the laws of motion,' a definition that also appeared in Johnson's dictionary. Specifically produced to be practical rather than theoretical ('not only of bare *Words* but *Things*'[33]), Harris's text did not expend much space quibbling with the finer points of first causes. He did, however, set out for his readers a succinct framework of the philosophy of matter that informed the chief principles of the study of mechanicks, or corpuscular philosophy: 'that way of Philosophizing which endeavours to explain things, and to give an Account of the *Phænomena* of Nature. by the Motion, Figure, Rest, Position, *&c.* of Corpuscles or Minute Particles of Matter.' Harris itemizes the key points of Boyle's mechanical hypothesis: that there is a single 'Universal Matter' that is impenetrable and divisible, common to all bodies, and capable of all forms; that this matter forms all natural bodies and therefore must have inherent motion given by God; that matter can be divided into atoms, which have their own particular sizes, figures, and shapes; and finally, that these different-sized and -shaped particles have many orders, positions, situations, or postures, from which arise the great variety in the composition of bodies.

The meaning, significance, and implications of the definition of matter, however, were as various as the writers attempting to define it, and accordingly so too with the man-machine. At its most basic, disputes over human engines tended towards attempts to come to terms with objections such as those voiced by Bentley in his Boyle Lectures. He asked, could self-consciousness, understanding, judgment, memory, wit, liberality, justice, prudence, magnanimity, charity, beneficence, the love and fear of God, insight, language, poetry, eloquence, lofty flights of thought, exalted discoveries of mathematical theorems, or divine contemplation be merely a product of the concussion and clashing of atoms one against another? Thus an orthodox philosopher might describe the human as a machine with an immortal and rational soul separate from the body and capable of all those elevated faculties; another might scorn any such terminology as machine to be belittling to spiritual humanity. A free thinker might either embrace pure mechanism or scorn the purely mechanical as lacking a convincing explanation for sensitivity, thought, or generation. In general, however, the man-machine can be understood as a renewed interest in how the material components of the human body and psyche function dynamically. As Boyle described it, philosophers needed to go beyond meticulous examinations of human physiology as if to study 'the barrel, wheels, string, ballance, index, and other parts of a watch, without examining the nature of the spring, that sets all these a moving.'[34] The question whether matter contained motion, or was given motion by an outside force was fiercely debated, but the eventual outcome was an essentially materialist premise – the foundation for cyborg technologies today that assume a material body in order to mathematically represent it: bodily and mental, organic and inorganic, or natural and artificial processes are described with incredible precision using abstract symbols that presuppose the sameness of basic atomic materials in both humans and machines.

Probably the most recognized and infamous image of the man-machine is Descartes' early hypothetical model of men composed of a soul and a body: 'I assume their body to be but a statue, an earthen machine,' he wrote in the *Traité de l'homme* (written ca. 1629–33 and unpublished until 1664, years after his death). By machine Descartes meant, according to Louis de La Forge, who annotated the first edition, any 'body composed of several organic parts which being united conspire to produce certain movements of which they would be incapable if separate.'[35] Descartes' famous comparison of the movements of living bodies to those of water automata seemed to his detractors too limited an explanation of human vitality, motion, and will (see figures 17 and 18). Descartes postulated that, like fountain automata, the body's involuntary movements were responses to stimuli, while other motions were determined and mechanically enacted by an intelligent principle residing in the body: 'One may compare the nerves of the machine I am describing,' Descartes wrote, 'with the pipes in the works of these fountains, its muscles and tendons with the various devices and springs which serve to set them in motion, its animal spirits with the water which drives them, the heart with the source of the water, and the cavities of the brain with the storage tanks. Moreover, breathing and other such activities which are normal and natural to this machine, and which depend on the flow of the spirits, are like the movements of a clock or mill, which the normal flow of water can render continuous.' When the rational soul is 'present in the machine,' it resides in the brain like a fountain-keeper who oversees the tanks of water used to 'produce, prevent, or change [the] movements.'[36] The Cartesian man-machine was an influential image throughout Europe, but from the very beginning this dualism of body and mind was contested, reframed, embellished, approved, and rejected by both its proponents and its opponents. Regardless of how one defined its nerves, fluids, and particles, however, the human-machine, a self-moving, complex engine made up of particles or corpuscles, announced a crucial shift from human identity as a vague balance of humours to a complex of mechanical-chemical reactions.

In light of the supposed newly indistinct boundaries epitomized by current cyborg theory, it is worth noting that there is no distinction made here between natural and mechanical, insofar as nature's laws are what dictate the mathematics of motion. The mechanical here *is* wholly natural. Descartes had stated in *Principles of Philosophy* (1644) that for the purposes of his philosophy the difference between machines and natural bodies was that the workings of machines are readily perceptible, whereas natural processes are not. Mechanics, however, as a subset of physics, is natural: it is as natural for a clock to tell the time as it is for a certain tree to produce a certain fruit, he explained in describing his attempt to investigate the imperceptible causes and particles of perceptible bodies: 'I do not recognize any difference between artefacts and natural bodies except that the operations of artefacts are for the most part performed by mechanisms which are large enough to be easily perceivable by the senses ... The effects produced in nature, by contrast, almost always depend on structures which are so minute that they completely elude our senses. Moreover, mechanics is a division or special case of physics, and all the explanations belonging to the former also belong to the latter.'[37] At the early modern origins of human sciences, the machine was simply the application of natural laws, and for some influential philosophers the human was no different.

A crucial point here is the notion that the smallest workings of nature and artifice – humans and machines – are not two separate categories, except as observed through the limited perceptions of human senses. Similarly, according to Boyle in 1665 the mechanical philosophy demonstrated that nature was different from art only in that its mechanisms were of a different size. Corpuscularian philosophy supposed

the whole universe (the soul of man excepted) to be but a great Automaton, or self-moving engine, wherein all things are performed by the bare motion (or rest) the size, the shape, and the situation, or texture of the parts of the universal matter it consists of; all the phænomena result from those few principles, single or combined ... So that the world being but, as it were, a great piece of clock-work, the naturalist, as such, is but a mechanician; however the parts of the engine, he considers, be some of them much larger, and some of them much minuter, than those of clocks or watches.[38]

Body components functioned, like machines, in calculable and predictable ways. As the Italian iatromechanist physician Giorgio Baglivi (1668–1707) wrote in 1696:

Whoever examines the bodily organism with attention will certainly not fail to discern pincers in the jaws and teeth; a container in the stomach; watermains in the veins, the arteries and the other ducts; a piston in the heart; sieves or filters in the bowels; in the lungs, bellows; in the muscles, the force of the lever; in the corner of the eye, a pulley, and so on ... all these phenomena must be seen in the forces of the wedge, of equilibrium, of the lever, of the spring, and of all the other principles of mechanics. In short, the natural functions of the living body can be explained in no other way so clearly and easily as by means of the experimental and mathematical principles with which nature herself speaks.[39]

Again, not only the obvious structures of the body's organs but also those of its smallest parts began to be interpreted as mechanical. 'The human body,' Baglivi wrote, 'is nothing more than a complex system of mechanical and chemical movements that obey mathematical laws.'[40]

As we have seen, in England, Boyle had revised and qualified the model of the human machine according to the chemical principles he and his colleagues were investigating. He had emphasized in 1663 in *Some Considerations Touching the Usefulness of Experimental Naturall Philosophy* that although there were some actions characteristic of man – namely, intellect and will – that could not be satisfactorily explained as 'the actings of meer corporeal agents,' most of natural phenomena could be explained by the motions of particles or corpuscles that made up all of matter.[41] In his later *Free Enquiry into the Vulgarly Receiv'd Notion of Nature* (1686), Boyle questions the old notion 'that the world was governed by a watchful being, called nature,' arguing that what were ambiguously called 'real qualities' or 'natural powers' or 'faculties' such as magnetism were not mysterious cogent powers but could more exactly be understood as material effects, 'produced by the texture, figure, and, in a word, mechanical disposition' of another agent. He describes organic life as a mechanical process of maintaining equilibrium in 'living automata' at the invisible level, where chemical components 'fermented' or reacted with one another to maintain stability. Much as the magnetic needle of a compass will settle again to its customary position pointing north, a diseased body will recover its former state 'by the bare mechanism of the instrument itself, and of the earth, and other bodies, within whose sphere of activity it is placed.' Another analogy is the scale or balance, he suggests, which upon being disturbed will eventually settle 'again in an exact *Aequilibrium*' without the action of an internal intelligent principle of nature overseeing the process.[42] Crucially, Boyle's view of the body was one that incorporated a system to maintain stability at a chemical-mechanical level, rather than representing the body as merely a clockwork mechanism of gears, pulleys, and hydraulics responding to stimuli, as in Descartes' model of living automata. 'I desire to have it taken Notice of,' he reiterates, echoing a point he had made numerous times,

> that I look not on a human body, as on a watch or a hand-mill, i.e. as a machine made up only of solid, or at least consistent, parts; but as an hydraulical, or rather hydraulopneumatical engine, that consists not only of solid and stable parts, but of fluids, and those in organical motion: and not only so, but I consider, that these fluids, the liquors and spirits, are in a living man so constituted, that in certain circumstances the liquors are disposed to be put into a fermentation or commotion, whereby either some depuration of themselves, or some discharge of hurtful matter by excretion, or both, are produced, so as, for the most part, to conduce to the recovery or welfare of the body.

Boyle is careful, however, to emphasize once again that the human engine, however mechanical, was guided by a high-order intelligence: 'though the body of a man be indeed an engine, yet there is united to it an intelligent being (the rational soul or mind).' A man, he concluded, 'is not like a watch, or an empty boat, where there is nothing but what is purely mechanical; but like a mann'd Boat, where, besides the mechanical part, (if I may so speak) there is an intelligent being that takes care of it, and both steers it, or otherwise guides it, and, when need requires, trims it; and, in a word, as occasion serves, does what he can to preserve it, and keep it fit for the purposes it is designed for.'[43] As we will see in later chapters, this image of the human-machine as a steered vessel is essential to the history of the cybernetic organism, or cyborg, as an organism that is mechanically and/or chemically steered to maintain equilibrium. To this extent, Boyle's human-engine is a cybernetic, homeostatic mechanism.

Despite their many disputes and controversies, what early philosophers from various locales across western Europe were reiterating was a similar theme: humans are machines created by God's art, and they are superior machines precisely because God's art is more intricate and minute than that of man. These assumptions appear commonly in the works of philosophers positing God as the engineer of the human-machine. And now, with the aid of electron microscopes and computer technologies – when these smallest of machines in their smallest parts are readily perceptible by our enhanced senses – when their manufacture 'is humanly possible' – Descartes' differentiation between machines and natural objects is rendered meaningless. The seventeenth-century image of God the Engineer is, at the very least, inapt when doctors use gene therapy to replace malfunctioning gene alleles with 'normal' ones, or when research scientists use gene splicing to mix the genetic materials of two organisms to create a new one. Today the categories organism as opposed to machine or artificial as opposed to natural do not sit well with physiological technologies that function at the level of molecules.

Natural philosophers in the great majority studied the mechanisms of the universe to reveal God's creative greatness. One can understand the concern (then as today) about reductionist views of the world, but for most of the early mechanical philosophers the representation of humans as mechanical entities was a celebration of God the Engineer's eternal power and wisdom. If Boyle agreed with Descartes that the body is a machine, he did so because he – like Descartes – revered God's workmanship and design. The image of the world as an elaborate engine created by God as engineer, or 'great Author and Governor,' said Boyle, is a better illustration of God's wisdom than to imagine nature as a mysterious force interposing upon natural events and human lives: 'it more sets off the wisdom of God in the fabric of the universe, that he can make so vast a machine perform all those many things, which he designed it should, by the meer contrivance of brute matter managed by certain laws of local motion and upheld by his ordinary and general concourse.'[44]

Boyle's colleague, the Oxford neuroanatomist Thomas Willis, also took great care to profess the 'great God, as the only Work-man' and the 'first Mover' who impressed the powers and faculties into matter necessary for the sensitive life of living beings. In his *Two Discourses Concerning the Soul of Brutes,* Willis negotiated a problematic explanation of the organic machine made active and vital by powers akin to the energetic powers of fire, air, and light that are used by men to make machines in human industries. To represent the body as a machinelike mechanism was not a particularly contentious issue, but to explain life, movement, reproduction, and instinctive knowledge as such risked displacing the positions of God and the soul in human lives. Willis explained that the subtle and highly active particles instilled from the blood into the brain are shut up in passages 'as it were Pipes and other Machines,' to effect both sensitive and 'loco-motive powers,' but he reiterated the crucial difference between machine and man-machine: if humans are created of invisible particles of matter, it only proves the superior craftsmanship of God. As human industry makes such devices as 'Dioptrick Glasses, Musical, Warlike, Mathematical Instruments, with many other Machines' as well as 'the Furnaces of Smiths, Chymists, and Glass-men, and of other boylers of several Kinds,' Willis explained, God 'the Great Workman' made the souls of living creatures out of material particles 'to which he gave also a greater, and as it were a supernatural Virtue and Efficacy; from the Excellent structure of the Organs, most Exquisitly laboured, beyond the Workmanship and artificialness of any other Machine.'[45]

For all his decorous rhetoric about spiritual matters, however, what Willis proposed was a model of human heat, motion, life, and even consciousness based on chemical reactions of material particles. Key to this system was the concept of fermentation, familiar enough in the making of bread, wine, or beer, but almost unknown, he pointed out, in the systems of 'Vulgar Philosophy' that explained nature as vaguely defined forms and qualities. Willis had noted, however, in his 'Of Fermentation or the Inorganical Motion of Natural Bodies' that in recent – and 'more sound' – philosophies that examined matter and motion, fermentation is a common object of study. Fermentation, or 'growing hot' by the agitation of particles in a body, was a process that occurred in a variety of natural bodies, whether 'Thin or Thick, Liquid or Solid, Animate or Inanimate, Natural or Artificial.' The process was characterized by what we would call chemical reactions that transform or purify compounds and create gases, liquids, and purified material residues: Willis described particles in all these substances that are light and want to fly away, while others that are thick and earthy, and want to 'detain' the lighter ones in their 'Embraces.' He proposed that the human brain and nervous system were comparable to an alembic, the apparatus used in the process of distillation. Typically the alembic consisted of a boiler and a covering called a head. The liquid was put into the boiler, and placed over a fire to raise vapours, which were then condensed by cooling the head with cold water. Willis favoured the chemical opinion of the nature of matter, which held that all bodies consist of particles of spirit, sulphur, salt, water, and earth. These elements are broken down by fire in the process of distillation, where vapours rise from a mixture and leave behind inert solids. Spirits, Willis explained in his tract on fermentation, are the aethereal active particles that constantly endeavour to fly away, unless bound to thicker particles by entering into them and 'subtilizing' them. They 'conserve the bonds of the mixture by their presence,' he explained, and they 'determinate the Form and Figure of every thing.' These are 'as it were the Instruments of Life and Soul, of Motion and Sense, of every thing.' They are responsible for the vegetation and fermentation of bodies, and they repeat the same motions in the bowels of living creatures to perform the 'works of Life, Sense, and Motion.' The crudest particles, at the other end of the scale, comprise earth, which is the dead or 'damned' substance incapable of change that remains after all else is removed from the mix. Willis's aethereal spirits here, like Newton's, are early descriptions of what we designate electronic activity in chemical reactions or in nervous tissue.

In terms of the man-machine image, Willis characterized the processes of life as a series of chemical reactions occurring within a vessel: 'As the Blood in the Heart, and appending Vessels, the Chyle in the Ventricle, so the Animal Spirit is wrought in the Brain,' he surmised.

> It seems to me that the Brain with Scull over it, and the appending Nerves, represent the little Head or Glassie Alembic, with a Spunge laid upon it, as we use to do for the highly rectifying of the Spirit of Wine: for truly the Blood when Rarified by Heat, is carried from the Chimny of the Heart, to the Head, even as the Spirit of Wine boyling in the Cucurbit, and being resolved into Vapour, is elevated into the Alembic; where the Spunge covering all the opening of the Hole, only transmits or suffers to pass though the more penetrating and very subtil Spirits, and carries them to the snout of the Alembic: in the mean time, the more thick Particles are stayed, and hindred from passing.[46]

The blood is purified in the head in this manner, and the subtle and active particles of the aethereal spirits are then, as in the alembic, 'drunk up by the spungy substance of the Brain, and there being made more noble or excellent, are derived into the Nerves.' The thicker particles in the blood are subsequently carried back to the body by the circulation of the blood. This description of bodily chemistry was reiterated in the massive three-volume *Bibliotheca Anatomica, Medica, Chirurgica* (1711–14) by Daniel Clericus and Jacob Mangetus, and although Willis's vision of human neurochemistry would be eventually replaced by better models, his work established essential principles in the model of the human nervous system and profoundly influenced the trajectory of the man-machine.

Newton, like Willis, surmised that the aether could comprise subtle but material bodies that permeated larger bodies, instigated both motion and vitality in living matter, and transformed bodies from one material to another. 'Are not the Rays of Light very small Bodies emitted from shining Substances?' Newton asked in his *Opticks* (1717). Not only might vision, hearing, and animal motion be performed through the vibrations of the aethereal 'Medium excited in the Brain by the power of the Will, and propagated from thence through the solid ... Nerves into the Muscles,' but material bodies and light may be the same substance transformed: 'Are not gross Bodies and Light convertible into one another, and may not Bodies receive much of their activity from the Particles of Light which enter their Composition?' Newton, too, speculated that the workings of the living organism – vitality and growth, reproduction, and decay – could be mechanisms of extremely small bodies. He described how nature delights in transmutation of material: 'Eggs grow from insensible Magnitudes, and change into Animals; Tadpoles into Frogs; and Worms into Flies. All Birds, Beasts and Fishes, Insects, Trees, and other Vegetables, with their several parts, grow out of Water and watry Tinctures and Salts, and by Putrefaction return again into watry Substances. And Water standing a few Days in the open Air, yields a Tincture, which ... by standing longer yields a Sediment and a Spirit ... And among such various and strange Transmutations, why may not Nature change Bodies into Light, and Light into Bodies?'[47] The relationship between light and mass would be described only in 1905 by Albert Einstein: $E = mc^2$ (energy, such as particles of light, is converted into mass at a fixed ratio). Electrons, both matter and energy, function to bond material together at the molecular level, impart energy in both chemical and electrical processes, and transform materials by their activity, sometimes giving off both heat and light in the reaction. In the time of Newton and Willis, such speculations hinted at an astonishing transformation in concepts of how the entire universe works, but without experimental confirmation, atomism would remain an acceptable hypothesis until the nineteenth century. Most significant for the man-machine figure was the realization that there might be only material substance in the world, modified to function in different ways. The danger of such a speculation, of course, was that it could be interpreted as removing a separate soul – and God himself – from the material world. No wonder then, Newton's cautious queries were phrased as questions rather than statements. Descartes and Willis, both orthodox men themselves, had both stopped short of describing the human being wholly as an automaton: animals, according to Descartes, were true automata while humans were mechanisms with souls; for Willis, the soul of animals was equivalent to the vital and sensitive soul of humans, which infused the body along with an immortal and non-corporeal soul.

Locke would later draw on Willis's mechanistic portrayal in his *Essay Concerning Human Understanding*, but the first sustained explication of the man-machine as a wholly material mechanism was La Mettrie's *L'homme machine*. La Mettrie's infamous claim that 'man is a machine and that there is in the whole universe only one diversely modified substance' united mind with body, interpreted matter as both sensitive and active, and denied the existence of spirit separate from the body's gross and corruptible matter. The rational mind or soul, however, responsible for knowledge, judgment, morality, and sensibility, was the most important characteristic separating man from beast. If it could be reduced to a function of mere particles in motion, then not only were traditional hierarchies, both secular and spiritual, threatened but so also was the very nature of human identity. 'Now that it has been clearly demonstrated, against the Cartesians, the Stahlians, the Malebranchians, and the theologians – who are hardly worthy of being mentioned here – that matter moves by itself not only when organised, as in a whole heart for example, but even when that organisation is destroyed,' wrote the scandalous La Mettrie, human curiosity would like to discover how a body finds itself adorned with the faculties of feeling and thought. We will never be able to determine the nature of matter, he concluded, nor of motion: we can grant only that everything depends upon the variety of organization of matter. Man is to the ape, he concluded, 'what Huygens's planetary clock is to one of Julien Leroy's watches. If it took more instruments, more cogs, more springs to show the movement of the planets than to show or tell the time, if it took Vaucanson more artistry to make his flautist than his duck, he would have needed even more to make a speaking machine, which can no longer be considered impossible, particularly at the hands of a new Prometheus.'[48] This was the disturbing and godless conclusion that the earlier philosophers had been so careful to circumvent in their writings. If there were only one diversely modified substance in the world – matter alone rather than matter and spirit – then human beings were merely more complicated mechanisms than either automata or animals. Taken to its most outrageous extreme, mechanism ultimately suggested that the soul or spirit was nothing more than an elaborate fiction.

To reiterate, *mechanical* simultaneously implied a lower order of humanity and the mathematical description and determination of material bodies in motion. The new mechanical philosophy, equally applicable to the motions of engines built by men and those built by nature or God, set up philosophical expressions of the mechanisms of human bodies, personalities, and consciousness that would make it increasingly difficult to accommodate an immaterial soul. The image of science would from here on walk a fine line between perceptions of progress and godlessness or base materialism, although the reality was, of course, much more complex. The variants of this trope throughout the period accentuate most emphatically that the Cartesian identity of automaton-with-soul did not establish supremacy over competing theories of human motion and consciousness. Indeed, the philosophy of particles from very early on seemed to exclude the immaterial spirit from the increasingly embodied soul.

Defining the Man-Machine II: From Aether to Ethernet?

The subtle aetherial fluid predominantly accepted as the medium containing the particles or atoms that made up all material hovered ambiguously between being material and immaterial; matter itself hovered ambiguously between vitality and passivity. Aether, once imagined as the upper region of space beyond the sphere of the moon, the stuff of stars and planets and their celestial spheres, was traditionally a substance different from terrestrial substances. As the element breathed by the gods, aether was a 'diviner air' (*Oxford English Dictionary*). Influenced by the Platonic idea of the soul, aether was presumed by some early philosophers to be the constituent substance of the soul, and it became the subtle fluid in the nerves giving the body motion and vitality. In natural philosophy it would become an elastic and subtle substance permeating all of space between the planets, between particles of air, and between the elementary particles of all matter. Aether was the medium through which Newton speculated that waves of light travelled, and it would also be associated later with the 'electrical fluid.' Descartes supposed the aether was very dense; Newton believed that it was very rare and tenuous, and speculated in the *Opticks* that it might contain particles exceedingly smaller than those of air or even of light. 'I do not know what this *Aether* is,' wrote Newton in Query 21, but its exceedingly small particles might recede from one another to make the subtle fluid both elastic and extremely rarefied. 'Is not vision perform'd chiefly by the Vibrations of this Medium, excited in the bottom of the Eye by the Rays of Light, and propagated through the solid, pellucid and uniform Capillamenta of the optick Nerves into the place of sensation?' he asked. Is not hearing performed by a similar vibration? Newton certainly did not advocate the image of the human as a purely mechanical machine. The long-range results of searching for the activities of such mysterious natural phenomena as light on the eye, in human perception, and through glass prisms, however, would be a mathematical conception of the unity of nature and the possibility of the mechanisms of a quasi-material substance permeating the world. The unsolved question throughout this period was what the universal element was that held bodies together, gave them life, and made them move and – dangerously – in the case of human bodies, perceive, feel, or think. 'Is not Animal Motion perform'd by the Vibrations of this Medium, excited in the Brain by the power of the Will, and propagated from thence through the solid, pellucid and uniform Capillamenta of the Nerves into the Muscles, for contracting and dilating them?'[49] Newton asked. In the General Scholium of *Principia* he elaborated:

And now we might add something concerning a certain most subtle spirit which pervades and lies hid in all gross bodies; by the force and action of which spirit the particles of bodies attract one another at near distances, and cohere, if contiguous; and electric bodies operate to greater distances, as well repelling as attracting the neighbouring corpuscles; and light is emitted, reflected, refracted, inflected, and heats bodies; and all sensation is excited, and the members of animal bodies move at the command of the will, namely, by the vibrations of this spirit, mutually propagated along the solid filaments of the nerves, from the outward organs of sense to the brain, and from the brain into the muscles. But these are things that cannot be explained in few words, nor are we furnished with that sufficiency of experiments which is required to an accurate determination and demonstration of the laws by which this electric and elastic spirit operates.[50]

These speculations would formulate a substantial foundation for subsequent physiological inquiries into human will, action, and motion. Whether the cause of the motions of the particles in matter was mechanical (attraction and repulsion of the particles caused by the action of the corporeal or incorporeal medium they were in) or non-mechanical (action at a distance, where the particles are attracted or repulsed by a force inherent in themselves) provoked considerable debate.

To suggest that the substance that makes humans live and move and think was solely these subtle bodies acting in predictable mechanical patterns was a contentious argument at the very least and, in certain political climates, quite dangerous. To materialize the soul ultimately would be to question established doctrines of faith, spirituality, and government of both body and body politic. And, despite the attempts of early philosophers to rationalize the mechanical bodies of humans with immortal spirit, the period is characterized by certain growing doubts about the relationships of matter to spirit, of God's artifice to that of humans. 'Our original ideas ... all might be reduced to these very few primary and original ones, viz. extension, solidity, mobility, or the power of being moved,' Locke suggested in 1690, stating in stronger terms Willis's representation of human knowledge as matter in motion.[51]

Matter acting upon matter 'can communicate nothing but Motion,'[52] Bentley countered in 1693, and he concluded that there had to be an incorporeal substance within human bodies that was created by an immaterial, intelligent, and eternal Being. This line of reasoning was why action at a distance was for some philosophers an appealing alternative to pure mechanism in describing the functions of both human and celestial bodies: gravity itself seemed to provide irrefutable evidence of an immaterial force throughout nature that could only be proof of God's existence. As Bentley argued in the fourth of his Boyle Lectures, gravity, which is the basis of all mechanism, may be proven not to be mechanical itself, but rather 'the immediate *Fiat* and Finger of God, and the Execution of the Divine Law.' Furthermore, if compound bodies cannot exist without gravity, and if gravity originates from 'a Divine Power and Energy,' then the atheists' arguments will be for naught, 'though they should be able to explain all the particular Effects, even the Origination of Animals, by mechanical principles,' he concluded with conviction.[53] Bentley's position in this matter, as in others more infamous, was not uncontroversial but it does summarize a key question posed by the man-machine or mechanistic hypotheses.

The question of whether matter could move without some other force communicating that motion – and whether the mysterious entity that caused matter to move was itself material – were questions Newton himself addressed in his letter to Bentley (25 February 1692) on the subject, although he did not care to answer the latter:

> It is inconceivable, that inanimate brute matter should, without the media-
> tion of something else, which is not material, operate upon, and affect other
> matter, without mutual contact; as it must do, if gravitation ... be essential
> and inherent in it ... That gravity should be innate, inherent, and essential
> to matter, so that one body may act upon another at a distance ... is to me so
> great an absurdity, that I believe no man who has in philosophical matters a
> competent faculty of thinking, can ever fall into it. Gravity must be caused
> by an agent acting constantly according to certain laws; but whether this
> agent be material or immaterial, I have left to the consideration of my
> readers.[54]

To reduce the motions of the material world – and thus human thought, judgment, volition, and ultimately will – to mechanism was problematic, to say the least. In its most negative interpretation, the organism as a mechanical entity threatens any sense of purpose in human lives; it undermines concepts of free will, autonomy, and moral-ity. If the body's motions can be explained as mechanical reactions to external circumstances, if inert mass can form the will to perform an action because of a predictable series of motions in its material particles, where do God and moral judgment fit in? Is the human being a mere automaton, purely material, acting and reacting based upon mechanical laws, or a spiritual creature with an immortal and rational soul? To use a common metaphor in eighteenth-century philosophy: if the body was a musical instrument, and the rational self or soul was making the music, what caused the musician to play as he did? Or, in the moral terms of motion, what moves us and how are we moved?

Many commentators reiterated the argument that the primary cause of certain forces was incomprehensible and therefore impossible to explain by mechanical philosophy. In 1711 Joseph Addison postulated that the care and 'natural Love' that even animal parents display for their young is a divine energy as innate and inexplicable as gravity:

> There is not, in my Opinion, any thing more mysterious in Nature than this
> Instinct in Animals, which thus rises above Reason, and falls infinitely short
> of it. It cannot be accounted for by any Properties in Matter, and at the same

time works after so odd a manner, that one cannot think it the Faculty of an intellectual Being. For my own Part, I look upon it as upon the Principle of Gravitation in Bodies, which is not to be explained by any known Qualities inherent in the Bodies themselves, nor from any Laws of Mechanism, but, according to the best Notions of the greatest Philosophers, is an immediate Impression from the first Mover, and the Divine Energy acting in the Creatures.[55]

A common conundrum, however, was even if motion was inherent in matter, and placed there by God, how was nature – and humanity – governed? In his first reply to Gottfried Wilhelm Leibniz (26 November 1715), for example, Samuel Clarke wrote: 'by the same reason that a philosopher can represent all things going on from the beginning of the creation, without any government or interposition of providence; a sceptic will easily argue still farther backwards, and suppose that things have from eternity gone on ... without any true creation or author at all, but only what such arguers call all-wise and eternal nature.' Furthermore: 'And as those men, who pretend that in an earthly government things may go on perfectly well without the king himself ordering or disposing of any thing, may reasonably be suspected that they would like very well to set the king aside.' As Clarke wrote in his fifth reply, whoever might think that the universe does not constantly need 'God's actual government' and that the laws of mechanism alone would allow it to continue in perpetual motion 'in effect tend to exclude God out of the world.'[56]

Others tried to incorporate the new philosophy into their spiritual lives, especially as the new philosophy seemed more and more irrefutable. 'The BODY OF MAN [is] ... a *Machine* of a most astonishing Workmanship and Contrivance!' the Massachusetts minister Cotton Mather would exclaim. 'My God, I will praise Thee, for I am strangely and wonderfully made!' But, he wondered, 'is it possible for me to consider this BODY as any other than a Temple of GOD!' The need for moral government became all the more pressing if humans were mere machines: 'Matter keeps [God's] Laws; but, O my Soul, wilt thou break 'em!' he lamented.[57] As aether became solidified in the popular culture, authors of fiction and poetry found new ways to revivify human spirit. As we will see in Chapter 4, one important response was the cult of sensibility, which positioned a new version of human sensitivity and morality in the material structures of the nerves: the alternative was simply too unpalatable.

Many readers will be familiar with the scabrous satire of contemporary government in *The History and Adventures of an Atom* (1769). It is attributed to Tobias Smollett, whose *Expedition of Humphry Clinker* (1771) is a defining text of eighteenth-century sensibility. The former book presented a far more cynical version of human spirit and government based on the then-new particulate theories than did the later novel. Here, an atom residing in the pineal gland (Descartes' seat of the soul) of one haberdasher and author of St Giles, Nathaniel Peacock, speaks aloud one day. 'What thou hearest is within thee – is part of thyself,' a shrill voice explains. 'I am one of those atoms, or constituent particles of matter, which can neither be annihilated, divided, nor impaired: the different arrangements of us atoms compose all the variety of objects and essences which nature exhibits, or art can obtain. Of the same shape, substance, and quality, are the component particles, that harden in rock, and flow in water; that blacken in the negro, and brighten in the diamond; that exhale from a rose, and steam from a dunghill.' The atom describes its journey from the toenail of an amoral and petulant king to the tail of a willingly abused underling, eventually being discharged in a scorbutic dysentery, taken up in a heap of soil to manure a garden, raised to vegetation and eaten in a salad, and so on until finally, residing in the animal spirits of Nathaniel Peacock's father, the sentient particle ends up in the seminal animalcule that eventually forms Nathaniel himself. The particle then recounts the remarkable history of its residence in the bodies of governmental officials in the fictionalized land of Japan (a little-disguised analogy for Great Britain). The atom resides in the toenail of the 'Japonese' king Got-hama-baba (George II), until it is dislodged by a violent kick upon the posterior of the king's Cuboy (chief-minister) Fika-kaka (Thomas Pellham-Holles, the duke of Newcastle). The atom finds itself 'divorced from the great toe of the sovereign, and lodged near the great gut of his minister' for a lengthy time until it is 'discharged in the perspiratory vapour from the perinæum of the Cuboy, and sucked into the lungs of Mura-clami' (a lawyer). Smollett unremittingly calls to the reader's attention not only the stupidity, ignorance, and avarice of the governors of his fictional Japan, but also notes the absurdity of philosophical reductions taken to their extreme:

[Fika-kaka] therefore assembled all those ... who had the reputation of being great philosophers and metaphysicians, in order to hear their opinions concerning the nature of the soul. The first reverend sage who delivered himself on this mysterious subject, having stroked his grey beard, and hemmed thrice with great solemnity, declared that the soul was an animal; a second pronounced it to be the number *three*, or proportion; a third contended for the number *seven*, or harmony; a fourth defined the soul the *universe*; a fifth affirmed it was a mixture of elements; a sixth asserted it was composed of *fire*; a seventh opined it was formed of *water*; an eighth called it an *essence*; a ninth, an idea; a tenth stickled for *substance without extension*; an eleventh, for *extension without substance*; a twelfth cried it was an *accident*; a thirteenth called it a *reflecting mirrour*; a fourteenth, *the image reflected*; a fifteenth insisted upon its being a *tune*; a sixteenth believed it was the instrument that played the tune; a seventeenth undertook to prove it was *material*; an eighteenth exclaimed it was *immaterial*; a nineteenth allowed it was *something*; and a twentieth swore it was *nothing*.

Fika-kaka is so confused he falls to the floor, deprived of sense and motion. When he recovers, he sends for his counsellor in great fear, for there seems to be no reason to believe in an immortal soul: those philosophers are 'good for nothing but to kiss my a—se;—a parcel of ignorant asses! – Pox on their philosophy! Instead of demonstrating the immortality of the soul, they have plainly proved the soul is a chimaera, a will o'the wisp, a bubble, a term, a word, a nothing! – My dear Mura! prove but that I have a soul, and I shall be contented to be damned to all eternity!'[58]

To reduce consciousness, and possibly the soul itself, to material particles comprised a complex shift in metaphorical structures. Smollett's analogy is one glimpse of an evolving and many-sided metaphorical tradition: depending on the author's stance, philosophical explanations of human identity, will, and rationality as material mechanisms either undermined or celebrated the human as rational and moral being. It either confirmed or destroyed government of the individual body and of the social body based on a religious and moral authority. The premise is clear in Smollett's merciless exaggeration. On the one hand, his work clearly satirizes a particular series of political events; on the other, surely, he decries a purely mechanical version of human spirit and moral government. If human rationality and vitality are merely the jostlings of base material particles, then generation, ensoulment, and judgment, Smollett implies, are equivalent to shit.

Smollett's satire is only one of many thorny articulations of human identity as mechanism during the Enlightenment, and it encapsulates a number of issues that defined resistance to the man-machine. The obvious criticism of mechanistic theories was that the human was a sensitive, thinking, feeling being, a spiritual and moral being: the human being had a soul and generated new human beings with souls. How could all these traits be accounted for if human spirit were made of the same stuff, as Smollett so repellently represented it, as fart particles? Theories of vitalism or animism, based on the premise that living bodies possess an active force different from that possessed by non-living matter, attempted to redress what seemed to be a negation of human spirit. Various philosophers such as Georg Ernst Stahl (1659?–1734), Théophile Bordeu (1722–1776), and Marie François Xavier Bichat (1771–1802), among many others throughout the eighteenth century, responded to mechanism by arguing that the living body possesses some unique property, a soul, force, or principle that transcends the realm of base inert matter. In some circles, vitalism and sensibility replaced mechanism as a ruling concept, but to overemphasize this as an overarching philosophical change distorts the fact that the philosophy of mechanics never disappeared. It has remained a foundational cultural influence in the physical and medical sciences, and in biomechanics and nanotechnology today.

For example, in 'From *Homme machine* to *Homme sensible*: Changing Eighteenth-Century Models of Man's Image,' Sergio Moravia has suggested, that iatromechanism was replaced by Bordeu's representation of life as sensitivity: 'For the iatromechanists life was movement; for Bordeu, life is sensitivity. For the iatromechanists, man was a mechanical apparatus; for Bordeu, man is an organism.' After Bordeu, Moravia argues, the man-machine was all but gone: 'As for the living individual, it is no longer (not even metaphorically) a *machine*, but an *être sensible*. It is an organic being made up of flesh, nerves, and muscles; possessing dynamic forces and impulses; and characterized by processes that have nothing to do with the working of a machine.'[59] The desire to portray Bordeu's theories as indicative of 'changing eighteenth-century models of man's image' betrays a problem of emphasis. Although individual philosophies may have invoked sensibility, irritability, or vitality as a specific contrast to mechanism, *vital* and *mechanical* are not mutually exclusive terms and both are used, frequently by the same author, to describe human physiology throughout the 1700s. Theodore M. Brown, in 'From Mechanism to Vitalism in Eighteenth-Century English Physiology,' has also argued that 'one of the most striking features of English physiology in the forty years between 1730 and 1770 was the dramatic, indeed precipitous decline of varieties of mechanism and the rapid rise to pre-eminence of alternate varieties of vitalism.' This he summarises, citing Robert

Schofield's *Mechanism and Materialism,* as the change in physiology 'from a mechanistic style to one in which fluid, spirits, ether, electricity, fire, and irritable or sensitive fibers were the principal explanatory entities.'[60] Forms of vitalism and animism characterized by irritability or sensitivity were indeed emphasized by Bordeu at Montpellier, Stahl in Halle, William Cullen in Edinburgh,[61] or Albrecht von Haller in Göttingen, and by all their followers. Mechanism nonetheless played a strong role in physiology, and this is particularly evident in Newtonian English physiology. Oliver Goldsmith matter-of-factly describes the sensitive machine in *An History of the Earth, and Animated Nature* (1774): 'It is well known in mechanics, that the finest and most complicated instruments are the most easily put out of order, and the most difficultly set right; the same also obtains in the animal machine. Man, the most complicated machine of all others, whose nerves are more numerous, and powers of action more various, is most easily destroyed.'[62]

The chemist and physician Thomas Beddoes similarly remarks in passing that the 'human machine, at all times frail, consists at first of parts peculiarly delicate and having an adjustment peculiarly liable to disorder,' in his 1797 publication entitled *A Lecture Introductory to a Course of Popular Instruction on the Constitution and Management of the Human Body.*[63] These were not atoms-as-billiard-balls or nerves-as-hydraulics mechanistic interpretations of human vitality, of course, but to imagine that the human-machine and mechanistic explanations for physiological functions were ousted from medical and popular discourse and replaced with aether, spirits, electricity, and so on ignores a terminology of material mechanism that includes chemical-electrical processes. It also ignores the tacit acceptance implied by the not uncommon use of the terms machine or engine to describe human bodies. The terminology of mechanism did not exclude vital fluids or electrical fire. Indeed, the self-taught natural philosopher Richard Lovett (1692–1780) had earlier suggested in his *Philosophical Essays* (1766) that 'electrical fluid' is the 'active mechanical Agent in Nature' that gives motion to both the world and to human bodies. 'In a Word,' he claims, 'this is the universal subtile Æther' of Newton. That this subtle material is active and vital does not preclude it from being mechanical. Lovett specifically distinguishes gross manual mechanics from the mathematical principles of 'accurate Mechanics.' Citing Newton, he explains: '"To practical Mechanics all the manual Arts belong, from which Mechanics took its name. But as artificers do not work with perfect accuracy, it comes to pass that Mechanics is so distinguish'd from *Geometry,* that what is perfectly accurate is call'd Geometrical and what is less so is call'd Mechanical: But the errors are not in the Art but in the Artificers." Hence Geometry, i.e. Mathematics, is mechanical in the strictest Sense, and *Mathematical* Philosophy and *Mechanical* Philosophy, are synonymous.'[64] The language of 'mechanical' philosophy would evolve into different mathematical expressions describing the material and forces of body structures, but fundamentally mechanism would not be eradicated from the body even as a vital, sensible, and irritable system. Moreover, the problematic issues regarding what the vital fluid or spirit was were hardly settled. Although Lovett's earlier publication, *The Subtil Medium Prov'd, or, That Wonderful Power of Nature ... Call'd Sometimes Aether, but Oftener Elementary Fire, Verify'd* (1756) had been heavily criticized, his *Philosophical Essays* – quoting extensively from Isaac Newton, Benjamin Franklin, and the *Philosophical Transactions* – was judged to be 'a work of genius' by the editor of the *Critical Review,* writing in October 1766.

There has been a tendency, however, to emphasize 'the triumph of vitalism over mechanism'[65] or the 'widespread rejection of mechanist physics'[66] as a new expression of Enlightenment sentiment or sensibility. More recently Peter Hanns Reill has presented an excellent summary of eighteenth-century mechanism and vitalism. His aim, however, to refute the critiques of the Enlightenment's hidden agenda of universalizing reason and totalitarianism such as expressed by Max Horkheimer, Theodor Adorno, Michel Foucault, or more recently Stephen Toulmin in *Cosmopolis: The Hidden Agenda of Modernity,* seems to result in the somewhat overgeneralized claim that 'the late Enlightenment' in contrast to the earlier period endeavoured to 'create a place for the soul and the individual, to avoid the rush to reductionism, and to recognize the epistemological value of ambiguity and paradox.'[67] While, as Reill argues, eminent vitalists such as Stahl and Johann Friedrich Blumenbach (1752–1840) in Germany, John Hunter in Great Britain, or Paul-Joseph Barthez (1734–1806) in France rejected mechanism, nonetheless mechanick philosophers had their own versions of vital forces. The tendency to claim that mechanistic philosophies supplanted entities such as soul and individuality, which were then reintroduced by vitalists simply does not take into account consistent speculations that matter in addition to having mechanical properties had vital or active properties, and indeed that vital-mechanical processes of matter could be seen as proof of God's presence and greatness: an unnecessarily narrow notion of mechanism seems to be distorting present-day assumptions of its inevitable demise.

Describing the forces of vitality, irritability, and sensibility was a predominant concern for natural philosophers throughout the eighteenth century, but to suggest that vitalism replaced mechanism as a physiological paradigm is to ignore the longevity of mechanism to describe physical processes of limbs, muscles, and molecules.[68] Rather than an *être sensible*, the period might be said to have produced an understanding of the human as *machine sensible*. Moreover, the mechanisms of bodies and machines even in early modernity are not completely distinct, as Moravia implies in the extracts cited above: the workings of machines were progressively characterized by dynamic forces and impulses – particularly in self-regulating furnaces, steam engines, and electrical mechanisms. Electrical demonstration machines incorporating the human body as part of the dynamic mechanism were popular throughout the eighteenth century.[69] Electricity itself was defined in mechanical terms until it was better understood, and it was defined in terms of animate motion. Indeed, in the 1740s the notion of a vitality inherent in the body's mechanical structure was given new emphasis by discoveries and experiments in electricity. Newton's publications had a lasting influence on speculations, examinations, and proofs of the mechanisms of active and vital bodies.

William Watson cautioned against such an interpretation in the 1746 publication of the *Philosophical Transactions*: 'Does not the Power we are now Masters of, of seeing the Separation of Fire from Bodies by Motion, and of seeing it restored to them again, and even after that Motion has ceased, cause us rather to incline to the Opinions of *Homberg, Lemery the younger, s'Gravesand* and *Boerhaave,* who held Fire to be an Original, a distinct Principle, formed by the Creator himself, than to those of our illustrious Countrymen, *Bacon, Boyle,* and *Newton,* who conceived it to be mechanically producible from other Bodies?' And further: 'Must we not be very cautious, how we connect the elementary Fire, which we see issue from a Man, with the vital Flame and *Calidum innatum* of the Ancients; when we find, that as much of this Fire is producible from a dead Animal as a living one, if both are equally replete with Fluids?'[70] Nevertheless, the article following Watson's in this volume of the journal is Browne Langrish's 'Croonean Lecture on Muscular Motion,' in which he concurs with 'the learned Dr *Mead*' that 'no Regard ought to be had to the immechanical Notions of those Authors, who imagine that there is no such thing as a nervous Fluid in an animal Body; and that muscular Motion and Sensation are performed only by the Vibrations of the Fibres of the Nerves.' Rather, Langrish notes that the 'surprising Discoveries' made of late with regard to electrical fluid travelling instantaneously through the bodies of men 'communicating with each other by a Cane, Sword' or other conductor 'do in some measure give us an Idea of the great Subtilty and Velocity of the nervous Fluid.' The 'electric Matter,' he concludes, 'may be very similar to the Motion and Action of the nervous *Æther.*'[71]

What had been missing from Descartes' body-machine was a credible explanation for the complex homeostatic motions of what was then the invisible world. Proponents of the man-machine or materialist or mechanist philosophy attempted to consolidate rather than separate human vitality from human mechanism. As Aram Vartanian has remarked, 'the psychophysiological doctrines of La Mettrie and Diderot alike would seem, in fact, to have sprung from a common desire to bridge the conventional gap between mechanism and vitalism by assuming that deeper investigation of the phenomena peculiar to each of those conceptions would, in the end, wipe out the dichotomy erected arbitrarily between them.'[72] Elizabeth Haigh has similarly commented that 'Diderot himself came to understand sensibility much as La Mettrie had done. He saw it as a property of matter which is released under certain circumstances of organization. Matter then can be sentient. The age-old mind-body problem had effectively ceased to exist.' Haigh remarks on the irony that 'irritability and sensibility as forces of living matter were being used by mechanists and vitalists alike to support their particular viewpoints. Though the dispute between the two groups continued to be expressed enthusiastically, the difference between their basic positions was persistently fading.'[73] Although sensibility is seen as a dominant influence in many literary studies of the Enlightenment, from a medical-scientific point of view, however, vital forces became part of a mechanistic interpretation of human physicality. For example, in his 1787 *A Philosophical and Medical Sketch of the Natural History of the Human Body and Mind*, the physician James Makittrick Adair (1728–1802) stated simply that the most probable medium through which nervous communication conveys an animating principle to the body and impressions of sensation to the brain is 'electric fluid ... as its rapid movements render it peculiarly fit for that almost instantaneous communication which takes place between the brain and the other organs.'[74] Once the 'mechanical Agent' of the electrical fluid was discovered to induce motion in both man and machine, the image of the human as animated or sensible machine would become even more credible: this is partly why years later Mary

Shelley's *Frankenstein* was – and has remained – such a powerful cultural image.

As Lovett explains in his *Philosophical Essays*: 'Before a subtile Medium or Æther was discovered, Motion was given up, along with other immechanical Causes, to be performed by the immediate influence of the first Cause. – Before such Discovery, we seemed very naturally to conclude that all Matter was inert, whereas we now find we were greatly mistaken, and that there are active Particles of Matter existing as well as passive ... But had we attended to Sir *Isaac*'s Principia, we might have learned better.' The 'subtile Fluid' that Newton had discovered, he goes on to say, 'was contemned and rejected as an imaginary Fluid only' but when 'they had set aside his mechanical Agent, the Matter was so far from being mended, that it was thenceforward rendered so unintelligible, as that Motion itself must either be the Cause of its own Motion, or else that the Power and influence of the Deity is more interested in it than in most other Things.' Lovett specifically argued that this active principle was both mechanical and material: that 'this Electric matter is not subject to the same rules or laws of other Matter is most certainly a Concession that will be very readily granted, and therefore, the common Postulatum, that Matter differs from Matter in form only, having all other properties in common, must consequently be absolutely false. For, let me ask, Can the primary Corpuscles of all Matter differ from each other in figure or form only, and yet not be subject to the same Laws?'[75] Lovett had been severely criticized as being ill-educated and ignorant of the subject, yet not five years later, the eminent Henry Cavendish also attempted to explain electrical fluid as 'matter of a contrary kind to other matter' using a mechanics of attraction and repulsion.[76]

The understanding of mechanics as the explication of natural processes made possible the inclusion of the human body in works that concentrated not upon physiology or experimental chemistry, but upon the philosophy of the practical functioning and construction of machines. *The Principles of Mechanics* (1758) by the mathematician William Emerson (1701–1782) is one such example. Emerson introduces his book on applied mechanics with a lengthy defence of its practical use. 'Upon mechanics,' he explains, 'is also founded the *Newtonian,* or only true philosophy in the world.' Mechanical philosophy describes not only the motions of visible bodies of the universe such as planets, comets and the moon but, according to Emerson, also those invisible forces belonging to small particles of matter, of which we are still ignorant and upon which depend both human will and thought, processes of death and life, light, cold, or heat, and chemical reactions:

> There are also certain forces belonging to the small particles of matter, which we are still ignorant of; by which they are either impelled towards one another, and cohere in regular figures; or are repelled and so recede from each other. For the particles of different sorts of bodies have different laws; since the small particles of some bodies attract one another, whilst those of other sorts repel each other; and that by forces almost infinitely various. Upon these forces the cohesion, solidity, and fluidity of bodies depend. The nature of elasticity, electricity, and magnetism. Upon these also depend the principles of fermentation, putrefaction, generation, vegetation, and dissolution of bodies; digestion and secretion in animal bodies; the motion of the blood and fluids in animals, and the moving of the members by the command of the will; the exciting sensations in the mind; the emission, reflection, refraction, and inflection of light; freezing by cold, burning by fire; all operations in chemistry, &c.

Emerson concludes that 'by mechanics we come to understand the motions of the parts of an animal body; the use of the nerves, muscles, bones, joints, and vessels. All which have been made so plain, as proves an animal body to be nothing but a mechanical engine.' In a later section Emerson includes the human body among descriptions of other compound engines such as a cheese press, a spinning wheel, 'a machine to raise a weight by the force of the running water,' a cart or carriage, a wagon, the sails of a windmill, a 'common sucking pump,' an artificial kite, a wooden bridge, a syphon or a crane, and barometers, to name only a few.[77]

Although Emerson's influence as a Newtonian mathematician was significant, his influence on Enlightenment medicine and physiology would not have been far reaching: according to his biographer W. Bowe, he did not wish to be admitted to the Royal Society 'because (he said) it was a d—n'd hard thing that a man should burn so many farthing candles as he had done, and then have to pay so much a year for the honor of F.R.S. after his name. D—n them and their F.R.S. too.'[78] If Emerson was somewhat marginal in the ongoing dialogue between the influential members of the Royal Society, his work is important insofar as it represents a historical record of the physiology of the man-machine as a subcategory of practical engineering. Emerson's text was absorbed fully with practical applications of Newtonian mechanics, and as such it emphasized the treatment of the body as a calculable working device. The mathematical description of the motions of bodies and machines is precisely the same language. The human body is just another engine, and for Emerson there is no hierarchy implied about its perfection – or its purpose – above either animals or human-made machines. The human body simply appears as number 38 and 39 of 47 complex engines (see figure 15):

Ex. XXXVII.

BC, CG are two *bones* of an animal, moveable about the joint *FK*, by help of the muscle *KD*. The joints of animals are either spherical or circular, and the cavity they move in, is accordingly either spherical or circular. And the center of motion is in the center of the sphere or circle, as at *C*. Let *W* be a weight hanging at *B*, and draw *CP*, *CK* perpendicular to *BW*, *KD*. Then if the weight *W* be suspended by the strength of the muscle *KD*; it will be as *CK* : *PC* : : *W* : tension of the muscle *KD*.

The bone *BC* is moved about the joint *FK*, by the strength of the muscle *KD*. For when the muscle is contracted, the point *K* is moved towards *D*; and the end *B* towards *E*, about the immoveable center of motion *C*.

... It is evident that all animal bodies are machines. For what are the bones but leavers, moved by a certain power placed in the muscles, which act as so many ropes, pulling at the bones, and moving them about the joints? Every joint representing the fulcrum or center of motion. What are all the vessels but tubes which contain fluids of different sorts, destined for the use or motion of the several parts of the machine? And which by opening or shutting certain valves; let out or retain their contents as occasion requires; or convey them to distant places, by other tubes communicating therewith. And therefore all these motions of an animal body, are subject to the general laws of mechanics.[79]

That is to say, there *were* certain vitalist or sensibilist reactions to the Cartesian image of the human as mechanical engine in the period, but the human-machine was never displaced by vitalism, animism, or sensibility. What we see, instead, is a gradual convergence of the scientific language used to describe the invisible and active structures and processes of nature, human, and machine to the point where inorganic matter and much of organic life can be expressed as equations of the subsets of mass and energy. Emerson's work is one such example, but the mathematical abstraction and description of powers and forces, as well as the increasing understanding of chemistry and electricity as 'mechanical' forces, was the beginning of that convergence of language and sign that we now take for granted where abstract symbols can describe all of nature.

At the end of the century *A System of Familiar Philosophy* (1799) by the popular scientific lecturer Adam Walker (1730/1–1821) also celebrated a mechanistic approach that derived from Newtonian philosophy. Walker begins his lectures with the following premise: 'Attraction and repulsion are the great acting principles of the universe. By attraction, I mean that tendency which the atoms of matter have to unite and form bodies; and by repulsion, the reaction of fire, in its combined, active, and latent state; or in its character of electricity, heat, or light. The experiments through the course of this work, will principally tend to the establishment of this persuasion.' He dismissed the need for any distinction between active and passive matter: 'The cavillers at the Newtonian doctrine, quarrel with the word attraction; and much waste paper has belumbered the world, in disputing whether it is a cause or an effect. It is natural to suppose that one body cannot act upon another, but by any means of some interposing medium; but we have never been able to discover what this medium is. We know that the particles of matter are held together, by some sort of influence they have upon each other; and why not call it attraction? What is it to us, whether it is inherent in matter, or a quality imparted to it by the Author of nature?' For Walker, *attraction* and *repulsion* are terms applicable to all matter: the terminology to describe the forces and energies within both living and inert bodies had begun to recognize that the origin of these energies may be either God-given or naturally inherent without consequence to the functioning of the mechanism itself. Moreover, not only does Walker explicate the arm as a lever among many other mechanical devices (see figure 16), he describes Boulton and Watt's steam engine – known for its innovation of a feedback device, the centrifugal governor, which reduces steam supply if the engine begins to run too fast – and his own improved steam engine, as approaching organic bodies: 'These kind of machines approach nearer to the animation and powers of animal mechanism, than any yet invented by man,' he suggested (see figure 19).[80] Whereas at the end of the seventeenth century Willis's steam-producing alembic had seemed to provide an ideal correspondence to the animal mechanism in his work on fermentation, at the end of the eighteenth the steam engine with its governor to maintain equilibrium provided a new means of understanding animate motion and homeostasis through what we now call feedback.

On animal electricity, Walker demonstrated that electricity applied to nerves of various animals results in muscular convulsion, that animals 'almost dead' can be revived using electricity, and that electrical current applied to points in the mouth results in 'a flash of lightning ... seen by the eyes, even though the person be in the dark, or hoodwinked.' Walker, who incidentally lectured to the poet Percy Shelley at Eton, postulated a connection between electricity and life as had been illustrated by such galvanic experiments: 'Can any doubt remain that the wonderful agent is a prime instrument in muscular motion?' he asks. 'Fire is the first mover in the animal machine,' he states unequivocally in the lecture on medical electricity. Moreover, the mechanism of electrical fluid was presented to his audience of enthusiastic amateur philosophers as a world soul animating all material: 'Its power of exciting muscular motion in apparently dead animals, as well as of increasing the growth, invigorating the stamina, and reviving diseased vegetation, prove its relationship or affinity to the *living principle*. Though Proteus-like, it eludes our grasp; plays with our curiosity; tempts enquiry by fallacious appearances, and attacks our weakness under so many perplexing subtilties; yet it is impossible not to believe it the soul of the material world, and the paragon of elements!'[81] Mechanism, electricity, and vitality: all were aspects of the man-machine as a material being made up of and animated by material particles.

Indeed, in an examination of the Croonian lectures on muscular motion, which were read annually before the Royal Society from 1738 onward, one can observe a diminishing interest in explaining the action or existence of the soul alongside a growing interest in explaining pure mechanism. By pure mechanism I mean the attempt to explain muscular motion without commentary on the soul or will, and focusing exclusively on the physical mechanism as a cause in itself. As an illustrative example, James Parsons's 1744 preface notes specifically:

What the *Soul* is, or in what Manner she makes her Impulse on those Parts of Animals that are the immediate Instruments of Motion, we dare not attempt to guess; these being wrapp'd up among those Secrets only known to HIM that order'd all Things: But, as the Bodies of Animals are mechanical, and therefore naturally fall within the Sphere of our Understanding, we may make some Attempts toward explaining the several *Phænomena* that belong to it; and therefore we can only consider how its Organs are actuated, and not what is the Cause of their Motion; and must take it for granted, that the *Soul* makes her Impulse on the Organs, and then endeavour to shew the

Nature of the several Consequences of that Impulse, as far as it relates to the Motion of the Muscles.[82]

Sixty years later, Anthony Carlisle introduced his lecture in 1804 with the observations that causes can be known and that life and death are defined by the motion of muscles:

> Muscular motion is the first sensible operation of animal life: the various combinations of it sustain and carry on the multiplied functions of the largest animals: the temporary cessation of this motive faculty is the suspension of the living powers, its total quiescence is death.
>
> By the continuance of patient, well directed researches, it is reasonable to expect much important evidence on this subject and, from the improved state of collateral branches of knowledge, together with the addition of new sources, and methods of investigation, it may not be unreasonable to hope for an ultimate solution of these phenomena, no less complete, and consistent, than that of any other desideratum in physical science.
>
> The present attempt to forward such designs is limited to circumstances which are connected with muscular motion, considered as causes, or rather as a series of events, all of which contribute, more or less, as conveniences, or essential requisites, to the phenomena.

Whether one is investigating a lower animal or 'the human frame,' comments Carlisle, 'the variety of accessory parts, and of organs by which a complicated machinery is operated, exhibit infinite marks of design, and of accommodations to the purposes which fix the order of nature.'[83] The human-machine is still a figure apparent in Carlisle's terminology; however, the soul is not present in his text. If for Parsons the mechanical human was constructed by 'HIM that order'd all Things,' for Carlisle the origin of the 'purposes which fix the order of nature' was more ambiguous. Even though in Carlisle's paper 'design' could imply God's presence in nature, the remark is vague and merely suggestive. Samuel Clarke's concern about the laws of mechanism alone governing the universe had indeed begun to 'exclude God out of the world.'

The extension of the soul – that is, giving it properties of material bodies – would in effect make it contiguous with the material electrical (and, eventually, electronic) world. As early as 1833 one commentator, A.P.W. Philip, would explain in the *Philosophical Transactions* that since the nervous influence 'has been found to perform its functions in the animal economy after it has been made to pass through a space not less than a quarter of an inch between the divided ends of a nerve, we must suppose either that, like magnetism or gravitation, it is capable of extending to a distance from the body in which it exists, or of passing through other conductors than the nerves.' The conclusion, then, is 'unavoidable,' namely:

> the nervous influence is capable of its functions after having passed through, and consequently existed in, other conductors than the nerves. It is therefore not peculiar to the nervous system, but capable of existing elsewhere, and consequently is not to be regarded as, strictly speaking, one of the vital powers of the animal body, but as an agent employed by them. On the other hand we find that voltaic electricity, applied under the same circumstances, is capable of all its functions, of exciting the muscles ... and maintaining all the other processes of assimilation on which the healthy structure of every part depends.
>
> A vital power has no existence except in the particular mechanism to which it belongs, and its functions are of a nature which admits of the substitution of no other power ... The nervous influence is therefore an inanimate agent ... one capable of existing in inanimate nature, and consequently, independently of the mechanism to which in the animal economy it belongs.[84]

The nervous influence, the idealized symbol of human vitality and human sensitivity through the latter part of the eighteenth century, the purveyor of ideas and judgment, here becomes a mere inanimate agent – it is a mechanism solely due to its materiality, and it is transferable to other conductors outside the body.

What, then, comprises the soul, human individuality and human consciousness? Such ideas have had clear implications for media theory since Marshall McLuhan first postulated the extension of the nervous system by electric media, as well as for fictional representations of the human identity escaping into the ethernet of global communications systems.

They have also had clear implications for the theory of thinking machines. Consider, for example, the strongly worded reductionist argument of Daniel C. Dennett, a member of the MIT team working to create a humanoid robot called Cog. In a 1994 article in *Philosophical Transactions* entitled 'The Practical Requirements for Making a Conscious Robot,' Dennett plainly scoffed at any suggestion that vitalism might have held sway over materialism in terms of explaining consciousness:

Let us briefly survey a nested series of reasons someone might advance for the impossibility of a conscious robot.

1. Robots are purely material things, and consciousness requires immaterial mind-stuff. (Old-fashioned dualism).

It continues to amaze me how attractive this position still is to many people. I would have thought a historical perspective alone would make this view seem ludicrous: over the centuries, every other phenomenon of initially 'supernatural' mysteriousness has succumbed to an uncontroversial explanation within the commodious fold of physical science. Thales, the pre-Socratic protoscientist, thought the lodestone had a soul, but we now know better; magnetism is one of the best understood of physical phenomena, strange though its manifestations are. The 'miracles' of life itself, and of reproduction, are now analysed into the well-known intricacies of molecular biology. Why should consciousness be any exception? Why should the brain be the only complex physical object in the world to have an interface with another realm of being? Besides, the notorious problems with the supposed transactions at that dualistic interface are as good as a *reductio ad absurdum* of the view. The phenomena of consciousness are an admittedly dazzling lot, but I suspect that dualism would never be seriously considered if there weren't such a strong undercurrent of desire to protect the mind from science, by supposing it composed of a stuff that is in principle uninvestigatable by the methods of the physical sciences.

But if you are willing to concede the hopelessness of dualism, and accept some version of materialism, you might still hold the following.

2. Robots are inorganic (by definition), and consciousness can exist only in an organic brain.

Why might this be? Instead of just hooting this view off the stage as an embarrassing throwback to old-fashioned vitalism, we might pause to note that there is a respectable, if not very interesting, way of defending this claim. Vitalism is deservedly dead; as biochemistry has shown in matchless detail, the powers of organic compounds are themselves all mechanistically reducible and hence mechanistically reproducible at one scale or another in alternative physical media.[85]

I do not quote Dennett at length to provide some definitive answer to the vitalist-mechanist debate but rather to indicate that it remains unresolved in modern-day discourses of machine-consciousness, and in part, too, because we continue to resist the notion that such human traits as thinking, judgment, or knowledge – the stuff of the soul – could be mechanically produced. Despite his off-putting condescension, Dennett makes some telling points. First is one that I have already made, that dualism as an explanation for consciousness was long ago rejected by many members of the scientific community. Accordingly, to suggest that the cyborg as a physical entity has anything to do with a Cartesian dualistic ideology seems merely anachronistic. Second, again emphasizing a point I have already made, a suggestion that vitalism replaced mechanism or materialism is in all probability coming from a perspective naive to some basic assumptions in the physical and computational sciences. Third, that there is an undercurrent of desire to protect the mind from science is undoubtedly an accurate observation: we can see it in the culture of sensibility in the eighteenth century, and we can see it in the culture of resistance to the cyborg in the twentieth and twenty-first. Fourth, Dennett's words reiterate a point I am making throughout this book: the material extension of the nervous system creates a new paradigm for the governing mechanism of the body – human consciousness – and as such generates tremendous unease that manifests in extravagant metaphorizing in the writings of both the critics and the proponents of cyborg culture. Finally, Cog, as an artificial consciousness, could theoretically have organic components[86] and could comprise a cyborg of sorts. As a networked computer complex (comprised in 1994 of four backplanes, each with sixteen nodes that were basically Mac II computers), Cog represents the material manifestation of aether to ethernet. Research on Cog continues at MIT today because analysts there continue to believe that cognition can be reproduced as machine computation. This ongoing research into how machines learn is defended by researchers as a means of understanding *human* learning and cognition: current studies of cognitive science and artificial intelligence include devising a way for Cog to learn the causal relationships between its own motions and input via vision and mechanical proprioception.[87] If Cog thinks, it is the ultimate and (for some) disturbing realization of the man-machine philosophy. The 'Blue Brain Project' is a collaborative research project of IBM and the Ecole Polytechnique Fédérale de Lausanne (EPFL) to create a three-dimensional digital model of the brain; it is also purposed to better comprehend the organic, living brain. The digital model, to be housed in a supercomputer system that will occupy a floor space that is approximately that of four refrigerators, will help scientists to 'understand more about how and why certain microcircuits in the brain malfunction – thought to be the cause of psychiatric disorders such as autism, schizophrenia and depression,' according to an IBM press release. The first phase of the project is to create a software replica of a column of the neocortex.

To conclude, the history of the cyborg in the early modern man-machine is based on these two related facets of techno-bodies. First, is the mechanistic view of the body as an engine composed of multiple parts, seen today as rather unremarkable: surgical procedures such as inserting stents in arteries to maintain unimpeded blood flow, or prosthetic limbs, or the replacement of organs, derive from a mechanistic understanding of the body's visible parts as mechanisms within a system or engine. Second, the view of the mind or consciousness as a series of electrochemical reactions derives from a mechanistic or materialist understanding of the fine particles within the body. This view has been significantly more problematic than the first, resulting in centuries of dispute about what makes us rational, spiritual, emotional, sensitive, and vital beings. How we think or feel is understood in the biosciences today as being governed to a great extent by mechanical causes – the molecular structure of the brain, or molecular reactions triggered in the medial and ventral prefrontal cortexes in the brain – but the mechanical directly opposed to the moral, thinking, or feeling being is still a common theme of cyborg literature. The cyborg figure, like the man-machine, is a metaphoric construct and a figure of language, and it is defined by its precursors in early modern definitions of and debates over *mechanic, machine, matter,* and *aether,* which so radically changed the understanding and eventually the industries, of human bodies and machines.

Like the man-machine of early modernity, the cyborg of fiction and film, theory, and cultural studies depends to a great extent upon these most disputed definitions that characterize our material world and our material embodiment – what may be called natural or artificial; living or non-living; organism or mechanism; mechanical or vital; material or immaterial; corporeal or aethereal; body or mind, or spirit, or soul. Despite an apparent logic of binaries implied by this series of dualisms, these terms are for the cyborg and were for the man-machine sites of multiple meanings, of complexity, query, and contest. To return to Richard Bentley in the seventeenth century for a moment, the issue of 'what prerogatives this Rational Machin' can demonstrate to raise humans 'above other parcels of Matter'[88] was one of central importance during this period when orthodox theological truths were being challenged by the new mechanical philosophy. Matter and soul were hard to distinguish in a theory of invisible atoms: what directed or governed human actions? Literature today carries a residual discomfort over the displacement of spirit by mere action and reaction at the scale of invisible atoms and electrons.

The soul as material mechanism in both early modern and postmodern science competes with metaphors and analogies that consistently attempt to reaffirm the spiritual soul. What I will be examining from here on is that tension between the practical man-machine or cyborg that exists as the manipulation of matter and energy in the human organism and the metaphoric man-machine or cyborg that carries with it the baggage of the soul. As a creature of the ethernet, of electrons and atoms, the cyborg is not just a person with a pacemaker or artificial heart valves; it is also a metaphorical creature, with a long tradition of associations with pre-electrical communications systems, individual and community government, and human spirit.

3 Some Contexts for Human Machines and the Body Politic: Early Modern / Postmodern Government and Feedback

In the upper region serving the animal faculties, the chief organ is the brain ... engendered of the purest part of seed and spirits ... and it is the most noble organ under heaven, the dwelling-house and seat of the soul, the habitation of wisdom, memory, judgment, reason, and in which man is most like unto God.

Robert Burton, *The Anatomy of Melancholy* (1621)

Why do people lose their tempers? What makes children bratty? Why do fools fall in love? What makes us laugh? And why do people believe in ghosts and spirits? ... I want to convince you that our minds are not animated by some godly vapor or single wonder principle. The mind, like the Apollo spacecraft, is designed to solve many engineering problems, and thus is packed with high-tech systems each contrived to overcome its own obstacles.

Steven Pinker, 'Standard Equipment,' *How the Mind Works* (1997)

Wisdom, memory, judgment, reason, morality, all these aspects of the soul or mind or identity were as vital to the earliest investigations of the man-machine as they are to cyborg literature and theory today. How does the body's material 'equipment' determine will, actions, or mental health? How does materially formed consciousness create the intangible combination of behaviours, intelligence, imagination, and morality or spirituality that comprises an individual self? We might be able to agree generally that the human body is a mechanical object, with the arms and legs as levers and pulleys, the heart and lungs as pumps, the circulatory system as pipes with valves, actin and myosin acting together within the muscle fibrils as ratchets, axons innervating sets of muscle fibres as motor units, or neurons as on/off switches transmitting impulses to and from the brain – but what makes this mechanism of flesh and bone ultimately do what it does? Perhaps more significantly, what makes it *decide* to do what it does? These are the questions that helped define the man-machine. If the muscles, tendons, and bones of the arm are a mechanical system of fulcrum and lever, then what is the force that makes one human-machine labour productively, an ideal citizen in the body politic, while another commits theft or murder? If knowledge, will, or even morality can be determined by physiology, how can we imagine or explain free will? What governs the human-machine? The discourse of human mechanism is also one concerning intellect and autonomy, morality and spirituality, and one of the most consistent metaphors used by philosophers, scientists, and writers of fiction alike to explore these issues in both man-machine and cyborg has been the image of a steers-man or governor within the body.

We imagine the human organism today as a complex communications system. Within the body, the nerves and brain communicate or trans-mit information through a series of electrical channels that can be compared to telephone wires or computer processors and, ever since McLuhan's pronouncement of a new global consciousness, media theo-rists have portrayed networked communications technologies as an ex-tension of the human nervous system. While the gradual transformation from humans as ensouled microcosms to humans as rational machines required an increasingly materialistic view of body-minds determined by molecular reactions and active forces, it simultaneously required an abstraction of the physical to represent the body-mind as numerical calculations, models, and programs, and new metaphors. Like the man-machine, the cyborg is also a textual construct, a mathematical figure, and a figure of language. However, many theorists have uncritically assumed that terms common to informatics and cybernetics or physiol-ogy and technology mean that the boundaries between human and machine have been eradicated. My hypothesis is that the figurative cyborg – the imagined techno-human upon which fiction and theory have imposed a radically altered spirit or consciousness – is the contem-porary iteration of metaphorical precursors to be found in the early modern man-machine. The early modern history of the human-machine as a communications system metaphorically manifested what would be-come a practical description of the cyborg from the 1960s onward. Then, as today, the material-electrical nervous system as analogous to a com-munications system represented not only physiological mechanisms but also moral and political ones.

I draw upon Norbert Wiener's term *cybernetics* (from the Greek *kybernetes*, or steersman), and Manfred Clynes and Nathan Kline's term *cyborg* from *cybernetic* and *organism*, to limit the seemingly ubiquitous cyborg to a more manageable object of study. To be quite specific, I demarcate the cyborg entity precisely by the derivation of the term from *cybernetic organism*. The origin of the term *cyborg* is well known in both science and humanities scholarship, but what is surprisingly neglected in many analyses of twentieth-century cyborgs is a characteristic that was originally their defining feature: the cyborg is an organism that is steered or governed by artificial feedback mechanisms. Wiener's invention of the field of cybernetics came about in part from his attempt to predict how pilots in the Second World War would control their airplanes as they flew in combat situations. Using terms from communications theory such as feedback, information, control, input, output, or homeostasis to describe the human nervous circuit and storage of information, Wiener suggested that at least some human voluntary behaviour could be characterized as machine feedback: 'It had ... become clear ... that the problems of control engineering and of communication engineering were inseparable ... and that they centered not around the technique of electrical engineering but around the much more fundamental notion of the message, whether this should be transmitted by electrical, mechanical, or nervous means.' Wiener explained that he and his colleague Arturo Rosenblueth had become aware of 'an essential unity of the set of problems centering about communication, control, and statistical mechanics, whether in the machine or in living tissue.' This realization necessitated a new term that encompassed this unity: 'We have decided,' wrote Wiener in an oft-reproduced statement, 'to call the entire field of control and communication theory, whether in the machine or in the animal, by the name *Cybernetics*, which we form from the Greek κυβερνήτης or *steersman*. In choosing this term, we wish to recognize that the first significant paper on feed-back mechanisms is an article on governors, which was published by Clerk Maxwell in 1868, and that *governor* is derived from a Latin corruption of κυβερνήτης. We also wish to refer to the fact that the steering engines of a ship are indeed one of the earliest and best developed forms of feed-back mechanisms.'[1]

From this definition we see that cybernetic devices are steered by a 'pilot' or 'governor'; they maintain homeostasis through the use of communications systems; they are governed by a feedback mechanism, which can either be informational or mechanical or both; and finally, in machines, animals, and humans these mechanisms may be computationally equivalent – and therefore potentially predictable and re-engineered. The equation of cybernetic human and machine in materi-ally and mathematically based analogies eventually would be posed in media and cultural studies as an actual synthesis of human being into cybernetic organism – a new being both human and machine. By the end of the twentieth century the cyborg, a self 'situated beyond the skin,' as the performance artist Stelarc has described it, would be imagined as a substantiated certainty of the radical evolution to post-humanity, where 'what it means to be human is no longer being immersed in genetic memory but in being reconfigured in the electromagnetic field of the circuit.'[2] Katherine Hayles wrote in 1993 of three-dimensional virtual reality (VR) environments about the illusory sensation of being *inside* the computer: 'I can attest,' she wrote, 'to the disorienting, exhilarating effect of feeling that subjectivity is dispersed throughout the cybernetic circuit. The user learns kinesthetically and proprioceptively in these systems that the boundaries of self are defined less by the skin than by the feedback loops connecting body and simulation in a techno-bio-integrated circuit.'[3] Writing in 1996, Hayles emphasized more strongly the necessity of embodiment to human subjectivity whether enhanced by technology or not: 'Far from being left behind when we enter cyberspace, our bodies are no less actively involved in the construction of virtuality than in the construction of real life.'[4] Nevertheless, the image of human identity extended into or even transferred over to the com-puter was a common feature of both theory and fiction throughout the 1980s and 1990s. Stelarc has proclaimed that 'subjectively, the body experiences itself as a more extruded system, rather than an enclosed structure. The self becomes situated beyond the skin. It is partly through this extrusion that the body becomes empty.'[5] Today, the joint research initiative of IBM and the Ecole Polytechnique Fédérale de Lausanne (EPFL) – dubbed the Blue Brain Project – fulfils earlier speculations that the human brain function corresponds to complex microcircuitry and that it can be modelled through man-made machines and mathematical computation (see figure 20).

These two general and often overlapping approaches to the character-ization of the human nervous system as a form of communications technology have complicated the study of cyborgs and cyber-identities. The model or analogy that is frequently adopted by engineers or physi-ologists to describe the function of humans and communications tech-nologies reappears in warped form in the notions of extension, as first popularized by Marshall McLuhan, where human consciousness is not

merely mathematically equivalent or analogous to communications technologies but in fact is continuous with them and accordingly modified by
them. 'The electric media,' McLuhan wrote, 'are a physical extension of
our own organic nervous system, which is literally constituted by electric
impulses. When you put an electric system, or field, around you, you
enlarge your own nervous system. Just as the wheel extends your foot,
the "wired planet" now extends our nerves. All technology is a physical
extension of man, and our bodies, said Emerson, are a "magazine" or
storehouse of all past, present, and future innovations.'[6]

In this particular vein of media or cultural studies, metaphor-cum-
reality has regularly translated into musing upon the morality and ethics
of the cyber-individual and of the networked cyberpopulace. The imagined melding of individual subjectivity with the network of electric and
electronic communications media symbolized for many the evolution to
a new technological or cyborg identity where technologized humanity
is characterized both as global citizens and as nodes in a global neural
net. And for many writers theorizing this new identity, the cyborg or
cybercitizen has signified either a more democratic future or an imminent disintegration of morals and community engagement. Why do
these metaphors so often assume the power of reality?

The question of what governs or steers human actions is the predominant theme for both cyborg studies and the philosophy of the
man-machine: communications in both the body and the body politic
thus become politically sensitive concerns. Such representations of
cyber-identity are contingent upon a history of metaphors, and their
apparent certainty has been reinforced by discursive tradition. The cyborg is a descendent of the man-machine, not merely because the
mechanistic philosophy made simplistic and ultimately unsatisfactory
comparisons of clockwork mechanisms to organic ones. The postmodern
cyborg and its historical counterpart in the early modern neurological
man-machine are organisms whose functions depend upon the transmission of messages. Contemporary speculations on the government of
cyborg individuals and cyborg society have a lexical and metaphorical
history in early modern representations of the man-machine's knowledge, judgment, and will as a function of control and communications. Because the man-machine, like the cyborg, is a communications
system, and because the nervous system was also characterized in late
seventeenth-century England by tensions surrounding the government
of those communications in the context of war, my suggestion is that
much of the proclaimed evolution to post-humanity in media and cul-

tural studies of the past decades is demonstrably a metaphoric tradition going back to discourse of the man-machine. Moreover, these analogies for systems of nervous communications and control in the body have also historically seemed to imply for other commentators some existent truth about the systems of communications and control in the body politic.

Needless to say, the interplays between contemporary analogies of communications, control, government, body, and body politic are complex and interrelated as it is without further complicating matters by venturing into an early modern history of these issues as associated metaphors. I would like to begin, accordingly, with three sets of contextualization before going on to explore the history of these analogies in the early modern man-machine and their implications for morality, ethics, and control in both the body and the body politic. First, I present a few of the countless analogies in natural philosophy, science, and medicine that compare the nervous system to technological communications systems. Second, I discuss how communications and control figure into twentieth-century configurations of media theory; third, I examine the significance and power of communications in the body from the perspective of an Aristotelian paradigm as Thomas Willis's predecessor at Oxford, William Harvey, imagined it. Finally, I discuss Willis's human engine as a machine characterized not by clockwork but by a form of feedback. These contextualizations function to demonstrate, first, that analogies for the human mind as a machine for communicating are an ancient practice that became problematized as the aethereal soul became materialized; second, that certain twentieth-century criticism has further problematized these analogies by mistakenly representing physically discrete systems of control and communications in human and machine as functionally equivalent and literally merged; third, the metaphorical language of control and communications must also be understood in terms of the long history of metaphors for the government of the human soul or spirit; and, finally, that the revolution in paradigms from humans as clockwork machines to humans as homeostatic machines does not originate with Wiener's cybernetics but is evident from around the time that feedback mechanisms were being developed and used for alchemical experimentation.

Cartesian mechanistic philosophy alone could not provide an answer to the problem of what makes matter move, and how that motion is governed. This governor, the soul or rational spirit, was represented as a monarch, pilot, steersman, or musician within the brain, but the analogy was potentially contentious, and carefully chosen. The governance of the body politic had been historically based upon a religious system: the accepted source of governance of the individual body was spiritual, and if human spirit could be reduced to mere matter in motion then not only morality, religion, and the immortal soul but also the authority of the Church, the monarchy, and their systems of government could be brought into question.

What is the history of the interplay of communications and feedback within the body, and their implications for both individual and community self-control? This investigation into what we might call the Enlightenment cyborg has in part been motivated by the confidence of many theorists that networked communications mean inevitable and profound changes to individual and social consciousness. The image of communications regulating the individual body and the supposed reconstitution of body and body politic due to ungoverned or networked communications systems are dominant metaphors in contemporary theories of postmodern transformations. Yet these are all to some extent constructs fashioned by early modern tropes for the human-machine.

Context 1: The Nervous System and Machines for Communicating

In *Are We Hardwired? The Role of Genes in Human Behavior*, researchers of molecular biology at the University of California, Los Angeles (UCLA), William R. Clark and Michael Grunstein, adopt a familiar model for the human organism as a system of information transmitted through a series of communications channels and devices. 'Neurotransmitters are the way nerve cells communicate with each other,' they explain,

> and with other cells in the body. We use neurotransmitters to relay information about the environment to the brain, to analyze that information, and to set in motion appropriate bodily responses ... The resulting cross-talk is necessary if the brain is to draw on all of its stored information in formulating responses to internal and external events. Interneurons connect nerve cells in all regions of the brain, from those regions in which experiences are first perceived, through extensive switching systems that route incoming signals through memory banks, and through the intellectual centers of the brain's cortex. The brain is literally bathed in information flowing in the form of neurotransmitters. Nerve cells in the brain and in the spinal cord also use these same neurotransmitters to communicate with other cells in the body, for example, to instruct a gland cell to release a hormone, or a muscle cell to contract in order to initiate movement.[7]

The analogies – bridges, cross-talk, stored information, switching systems routing incoming signals, flow of information, memory banks, communication – all describe a system of control for the human body as a computer system. They also recall ancient images of the brain as a storehouse. Steven Pinker effectively suggests in *How the Mind Works* that what was once called spirit or soul or consciousness is actually physically stored information that is analogous to computer data: 'The mind is what the brain does; specifically, the brain processes information, and thinking is a kind of computation. ... The computational theory of mind resolves the paradox [of the 'mind-body' problem]. It says that beliefs and desires are information, incarnated as configurations of symbols. The symbols are physical states of bits of matter, like chips in a computer or neurons in the brain ... Eventually the bits of matter constituting a symbol bump into bits of matter connected to the muscles, and behavior happens.'[8]

Such models or analogies of brain to communications technology reiterate a conventional trope of comparing the mind to the most sophisticated communications system available. The nervous system 'is more than an apparatus for receiving and transmitting signals,' Wiener wrote in his 1948 article, intended for a popular audience. 'It is a circuit in which certain impulses enter muscles and re-enter the nervous system through the senses.'[9] Dean E. Wooldridge compared neurons to on/off switches and input/output devices, the spinal cord to an input/output cable, and the brain to a central data processor in *The Machinery of the Brain* (1963).[10] And, as the Massachussetts Institute of Technology (MIT) professor Victor Weisskopf noted in 1979, 'nerves are long strands of special cells that, like telephone wires, transmit messages from one place to another.' The 'messages received are transmitted through connecting nerve cells to other parts of the unit so that the unit is able to coordinate locomotion and other reactions to the outside conditions.' The exchanges of information are also like those of a computer: 'A system of interlocking nerve cells is in many ways equivalent to a system of interconnected electronic vacuum tubes or transistors' albeit 'the brain of even an insect is a more complicated device.'[11]

Earlier analogies were to telephones and before that to telegraph wires. In 1892 Karl Pearson, writing of reflex action, thought, and consciousness, under the sub-heading 'The Brain as a Central Telephone Exchange,' stated that 'the view of brain activity here discussed may perhaps be elucidated by comparing the brain to the central office of a telephone exchange, from which wires radiate to the subscribers

A,B,C,D,E,F, etc., who are senders, and to W,X,Y,Z, etc., who are receivers of messages.'[12] So seemingly obvious were the similarities between transmissions over nerves and wires that Emil duBois-Reymond frequently compared the nervous system to a communications network. In a public lecture delivered in 1851, he reasoned that 'the wonder of our time, electrical telegraphy, was long ago modeled in the animal machine. But the similarity between the two apparatus, the nervous system and the electric telegraph, has a much deeper foundation. It is more than similarity; it is a kinship between the two, an agreement not merely of the effects, but also perhaps of the causes.' Seventeen years later, he opened his Croonian Lecture on Muscular Motion by explaining that 'just as little as telegraph-wires, do the nerves betray by any external symptom that any or what news is speeding along them.'[13]

Even much earlier in the eighteenth century, the mind was defined in accordance with technologies of representation or of communications. To La Mettrie 'judgement, reason and memory are only parts of the soul which are ... veritable modifications of that sort of medullary screen on which the objects painted in the eye are projected as in a magic lantern.'[14] John Locke described the original state of human understanding – the 'most elevated Faculty of the Soul' – as a surface clear of writing: 'Let us then suppose the Mind to be, as we say, white Paper, void of all Characters, without any *Ideas*.' Locke used the page as trope for the inscription of both rational thought and morality. In describing how seemingly self-evident or innate moral and religious beliefs are formed, he suggested that children '*grow up to the dignity of Principles* in Religion or Morality' through the doctrines professed by their caregivers, instilled into the unprejudiced understanding, 'for white Paper receives any Characters.'[15] The imagery recalls Hamlet's promise in Shakespeare's play, written a century earlier: 'I'll wipe away all trivial fond records' from memory, vows Hamlet, to better remember the words of the ghost 'within the book and volume of my brain.' Immanuel Kant compared the memory to an indexed library when he wrote: 'Most of all, the use of *topics* – that is, of a framework for universal concepts, called *general headings (loci topici)* – makes remembering easier, by dividing the material into classes, as when we arrange the books in a library on shelves with different labels.'[16]

Writers such as Pierre Gassendi (1592–1655), Robert Burton (1577–1640), or Thomas Hobbes (1588–1679) had restated Aristotle's description of the mind (in *De anima*) as being a blank writing tablet or *tabula rasa*. We might also recall Socrates's discarded analogy for the memory of the mind as a block of wax, like a wax writing tablet, that holds the impressions of sensory experience (recorded in *Theatetus*). As Frances Yates has reminded us, Cicero's *De oratore* placed the invention of the technology of memory much earlier with the poet Simonides (ca. 556–468 BC) who suggested the mnemonics of using 'places and images respectively as a wax writing tablet and the letters written on it.'[17] All this is to state the obvious: technologies of representation and communication have throughout human history tended to inspire analogies for the mind. As the technology changed, so did the analogy. Why, then, has the image of the mind as computer network inspired such vehement critique and such elated endorsement?

The analogies of the human mind as cybernetics communications device carry on what is in effect a traditional trope. What interests me here specifically is the 'cyber' genre of theory that has accepted as literal truth the sometimes fantastical images of a technologized human intelligence increasingly subsumed and driven by computer technologies – with heady repercussions predicted for both social and individual morality and spirituality. Human identity actuated by medical technology in the form of drugs or stimulants has been a regular feature of human life for centuries, but theories that a post-human identity can be actuated by new forms of communications technology cry out for historical contextualization.

Context 2: Communications and Control in the Cyborg

In suggesting a distinct correlation between the human nervous system and a computer system transferring information throughout the body in a feedback loop, Wiener set the groundwork for what would become the defining characteristic of the cyborg: not only is the cyborg a communications device, but it is a communications device that is controlled by a feedback circuit. Moreover, the relationship between human and machine for perhaps the first time in human history thereby was not merely analogical: the human nervous system and 'computing machine' were perceived to be not only mathematically equivalent but also physically compatible. Throughout the 1950s and 1960s, when cybernetics and cyborgs were first envisioned, abstract mathematical models of equilibrium and self-regulating systems became a prominent subject of study. These models yielded innumerable practical technologies. They also brought along with them a metaphoric system based on a history of ideologies of human government and communications. Clynes and Kline envisioned the cybernetic organism, an integrated system of living organism and exogenous regulatory components, thus: 'for the exogenously extended organizational complex functioning as an integrated homeostatic system unconsciously, we propose the term "Cyborg." The Cyborg deliberately incorporates exogenous components extending the self-regulatory control function of the organism in order to adapt it to new environments.'[18] The steering component was a material extension of the organic body's own regulatory systems to achieve homeostasis through a programmed feedback loop, a communications circuit.

But a few years after Clynes and Kline defined the cyborg, Marshall McLuhan incorporated those key components – communications, homeostasis, and a steering metaphor – to describe the revolutionary effects of media upon the nervous system, individual sensory perception, human souls, and the global society as a whole: 'The electric extension of the central nervous system has never occurred before, and therefore, we are without any navigational chart whatsoever. Mechanical systems are not closed in the way that electric ones are ... An electric system is very much like a small tribal society – that is, ecological and homeostatic,'[19] McLuhan wrote in a letter to John Snyder in 1963. McLuhan adopted the language used for medical-scientific engineering of the physical nervous system and transformed it to proclaim that communications technologies were literally and seamlessly conjoined to the nervous system of anyone using them. In his *Gutenberg Galaxy* (1962) McLuhan explained that our electric technology, 'has consequences for our most ordinary perceptions and habits of action which are quickly recreating in us the mental processes of the most primitive men. These consequences occur, not in our thoughts or opinions, where we are trained to be critical, but in our most ordinary sense life, which creates the vortices and the matrices of thought and action.'[20] That electrical or electronic technology affects not only our rational thought but also the sensory perception of the world and other physiological processes was a presumptuous and unproven contention that continues to reverberate in literary criticism and critical theory: more recent critics might speak of the mind uploaded to the computer, virtual identity or terminal identity, cybersubjectivity, or 'technopsychotic annihilation via cyborg,'[21] all suggesting a transformation of minds and society through integration with the machine.

In communications and cyberculture studies a commonplace claim has been that any change in our mental processes due to our technology (the metaphoric extension of the nervous system into or by the electronic medium) would affect individual human morals, the nature of humanity, or the indefinable and immaterial soul itself: 'Uncertainty in the strivings of the soul,' McLuhan wrote, 'is perhaps one of the aptest descriptions of man's condition in our modern crisis [wrought by] technology, the machine,' which has spread so ubiquitously through the world.[22] McLuhan's implicitly religious construct of technology's effect upon the human soul has formed a significant basis for theorizing the body and technology in the newly formed genre of cultural studies of the body – much of which makes claims to postmodernity but, in effect, has very close literary ties to the mythos of a much older Christian tradition. The cyborg merely reiterates well-established clichés about the much-contested grounds of identity and human nature. It is striking how so many media theorists took the ubiquitous analogies of nervous and computer communications to mean that electrical communications devices literally have altered our nervous system, and thus also foreshadow new forms of social decay or democracy.

By the end of the twentieth century, the cyborg – Stelarc's self 'situated beyond the skin' – would be imagined as a material and substantiated possibility of the radical evolution from human to post-human, an 'extruding of awareness,' where one's humanity is configured in the electronic circuit. This extruding of awareness frequently has been imagined as having moral and spiritual implications, the most extreme fears of which have been voiced by Jean Baudrillard and Arthur Kroker along with various co-authors. In 1983, for example, Baudrillard described the 'sovereign' human consciousness united to communications systems as 'ecstatic,' a lone pilot disengaged from the grounding of body, community and reality: 'at the controls of a hypothetical machine, isolated in a position of perfect and remote sovereignty, at an infinite distance from his universe of origin. Which is to say, in the exact position of an astronaut in his capsule, in a state of weightlessness that necessitates a perpetual orbital flight and a speed sufficient to keep him from crashing back to his planet of origin.' Left behind, he claimed, was 'a large useless body, deserted and condemned.' This imagined status of the mind's steersman abandoning the individual body also defined the body politic for Baudrillard as being 'not a public scene or true public space but gigantic spaces of circulation, ventilation and ephemeral connections.'[23]

Drawing on the same metaphoric tradition, Kroker and Weinstein consistently have imagined a dire future for society due to the escape of consciousness into the machine network: 'What does it mean when the body is virtualized without a sustaining ethical vision? Can anyone be strong enough for this? What results is rage against the body: a hatred of existence that can only be satisfied by an abandonment of flesh and subjectivity ... Virtuality without ethics is a primal scene of social suicide ... it is already the after-shock of the living dead: body vivisectionists and early (mind) abandoners surfing the Net on a road trip to the virtual Inferno.' In the world of virtual reality, 'the fantastic galaxy of the perverted image,' they claim, 'there are no sight lines, no solitary (sovereign) subject as the privileged locus ... of perspective.' The so-called sovereign subject of consciousness is a significant metaphor. What do Kroker and Weinstein suggest we are losing, should we understand the nerves to be equivalent to networked machines of communication? The long literary tradition points to an image of sovereign soul or rational consciousness with distinctly Christian overtones. Mistaking mechanical analogies used in describing human consciousness for veritable truth results in the assumption that, as our technologies become more sophisticated approximations of humans as communications machines, we become spiritually altered. 'The Net is the nervous system of the electronic body,' Kroker and Weinstein announce. Further, the ideal 'sovereign subject' vanished from the body transforms into the transgressive pilot in the amoral void of cyberspace: 'Virtual evil? That is cybernauts as the sign of the beast with two easily identifiable marks burnt on their electronic flesh. First, the mark of *forgetfulness*, as cybernauts systematically expunge from their world-view any account of the human costs associated with the coming to be of the technological dynamo. And secondly, the mark of *techno-fetishism*, as cybernauts transform their cyber-bodies and cyber-consciousness into living registers of emergent technologies. Total repression and total valorization, then, as the twin signs of virtual evil.'[24]

Such pronouncements of radical change in the last decades of the twentieth century were made in apparent unawareness of their fundamental reliance on three-hundred-year-old metaphors for human consciousness as pilot, steersman, or monarch in the brain communicating via material mechanisms with body systems and external environment. From very early in the history of the human-machine, the mechanism of body government was imagined in terms of signals, transmission and reception of commands or messages, and a circuit of communications between the governor in the brain and the active muscles of the body. To summarize, the governor trope in both cyborg and man-machine is characterized by the image of body as mechanism, the interplay of communications systems with body systems, and the assumption that the government of such a body is equivalent to or continuous with that of the larger body politic.

These metaphorical versions of the cyborg tended to postulate a separation of immaterial mind from the material body, even though the image of the mind's extension into the communications media paradoxically depended upon a materialistic understanding of consciousness: for the consciousness to be able to extend out of the body into the network, or to be influenced in any way by material communications technologies contiguous with the body, it must have quasi-material properties such as electronic flux; at the same time, the notion that the consciousness can escape the confines of the body posits a dualistic demarcation between physical body and non-physical mind. The sensationalistic temptations of the metaphor, however, proved powerful and pervasive. Others embraced a future where human consciousness is imagined as extruded from the body. Claudia Springer, for example, wrote of networked communications from a feminist perspective as 'a microelectronic Imaginary where our bodies are obliterated and our consciousnesses are integrated into the matrix.' She concluded that, with regard to sexuality and gender roles, 'depending on one's stake in the outcome, one can look to the cyborg to provide either liberation or annihilation.'[25] For an astonishingly long period the cyborg identity as an ungoverned cyberconsciousness was used carelessly, either to represent how communications technologies compromise both the health of the living body and the spirituality or morality of society, or more positively to represent the progressively liberating evolution of humanity.

The cyborg steersman released into or absorbed by the matrix, however, is a textual sleight of hand made possible only through a tradition of metaphors for the imagined opposition of mind and body, spirit and matter, form and substance. Vivian Sobchack, for example, has argued that 'the two-dimensional, binary superficiality of electronic space at once disorients and liberates the activity of consciousness from the gravitational pull and orientation of its hitherto embodied and grounded existence ... Subjectivity is at once decentered and completely extroverted ... [This euphoric presence's] lack of specific interest and grounded investment in the human body and enworlded action, its saturation with the present instant, could well cost us a future.'[26] Our language describing that steersman or governor in the mind has evolved so that human and machine both can be described as the same continuous cybernetic essence. An isomorphic terminology for human and machine does not mean, however, that we can presume our inherent subjectivity has been lost to the machine.

Baudrillard based his more recent collection of essays, *The Transparency of Evil* (1993), on this assumption that human and existing technological communications systems are transposable, permeable, and interchangeable entities. The diseased body and our diseased society both are the result of a virulent technology: 'Virulence takes hold of a body, a network or other system when that system rejects all its negative components and resolves itself into a combinatorial system of simple elements. It is because a circuit or a network has thus become a *virtual* being, a non-body, that viruses can run riot within it; hence too the much greater vulnerability of "immaterial" machines as compared with traditional mechanical devices. Virtual and viral go hand in hand. It is because the body itself has become a non-body, a virtual machine, that viruses are taking it over.' The diseases of the body and the body politic thus become equivalent outbreaks of a human 'vulnerability' to science and technology. 'AIDS, crack and computer viruses are merely outcroppings of the catastrophe,' Baudrillard portends. 'The full-blown, the absolute catastrophe would be a true omnipresence of all networks, a total transparency of all data ... [T]he culminating point of the development of information and communications ... [is] death.' In another essay in this collection, Baudrillard claims that although mechanical automata once 'played on the difference between man and machine,' they have now 'become isomorphic and indifferent to each other: neither is other to the other.' With a logic that makes sense only in light of the history of the pilot metaphor, Baudrillard conflates human body, computer network, and society, in describing the imminent apocalypse of our global '*bricolage* with the broken bits and pieces of the Enlightenment, of "progress."' Enigmatically, he contends that 'Aboriginals' have the 'power to destabilize Western rule,' that because of 'their viral, spectral presence in the synapses of our brains, in the circuitry of our rocketship.' Presumably the implication here is that communications technologies have rendered us as overtly technological vessels piloted by diseased consciousness, which might be healed only by a sort of preliterate or prelapsarian vestigial primitive. Obfuscating once again any physical and actual distinction between body, social order, and computer network, Baudrillard prophesies that at that point 'our ultra-sophisticated but ultra-vulnerable systems ... will ... convulse from within.'[27]

In a more recent (and more rational) article on cyberdemocracy,

Mark Poster also imagines that networked computer communications impose 'a transformation of the subject position of the individual who engages within it.' He speculates more optimistically that the decentralization and dematerialization of communications through postmodern Internet technology will transform the 'modern category' of democracy, the 'grand narrative of liberation,' or 'Enlightenment narrative.' Finally, 'because the Internet inscribes the new social figure of the cyborg and institutes a communicative practice of self-constitution, the political as we have known it is reconfigured.' Accordingly, Poster asks, 'How will electronic beings be governed?'[28] What is crucial here is the terminology that suggests that ungoverned networked communications systems allow seemingly chaotic and tumultuous – viral – communications to potentially infect and disease the social as well as the individual body. Even optimistic representations of the Internet frequently call for or emphasize new systems of government.[29] The trope of a centralized governor receiving and responding to the body's communications has been modified by our understanding of both machines and human systems operating at the level of molecular-electronic activity.

Notwithstanding their opposing views of the outcome of the Enlightenment narrative in postmodernity, these arguments encapsulate the prevalent assumption that technological communications networks will reconfigure the government and health of both the individual and the social body – based solely upon the government of networked body, networked machine, and body politic as semantically interchangeable entities. Despite the supposedly radical shifts in human identity and consciousness, these versions of human mind/soul and society continue to use a metaphoric language that is anticipated and to a great extent prefigured by the early modern discourses of mechanist and materialist philosophy. For centuries tropes of control in the human mind (particularly, consciousness as steersman, governor, or pilot of a mechanical device) have been related to communications metaphors and applied to social bodies. The conceptualization of the human body as a mechanistic communications system that governs individual actions and also extends to affect (or effect) the health or disease of the body politic appears in the works of such diverse writers as Hobbes and Willis in the seventeenth century, and Swift and La Mettrie in the eighteenth.

The body's government was a figurative device employed by Harvey, Descartes, and Willis. Thomas Willis described the human engine as a 'cybernetic' organism that is governed by a rational soul, which itself is to some degree regulated through the communications system of the body's citizens. The words that would eventually come to characterize cyborg communications systems – transmit and receive, communicate, signal – were all used centuries ago by Willis to describe the self-regulatory circuit of communications and commerce in the body's 'œconomy or government.'[30]

Context 3: Communications and Control in the Man-Machine

Particularly important to the early modern period was the modification of the man-machine as automaton with a non-physical soul to the man-machine as a physical entity animated by a motive energy comparable to other material forces and fluids in nature.[31] In the century roughly between Descartes' *Meditations* (1641), which described humans as mechanical assemblages with incorporeal souls, and La Mettrie's *L'homme machine* (1747), which flatly declared such a belief to be mistaken, the status of humanity began to change (in medical-physiological terms) from the spiritual to the secular. Descartes' dualism was one attempt to explain how matter is imbued not only with vitality but also with virtue, while La Mettrie argued that morality is mechanical, and solely due to the conformation of the body. Few philosophers fully endorsed La Mettrie's claims. Nevertheless, an increasingly material and mechanistic approach to the study of human physiology and 'psycheology,' to use Willis's term, would displace the inexplicable, immaterial, and immortal soul from the realm of experimental and analytical studies. Even as they regularly acknowledged God's presence, medical physicians and philosophers increasingly explained human motivation, motion, and vitality not by some undefined pneuma, divine breath, or vital spirit but as empirically verifiable functions of the body's conformation. How – or whether – the mechanisms of the infinitesimal particles of matter confer not merely life but also reason, morality, and identity was left unresolved. To some extent, this remains an unresolved issue to the present day. Today's terminology used in describing cybernetic-organic systems has indeed evolved far beyond that of seventeenth- and eighteenth-century physiologists; all the same, contemporary speculations upon the government of cyborg individuals and society have a history in early representations of knowledge, judgment, and will as a function of a governor in the brain directing the body's civilians (muscles, nerves, organs, and so on). The history of the cyborg is the history of this governor, its position within the body, and its relationship to human spirit.

In part, this image arises from the old analogy of the body politic, which is found in writings as early as the *Rig Veda* and Plato's *Republic*. The body politic analogy builds upon two principles that were described by the Greeks: hylozoism and the body and natural world as microcosm and macrocosm. Hylozoism claims that mind or life permeates and is inseparable from the natural world. The life or soul or mind of the individual is the same as the force that animates, unifies, and directs all other aspects of the world, including the state. The second principle asserts that a single pattern or identity exists at many levels of being. The individual, therefore, is a microcosm of the state or world both in body and mind.[32]

Plato and Hippocrates both placed the rational soul in the head, with the ventricles as the vessel for the spirit. Plato emphasized that the flesh of the body was inferior and the soul both incorporeal and immortal. In his *Republic* (360 BC), Plato imagined both the state and the individual citizen to function best with a centralized rational control and appropriate division of labour by the members of a lower order. An unhealthy state, like an unhealthy body, was characterized by 'their meddlesomeness and interference with one another's functions, and the revolt of one part against the whole of the soul that it may hold therein a rule which does not belong to it, since its nature is such that it befits it to serve as a slave to the ruling principle?' On the other hand, 'to produce health is to establish the elements in a body in the natural relation of dominating and being dominated by one another, while to cause disease is to bring it about that one rules or is ruled by the other contrary to nature ... and is it not likewise the production of justice in the soul to establish its principles in the natural relation of controlling and being controlled by one another, while injustice is to cause the one to rule or be ruled by the other contrary to nature?'[33] In another of his dialogues, Plato has Timaeus place the soul, the greatest human faculty, in the head – 'being the most divine part of us and the lord of all that is in us; to this the gods, when they put together the body, gave all the other members to be servants, considering that it must partake of every sort of motion.' The brain's ventricles served as a vessel for the immortal soul, which had been formed out of the seed created by God from the soul of the universe mingled with the elements: 'When he had sown them he committed to the younger gods the fashioning of their mortal bodies, and desired them to furnish what was still lacking to the human soul, and having made all the suitable additions, to rule over them, and to pilot the mortal animal in the best and wisest manner which they could and avert from him all but self-inflicted evils.'[34] In this version of the human body, the incorporeal soul is a pilot or ruler and all other members of the body are its servants. The system that Plato described was essentially hierarchical, that is, ruled from the top down.

Conversely, Aristotle proposed three souls in *De anima*: the vegetative soul responsible for growth and decay, possessed by plants, animals, and humans; the animal soul, which confers motion and sensation upon animals and humans; and the rational soul, which is the conscious and intellectual soul in the heart, and is possessed solely by humans. Galen (AD 129–199), whose theories of physiology and anatomy predominated in Western Europe for centuries, retained the three Aristotelian souls, but understood that mental faculties arose from the brain: the liver was the seat of the vegetative soul, the heart was the seat of the animal soul, and the brain was the seat of the rational soul. The brain distilled the ethereal animal spirits from the vital spirits, stored them in the ventricles, and distributed them through the hollow nerves. The hollow nerve theory did not finally recede until well into the eighteenth century: part of its attraction was that it supported the humoral theory of diverse liquid and aethereal substances circulating throughout the hollow spaces of the nerves and arterial-venous system, separate from the solid tissues, and mediating between corruptible body and incorruptible spirit. Willis noted that he could see no evidence of a hollow nerve in his observations using the microscope; this marked a turning point from this older system: for Willis, humans and animals both had a corporeal soul comprised of 'very subtil little Bodies,' which resided throughout the metropolis of brain, nerves, and fibres.

The analogy of government in the human body was treacherous ground. The figure of the body politic had been adopted by the heads of church and state throughout the Middle Ages, and the image reinforced structures of entrenched hierarchy and authority as natural and vital systems well into the sixteenth and seventeenth centuries. As Saint Paul's epistle to the Corinthians had explained, 'for as the body is one, and hath many members ... so also is Christ ... And the eye cannot say unto the hand, I have no need of thee: nor again the head to the feet ... Now ye are the body of Christ and members in particular'. (I Corinthians 12: 12–27). These words, adopted and developed by the Church, supported the Pope as the head of a communal body, with kings, emperors, and common people as citizens governed by that authority. By the sixteenth and seventeenth centuries challenges to the validity of this system resulted in secular and spiritual rulers competing for the title of head. With increasing scepticism and changes in governmental systems, as well as the discoveries of the new science, organic analogies would become less meaningful ways to describe political issues,[35] but in the seventeenth century, defining an apposite system for the government of both individual bodies and the body politic was imperative. For Willis working out his system of neuroanatomy through the turbulent years leading up to and during the interregnum following the execution of Charles I – who had advocated the divine right of kings and whose struggle for power with Parliament threw the English into civil war – the analogy must have been particularly expressive.

A generation before Willis, in his unpublished *De motu locali animalium*, William Harvey had represented the mechanical body as a political microcosm in which centralized communications and command from the brain were prominent as the mechanism of control. As in Aristotle's *De motu animalium*, Harvey observed, the body is like an automaton: 'The movements of animals may be compared with those of automatic puppets, which are set going on the occasion of a tiny movement; the levers are released, and strike the twisted strings against one another ... Animals have parts of a similar kind, their organs, the sinewy tendons to and from the bones; the bones are like the wooden levers in the automaton, and the iron; the tendons are like the strings, etc.' As early as this, the automaton as microcosm functions by the transmission of signals: observing Fabricius's dictum that 'nerve transmits the command from the brain,' Harvey conjectures that 'the use of nerve is to communicate to the brain that which is perceived by the senses that a judgement may be made.' Harvey's notes suggest a system of government where communications might go back and forth between citizens and ruler, but commands come from the top down – although he was unsure whether the centralized control and motive and sensory spirits came from the brain or the heart. He notes that 'there is no need of a soul or appetite in each part but in a central governing place,' and postulates several such possibilities for a suitable representation of that governing mechanism. Harvey acknowledges that the sensitive and motive soul was in the brain: But was the brain a general, the ruler of a senate, a choir-master, an architect, the master of a ship, or a prime mover? If the heart instead was the general or ruler, the brain was the judge, sergeant major, or marching overseer. If the heart was the musician or architect, the brain was the choir-master or surveyor. In each of these cases, the nerves held offices subservient to the heart or brain and in positions of authority over the muscles; the muscles were the citizens, the workers, sailors, dancers, singers, or soldiers. Harvey's treatise comprises more questions than answers about what governs the body's movement, and the frequent contradictions in his notes emphasize the importance of finding a fit metaphor for the controller of the body's 'citizens' or muscles. The analogy chosen for the body's governor is meaningful not only within the context of control over citizens' offices within the individual body; it also reiterates and reinscribes those of the body politic. Order is achieved, therefore, at the direction of the soul, as in Aristotle's 'well governed State': movements are 'commanded as one whole and, moreover, as with a given signal.' On the importance of the body's regulation for the purposes of order within the body (and body politic), Harvey comments, 'another point to be considered is its [the body's motion] regulation and its motor faculty. As a result of which chickens with their heads cut off move ... as do also men in delirium and drunkards, but they move with a disorderly action and not with the harmony and rhythm necessary for work.' Since the use of the nerves was to communicate to the brain so that a judgment may be made, 'when nerves are cut movement is vigorous but purposeless; when the head is cut off there are movements but they are uncoordinated.'[36]

Given his own governmental analogies and the civil turmoil at the time, the loyalist Harvey seems to be emphasizing an allegiance to the monarchy as a source of stable and undivided authority, as he did in his 1628 *Exercitatio anatomica de motu cordis et sanguinis in animalibus* (*An Anatomical Disputation on the Movement of the Heart and Blood in Animals*). In 1618 Harvey was appointed physician extraordinary to James I and physician to Charles I in 1630. In his much-quoted dedication of *De motu cordis* to Charles, Harvey used the Aristotelian notion of the centrality of the heart – the basis of life and chief member from whom the body's life, vigour, and strength arise – to glorify Charles's position in the national body: 'The animal's heart is the basis of its life, its chief member, the sun of its microcosm; on the heart all its activity depends, from the heart all its liveliness and strength arise. Equally is the king the basis of his kingdoms, the sun of his microcosm, the heart of the state; from him all power arises and all grace stems ... Placed, best of kings, as you are at the summit of human affairs, you will at least be able to contemplate simultaneously both the central organ of man's body and the likeness of your own royal power.'[37] Christopher Hill, however, has suggested that Harvey clearly refuted the analogy of the heart's sovereignty in his *De circulatione sanguinis* (1649) and *De generatione animalium* (1651). Hill argues that in these later texts the heart was no longer the source of heat, whereas the blood was; the heart no longer drove the blood but rather the pulse was derived from the blood. The heart was no longer the 'chiefest of all the parts of the body,' or the 'residing god of this edifice, the fountain and conduit head'; now the 'blood lives and is nourished of itself, no way depending upon any other part of the body as elder or worthier than itself.'[38] Furthermore, Hill notes that Harvey had 'dethroned the heart,' in the words of his contemporaries, in the same year that King Charles was executed and the English republic proclaimed. The implications of these texts, Hill proposes, 'can only be described as republican – or at best they suggest a monarchy based on popular consent.'[39] The supremacy of the blood, he notes, cannot support the analogy of absolute monarchy. This view has been contested by Gweneth Whitteridge, since Harvey was no parliamentarian.[40] Harvey's personal political views aside, what is worth noting is the necessary *caution* used in describing the body's government. The correlation of individual human bodies to the national body meant that the man-machine so perilously approaching autonomy – the individual who was mechanically driven by natural forces and energies communicating throughout the body, rather than spiritually governed by a centralized monarch – carried with it a dangerous aura of validity for the decentralized control of the body politic by its citizens.

William Harvey was an advocate of the new philosophy that emphasized observation and experiment – reading what was commonly called The Book of Nature or God's Book – rather than solely deriving knowledge from the texts of ancient authority: he practised vivisection on a number of species to conduct comparative anatomy, and used quantitative experiments to determine the amount of blood leaving the heart. His friend Thomas Hobbes' proposed image of the state as an 'Artificiall Man,' where the analogy of body politic was not determined by any absolute need for anatomical accuracy; the analogy of the body politic in Harvey's work, however was presumably of secondary importance to his accurate observations of the body itself. Harvey obviously recognized the value of the heart-sun-king analogy in gaining Charles's sanction for a theory of circulation that many of his contemporaries found outrageous and implausible. However, Harvey's later suggestion that the blood rather than the heart is analogous to 'the superior orbs, (but especially the sun and moon),' which 'do by their continual motions quicken and preserve the inferior world'[41] may have been written with the knowledge that the heart-sun-king metaphor was no longer necessary. His views may have changed in the two decades between these publications. Indeed, Harvey may have been inscribing republican leanings in his old age and following Charles's flight from Oxford, or he may have felt freer to express, with the authority of his own age and stature within that particular chaotic political milieu, what his experiments and observations had demonstrated to him; he may even have observed that the 'sovereign' of human action and vitality was not a single locus. Certainly, Harvey was deferential to Aristotle's position in the philosophers' canon in the *De motu animalium* of 1628, where he introduced his theory of the circulation of the blood with an anxious apology. Such matters, he explains, 'though well worthy of consideration, are so novel and hitherto unmentioned that, in speaking of them, I not only fear that I may suffer from the ill-will of a few, but dread lest all men turn against me. To such an extent it is virtually second nature for all to follow accepted usage and teaching which, since its first implanting, has become deep-rooted; to such extent are men swayed by a pardonable respect for the ancient authors. However, the die has now been cast, and my hope lies in the love of truth and the clear-sightedness of the trained mind.'[42]

By the time he wrote *De generatione animalium*, Harvey could confidently state that both Aristotle's and Galen's theories of generation 'are false and rash assertions [that] will instantly vanish like phantoms of the night when the light of anatomical dissection dawns upon them.' The unimpeachable authority now was the body's own text: with dissection 'the Reader himself, with his own eyes shall discover [the truth]. ... Then will he at the same time understand how unsafe, nay base, a thing it is to be tutored by other men's commentaries without making trial of the things themselves, especially since Nature's book is so open and so legible.'[43] Harvey's earlier metaphors of the heart could have been used precisely because they were acceptable, both as Aristotelian images[44] and as flattering analogies to the established hierarchy of monarchy. They clearly conferred greater credibility upon his claim for the blood's circulation: Circulation was a fact of observation supported by acceptable metaphor.

Thomas Willis's publications show wary usage of the body politic analogy. *Anatomy of the Brain* (1664) and *Two Discourses* (1672) were dedicated not to the king but to Gilbert Sheldon, Archbishop of Canterbury (1663–1677), with whose patronage after the Restoration of 1660 Willis had been appointed Sedleian Professor of Natural Philosophy at Oxford University. Sheldon had been a supporter of Charles I and constitutional royalism in the 1640s, but by the 1660s he was an active and powerful Parliamentarian in defending the Church of England from the tolerant settlements that Charles II was considering in order to conciliate Catholics and nonconformists. The growing power of Parliament was evident in its ability to force its ecclesiastical policies upon the king: by 1670 Parliament had passed the Conventicle Act to limit the assembly of dissenters, and Sheldon was writing that 'all the disorders have arisen from the King's family and servants.'[45] Importantly, Willis's king seated in the brain did not have absolute power over the body: 'Concerning the Soul,' wrote Willis in the epistle dedicatory to the latter work,

> I have enter'd upon a great and difficult thing, and full of hazard; where we may equally fear the Censures of the Church, as the Schools. For that I assert a Man (as the Mad-man in the Gospel posses't with a Legion) to be indued with many distinct Souls, and design sometimes a legitimate Subordination of them, and sometimes wicked Combinations, troublesom Contests, and more than Civil Wars ... And altho I have a place of Safety, in that the Arguments and Reasons fight on my side, and that I have got the Suffrages of the ancient Philosophers, and the holy Fathers (and especially of St. *Hierome* and *Augustine*, and among the Moderns of *Gassendus* and our *Hammond*) yet suffer your Grace for my greater Safety, to extend your help to me, and grant that I may profess in the Entrance to this Discourse, that I am Your Graces Most humble and devoted Servant.[46]

The 'legitimate subordination' of the souls included a limitation of the powers of the monarchical rational soul to govern the body. The body's citizens comprising the material souls also held a degree of power to command and govern the body's states. That both Harvey and Willis felt the need so cautiously to introduce their own publications seems indicative of the power of communications in terms of both the actual body politic and of the body's analogical government.

Context 4: Clockwork versus Feedback in Human Machines

Otto Mayr's book *Authority, Liberty, and Automatic Machinery in Early Modern Europe* (1986) remains the best examination of the emerging concept of governmental self-regulation in seventeenth- and eighteenth-century Britain and its influence and dependence on the new technology of feedback control. The convergence of metaphors for machine and national government is one side of the cyborg's history; the other is the convergence of these metaphors with physiological ones for government of the individual human body. Although Mayr refers to Descartes' clockwork metaphors for the body as automaton in his first section on Authoritarian Systems, but does not recognize the early metaphoric treatment of bodies as feedback mechanisms. Early anatomists also conspicuously reinforced the governance of the body as a communications system. As we have seen, the analogy chosen for the body's governor is meaningful within the context not only of control over citizens' offices within the individual body; it also reiterates and reinscribes those of the body politic – often, as Mayr has pointed out, in describing a clockwork system as opposed to a self-regulating one: authoritarian and hierarchical, emphasizing social order, obedience, centralized rational control, and apposite division of labour.

Descartes also wrote of the human soul as a pilot or an engineer within and united to the mechanical body. As he muses in the sixth of his *Meditations on First Philosophy*, entitled 'The Existence of Material Things, and the Real Distinction Between Mind and Body' (1641) experience suggests, that 'I am not merely present in my body as a sailor is present in a ship, but ... I am very closely joined and, as it were, intermingled with it, so that I and the body form a unit.'[47] (In the 1647 translation by Louis-Charles d'Albert, this passage was rendered as 'a pilot in his ship.') In his posthumously published *Treatise of Man* (1664), Descartes describes the mechanism of muscular action as a series of hydraulic effects of the animal spirits (fluids in the nerves responsible for animating the body) and the soul residing in the brain as a fountain-keeper who oversees the movements of fountain automata: 'When a *rational soul* is present in this machine it will have its principal seat in the brain, and reside there like a fountain-keeper who must be stationed at the tanks to which the fountain's pipes return if he wants to produce, or prevent, or change their movements in some way.'[48] The main difference, then, between the automaton and the human is the presence of the steersman or rational soul.

Descartes' steersman was incorporeal and immortal, and it owed its allegiance to God, but the man-machine could nevertheless have imperfect judgment. Problematic human will and actions could be compared to an ill-made mechanism of the body-machine, as in a faulty clock, rather than being the fault of the soul:

> Perhaps it may be said that [people] go wrong because their nature is disordered, but this does not remove the difficulty. A sick man is no less one of God's creatures than a healthy one, and it seems no less a contradiction to suppose that he has received from God a nature which deceives him. Yet a clock constructed with wheels and weights observes all the laws of its nature

just as closely when it is badly made and tells the wrong time as when it completely fulfils the wishes of the clockmaker. In the same way, I might consider the body of a man as a kind of machine equipped with and made up of bones, nerves, muscles, veins, blood and skin in such a way that, even if there were no mind in it, it would still perform all the same movements as it now does in those cases where movement is not under the control of the will or, consequently, of the mind.

The senses then, 'notwithstanding the immense goodness of God,'[49] might deceive the combined unit of body and mind, and result in faulty will or desires while the soul itself remains perfect.

The communications system in the man-machine is dependent upon the type of machine serving as analogy. For Descartes in France, the animal and the human body are machines like a clock or automaton but the immortal soul is distinct from the corruptible flesh. The machine has been created by God, and the separate soul confirms the traditional undivided and supreme authority of spiritual as opposed to material powers. We have seen Descartes' model (Figures 2 and 17) of mechanical body and animal spirits flowing from the ventricles through pores in the brain to the pineal body, which in turn, regulated their flow into the hollow nerves. The immaterial soul animating the mechanical human body did not mingle with the body, and accordingly Descartes argued that there were no conflicts as are 'usually supposed to occur between the lower part of the soul, which we call 'sensitive,' and the higher or 'rational' part of the soul – or between the natural appetites and the will – consist simply in the opposition between the movements which the body (by means of its spirits) and the soul (by means of its will) tend to produce at the same time in the [pineal] gland. For there is within us but one soul, and this soul has within it no diversity of parts: it is at once sensitive and rational too, and all its appetites are volitions.'[50]

Descartes' solution to the conundrum of how the body mechanism moves and lives was thoroughly mechanistic in that it requires no vital soul but relies solely on the mechanisms of the material itself: 'I should like you to consider that these functions follow from the mere arrangement of the machine's organs every bit as naturally as the movements of a clock or other automaton follow from the arrangement of its counter-weights and wheels. In order to explain these functions, then, it is not necessary to conceive of this machine as having any vegetative or sensitive soul or other principle of movement and life, apart from its blood and its spirits, which are agitated by the heat of the fire burning continuously in its heart – a fire which has the same nature as all the fires that occur in inanimate bodies.'[51] The hydraulic operations of nerves and muscles in Descartes' machine were preset like those of automata or clocks, that is, their capacity to self-regulate was limited. The fire that agitates the spirits is not the soul but rather a constant energy intrinsic to all bodies whether animate or inanimate.

In contrast, Willis, in England, cautions against comparing the animal spirits, 'the Authors of the Animal function,' which 'constitute the *Hypostasis* of the Soul it self,' to liquid spirits. It is better to imagine them as 'Rays of Light interwoven with the air' because 'as Fire and Light in Mechanical things so in Animals, they are chiefly Energetical.'[52] Indeed, this energy with properties like fire or light is the mechanism of a circuit of information exchange (what Willis calls 'news'); the animal spirits have properties both of particle and wave, characterized as subtle particles that are 'transmitted' or 'communicated' to the brain in a fluctuation like waves from a fountain or like light.[53]

Willis, as founder of modern neurology, serves as a particularly useful case study to examine the history of the analogy that portrays the organism as a governmental system of communications and control. His *Cerebri anatome* (1664)[54] and *De anima brutorum* (1672)[55] not only reinforced the old image of the human body as microcosm of the body politic, but also reinvented it as a circuit of communications and commerce. Notably, these communications create a system of government that is like a self-regulatory machine. The brain, seat of the rational soul in man and the sensitive soul in beasts, is 'the origine and fountain of all motions and conceptions': the functions of imagination, memory, and appetite depend upon the brain. The sense and motion, passions and natural instincts are functions of the medulla oblongata and cerebellum; the cerebellum, with functions common to man and beast, is responsible for involuntary motion such as the beating of the heart, respiration, or digestive functions. Here the animal spirits 'as it were in a certain artificial Machine or Clock' go about their offices 'without any driver, which may govern or moderate their motions'; the brain or cerebrum is the seat of the incorporeal rational soul, the monarch, and 'the chief mover in the animal Machine ... the origine and fountain of all motions and conceptions.'[56] Notably, the cerebellum responsible for involuntary motion is imagined as a clock 'without a driver,' while the cerebrum is imagined in terms of a governing monarch, and both conduct their operations through a communications system.

Otto Mayr has traced the shift in imagery to describe systems of government in seventeenth- and early eighteenth-century Britain from purely mechanistic ('authoritarian') clockwork to self-regulating ('liberal') systems characterized by balance and equilibrium of both power and commerce. Mayr suggests that with the rejection of the authoritarian clock paradigm there was no other equivalent but distinct mechanism to demonstrate the desired balanced system of government, and 'the earliest really clear statements of the concept of self-regulation can be found in economic theory, specifically in the doctrine of economic liberalism.' He cites as 'remarkable for having recognized self-regulation ... at an early point' the London businessman Isaac Gervaise, whose *The System or Theory of the Trade of the World* was published in 1720.[57] In Willis, however, we see a comparatively earlier image of the regulation of the self through communications or, to use an anachronistic term, *feedback* within the body as machine and microcosm of body politic.

One example of automatic self-regulation occurs in Willis's analysis of involuntary function. In the healthy state, the animal spirits are peaceful and even complacent, 'so long as the tranquil region of the Cerebel, like a serene and fair Heaven, is free from all perturbation, the Spirits its inhabitants, being poured out with a pleasing sense, or as it were a certain complacency, flow with their proper habitations, both with a gentle circulation, and also with an equal flowing out enter the beginnings of the Nerves serving to the Functions both vital and natural: by which indeed easie Respiration, the Pulse, Chylification, and other offices of the same nature, are performed peaceably.' In the case of a fever, however, an automatic regulating system is set in motion by communication from the animal spirits below the cerebellum: 'If at any time the *Præcordia* grow too hot, and are burnt with a feaverish heat, presently by the passage of the Intercostals and the wandring pair of Nerves, the Spirits residing in the Cerebel, being warned of this evil, institute more frequent and stronger acts both of the Pulse and of Respiration. In like manner, if by chance the humors and sharp Juyces irritate or greatly trouble or afflict the Coats of the Ventricles or Intestines through the sense of this affection communicated to the Cerebel, the instinct of performing the motion is reciprocated, whereby the fibres of the parts, being contracted and wrinkled together, endeavour the shaking off of the hurtful matter.' This process occurs 'without our knowledge,' but when the motion or heat becomes too great the higher government in the cerebrum is set in motion: '[When] that sense of the trouble being transmitted to the Cerebel, for that it is more vehement, it unfolds it self more largely, and like a stronger waving of waters, passing through the

Cerebel, goes forward further even to the Brain, and warns its inhabitants of the evil; by which they being incited to oppose the enemy, cause a motion of another kind.' Thus, in order to reduce the fever, the cerebel will increase pulse and respiration as an involuntary mechanism. In the case that the fever is too high, 'the Brain, being warned of the same trouble, seeks and diligently requires cold drink and other remedies to moderate the heat.'[58] This system of 'self-regulation' is analogous not merely to a fire burning but to the thermostatically controlled alchemical furnace such as the one invented by Cornelis Drebbel, who served both James I and Charles I. Drebbel's thermostatically controlled furnace (ca. 1620) was the first self-regulating system invented in modern Europe, according to Mayr. Heat produced by the furnace determined the expansion or reduction of alcohol vapor, which in turn resulted in mechanical adjustment of the air supply to the fire.[59]

When Norbert Wiener described the phases of understanding and interpreting human bodies, he emphasized that one of the most important basic ideas in cybernetics is feedback: 'Here, then, is a significant parallel between the workings of the nervous system and of certain machines. The feedback principle introduces an important new idea in nerve physiology. The central nervous system no longer appears to be a self-contained organ receiving signals from the senses and discharging into the muscles. On the contrary, some of its most characteristic activities are explainable only as circular processes, travelling from the nervous system into the muscles and re-entering the nervous system through the sense organs.' Noting that working simulacra of the body have for centuries reflected technology, Wiener categorized Isaac Newton's time as the age of the clock, where the automaton functioned as a clockwork music box; in the nineteenth century, the age of steam engines, the automaton became a heat engine that burned fuel. The twentieth century, he concluded, was the age of communication and control: thus, the nineteenth-century paradigm of the engineering of the body as a 'branch of power engineering' was being replaced. 'What distinguishes communication engineering from power engineering,' he wrote,

> is that the main interest of the former is not the economy of energy but the accurate reproduction of a signal ...
>
> We are beginning to see that such important elements as the neurones ... do their work under much the same conditions as vacuum tubes, their relatively small power being supplied from outside by the body's circulation,

and that the bookkeeping which is most essential to describe their function is not one of energy.

In short, the newer study of automata, whether in the metal or in the flesh, is a branch of communications engineering.[60]

I wish to call attention to these images in terms of the early modern man-machine: the circular processes of nervous activity; the feedback mechanism or governor; the image of the body as both a machine and a text (here described as bookkeeping); and the predominant concern with signals or transmission of communications. Automata of the seventeenth and eighteenth centuries are only a small part of cyborg history. If we position the history of cyborgs instead in the history of physiological mechanisms – the nascent theories of atoms not only producing energy to animate both humans and machines, but also as means of transferring information and feedback – we can see the cyborg not as Cartesian automaton with a soul but more accurately as a human-machine-text moved by energy and controlled by a circuit of communications. Thomas Willis's man-machine comprises a tripartite metaphor that distinguishes it as a precursor of the twentieth-century cyborg: the body and corporeal soul are a mechanical entity; the body and corporeal soul are a governed body; and, finally, government is effected through a circuit of communications to and from the governor.

The inferior (corporeal) soul of man, Willis claimed, was material and divisible, co-extended with the body, and its faculties of motion and sense were to some extent independent of the brain and nervous system within the various parts of the body, as demonstrated by cutting worms, eels, or vipers into pieces that continue to move and show sensitivity when pricked. It was diffused through both the blood of the circulatory system, and the nervous juices of the brain and its appendices. While it was not unique to characterize the soul as being material, and infused through the material tissues, Willis's depiction of animal spirits as material bodies sharing physical properties with other natural substances – moving as both particles and waves – was an important contribution to our own understanding of the mechanisms of human will and action. Willis suggested that these subtle bodies were capable of 'transmitting' communications in a circuit typified as commerce to and from various regions in the brain and body. Significantly, these communiqués influenced the body 'oeconomy' as much as did commands from the soul-monarch in the brain, whereby what we anachronistically might call feedback from the animal spirits influenced, modified, and checked the power of command held by the sovereign rational soul. Willis's discoveries and speculations inform our own metaphorical constructions of cyborg consciousness, morality, and society – not merely because his influential anatomy described the human as comparable to a machine of unsurpassed complexity created by the 'Great Workman' – but because he understood the mechanisms of the governor of the human body in terms of a material system of communications and feedback.[61]

4 The Man-Machine: Communications, Circulations, and Commerce

Cartesian mechanistic philosophy alone could not provide an answer to the problem of what made matter move, or how that motion was governed. Norbert Wiener's steering mechanism in the mind was material, prone to dysfunction, and reprogrammable through human intelligence (psychoanalysis) and man-made cybernetic mechanisms. More recently, when Steven Pinker wrote of the mind as being like the Apollo space-craft,[1] he emphasized that it may be steered by body systems alone (several Apollo flights were 'unmanned'). The body in this analogy is self-powered and self-controlled through complex communications of the central and peripheral nervous systems via such feedback mechanisms as hormonal, electrochemical, and mechanical-physical reactions. In the early modern era of Descartes, it was God that was believed to ultimately power the vessel, and under God's authority was that vessel steered by the soul-inhabitant. This chapter examines the gradual trans-formation from control by an immaterial immortal soul to control by a material, machinelike bodily feedback system. Thomas Willis's represen-tations of the body's government in *Anatomy of the Brain* and *Two Dis-courses Concerning the Soul of Brutes* were cautious explications of the material components of the body itself as steering mechanisms occasion-ally at war with a governing monarch in the brain. A close study of these works provides considerable evidence for the textual heritage of the cybernetic organism as an embodied, physiological mechanism. Willis's works, along with Hobbes's analogies for the restriction of communica-tions within the mechanical body and body politic also provide consider-able evidence for the cyborg as a metaphorical union of uncontrolled 'inner' and 'outer' communications that constantly threaten anarchy and chaos. Apprehensions concerning uncontrolled mechanical bodies and uncontrolled mechanical minds continued to be reflected long after Willis and Hobbes – particularly in the works of Jonathan Swift and the Scriblerians but also in a culture of sensibility that established a new hierarchy for new physiological mechanisms of thought and sensitivity – and new mechanisms for rampantly commercial and ill-controlled publications of thoughts. As Willis's aether or animal spirits gradually changed in medical descriptions to electrical fluid and eventually to electrochemical fluxes of charged ions, the nervous system has consistently provided a metaphoric predictor of the conditions of the larger body politic. And as the powers of a metaphorical steersman or pilot or sovereign subject in the brain have been usurped by the network of exchange within the body, so have various theorists transposed the image to celebrate or condemn an ungoverned 'networked' society at large.

Thomas Willis's Nervous Government

What will this sustained examination of the works of a solitary man in the seventeenth century accomplish? I do not mean to infer a direct causal connection between Willis's energetical human machine and our own, but as a case study in the history of mechanical consciousness, Willis provides an astonishingly rich set of metaphorical-cybernetical germs. First, Willis emphatically made the soul material, and although he assiduously reiterated the immortality of the higher rational soul as monarch, his works ultimately diminished its authority and power. Second, he added 'energetical' forces to the mechanical body, and in so doing described a newer machine powered by an energy that was both waves and particles. Third, Willis's man-machine compounded astute anatomical observations and shrewd chemical surmises of the body's mechanisms with political commentary on the body politic; moreover, it challenged Descartes' version of the unity of mind and the duality of body and spirit. Finally, Willis's man-machine was governed by mechanical feedback: it was a cybernetic machine controlled by communications, and one not ruled, as a clock, solely by automatic mechanisms that reacted in certain ways to certain stimuli and ruled from the top down by a governing monarch. Rather, Willis's engine with its 'locomotive' powers was regulated by a system of governing bodies that reciprocally exchanged a multiplicity of circulating news and commercial goods to maintain order and health.

How was the early modern human-machine governed? Willis's man-machine was a child of civil unrest and conflict: Willis's experiences as a soldier in the royalist army and as a royalist supporter during the Puritan interregnum no doubt coloured his metaphoric representation of the nervous body politic, and his nervous body politic as a function of control and communications is in significant contrast to Thomas Hobbes's more infamous version. *Leviathan* was published two years after the public execution of Charles I, and it caused both religious and political offence – royalists regarded it as designed to encourage Oliver Cromwell to take the crown. Willis, the devout Anglican royalist, published his significant anatomical works after the Restoration of Charles II. Hobbes (in 1651) and Willis (in 1664) mapped the functions of the body and nervous system in terms of the mechanistic philosophy of their day, and while Hobbes's body was of course the artificial man of the common-wealth, it bore metaphorical similarities to Willis's later anatomical descriptions. Essentially, to simplify matters drastically, if the materialist philosophy suggested that all of nature – including humanity – could be described according to mechanical laws of motion, then morality, religion, the soul itself, and the civil hierarchies based upon those premises were brought into question. In the tumultuous period in which Hobbes and Willis worked, it is not surprising that theories of the motions of tiny corpuscular bodies circulating within the body and instigating motion as well as health and sickness were used metaphorically to describe the effects of civic movements and communications on the healthy operation of England's social body. Historians and theorists have not yet begun to explore cyborg history in metaphors of these early mechanical bodies as communications systems; Hobbes to some extent and Willis to a much greater extent are exemplary in their descriptions of civic governance, communications, and control as mechanical systems. Later studies of muscular motion provide less figurative bodies politic, but provide a valuable record of the changing values of the material-aetherial spirits animating the body.

To reiterate, the term *mechanical* was associated with the geometry of motion; it was also associated with lower status labourers. Thomas Blount's *Glossographia* (1656) defines *mechanick* as 'a Handicrafts man, a man of Occupation, a Trades-man' and lists the seven mechanical arts as agriculture or husbandry, clothing, navigation, hunting, architecture, medicine, and military discipline. Samuel Johnson's first definition for *mechanical* or *mechanick* a century later was more disparaging: 'Mean; servile; of mean occupation.' The second definition Johnson provided was 'constructed by the laws of mechanicks' such as those that described Newton's mechanical universe. Mechanical thus implies not only the mechanism of moving parts as it relates to both human and machine but also a lower-order human function or station. If the human body was merely mechanical, for many that implied that natural philosophy had demeaned a higher-order function of soul – intelligence, creativity, or spirituality. In terms of the commonwealth body, a mechanistic and materialistic view meant that the higher-order minds of society also risked being reduced to the same status as society's muscle. In the philosophies of Hobbes and Willis both, free will may be possible but a healthy body, like a healthy society, requires an orderly system of designed offices decidedly situated in a system of hierarchical rule by an intellectual elite, commercial interests, and armed enforcement. It was a metaphorical inscription of order with a specific system of communication privileging the brain as sovereign over the muscles. What is key – and still in need of further investigation – is that in their view, the mechanical nerves were the communications system of the body politic.

Communications and the Sovereignty of the Soul in *The Anatomy of the Brain*

Willis painstakingly explained that his mechanical interpretation of the soul should not be interpreted as anything approaching atheism. The epistle dedicatory to Archbishop Sheldon in *The Anatomy of the Brain* is as cautiously defensive as the later one in *The Soul of Brutes*: 'I am not ignorant,' he wrote,

> how great the labour is that I undertake: For it hath been a long while accounted as a certain Mystery and School-house of Atheism to search into Nature, as if whatever Reasons we grant to Philosophy, should derogate from Religion, and all that should be attributed to second Causes, did take away from the first.
>
> But truly, he doth too much abuse the Name of Philosophy, who considers the wheels, curious frame, setting together, small pins, and all the make and provision of a Clock, by which invented Machine the course of the Time, the orders of the Months, the changes of the Sea, and other things of that kind, may be exactly known and measured, if that at length, when by this his search and consideration, he hath profited himself so much, he should not acknowledge the Artist, to whose Labour and Wit he owes all those things.[2]

His study and explication of brains, he explains artfully, is as one who reads the 'Pandects of Nature [for] there is no Page certainly which shews not the Author, and his Power, Goodness, Trust, and Wisdom.' Furthermore, no right judge would fault him, he argues, for having studied 'these Rolls of Nature, because some Atheists may be made thereby.' At the conclusion of his instructions for dissection of the brain, Willis defends his many animal dissections as a means of discovering and demonstrating God's 'wonderful artifice,' whereby 'shewing the finger and Divine workmanship of the Deity, a most strong and invincible Argument may be opposed to the most perverse Atheist.' Willis's caution in both books was prudent, since his depiction of the human being as machine and corporeal soul came dangerously close to undermining the rule of the body's monarch, the immortal and incorporeal soul. Indeed, Willis's text adopts and refashions William Harvey's sun-king analogy – not, as one might expect, for a rational and immortal soul ruling over all the body from the head, but for the sensitive soul extending throughout the body: 'if the Head containing it self the chief part and power of the sensitive Soul, be taken for the body of some Luminary, as of the Sun or a Star,' he wrote, 'the nervous System shall be that radiant or beamy concretion compassing it about.' The brain and cerebel thus send out animal spirits as from a 'double Luminary' to 'irradiate the nervous System.'[3] If Harvey had indeed sanctioned a 'natural' system of body government by the blood rather than the heart, then Willis reinforces this imagery with a government comprised of animal spirits extending throughout the body and illuminated by *two* sources of light.

The *Anatomy* specifically rejected several Cartesian premises. As to whether spirits generated in the ventricles are deposited in the pituitary gland and pass through the hollow nerves, Willis's dissections had demonstrated otherwise. He notes dismissively that the 'Moderns' have esteemed the ventricles 'so vile, that they have affirmed the same to be mere sinks for [...] carrying out the excrementious matter.' Indeed, 'almost all Anatomists, who are of a later Age, have attributed that vile office of a Jakes or a sink to this more inward chamber of the Brain.' Willis points out that the nerves do not appear to be hollow; rather, they have 'solid smooth bodies' through which humour or spirits might seep as fluids through fine cloth or sponge. He had determined, by microscopic examination of their structure, that the nerves are 'not bored through,' as are the veins and arteries, and thus he speculates that the subtle humour which carries the spirits is instead diffused through their porous structure. Thus, Willis reiterates, 'because the animal Spirits require no manifest cavity within the Nerves for their expansion; neither is there need of the like for them' in the brain. The ventricles, therefore, 'ought to be deputed to some other office than this.' Willis also speculates here that the animal spirits move in waves, like 'so many lucid particles' or 'so many diverse rays of light' through the bodies of the nerves. As for the pineal gland, 'we can scarce believe this to be the seat of the Soul,' he contends, since it is not only found in man but also in four-footed beasts, and even fish and fowl.[4]

Most significantly, Willis proposed that humans are imbued with many souls that are material and united to the body. 'Humane government' was by two distinct souls in mankind, which Willis called the rational and sensitive in the *Anatomy* and the rational and the corporeal in *Two Discourses Concerning the Soul of Brutes*. In the latter text, even the rational soul seems to reside in the body rather than only in the head: the corporeal soul is 'joyned immediately to the Body [and] intimately united' with the immaterial and immortal rational soul, which 'residing in its [the Corporeal Soul's] Bosom, inhabits the Body.'[5] Willis's cautious modification to emphasize the soul's corporeality was a radical proposal in a period when the predominant Christian faith held that the immortal soul was not material. After the Restoration the Anglican Church was given back most of its old privileges, and it re-established close ties to the Crown: to seem to undermine the Church's authority over the soul might not be prudent – besides, Willis was a devout man. The sustained analogy throughout the *Anatomy* and reappearing in *Two Discourses* represented solely secular politics, however: the materialization of the soul granted certain powers to the individual citizens in the body's government. Whether his dedication to the Archbishop of Canterbury and the 'I'm just reading God's Book' defence was to shield *Two Discourses* from objections to his political analogy of the body politic or to Willis's discoveries about human anatomy and identity is hard to say. In any case, Willis's corporeal soul provided a new paradigm that would eventually come to dominate medicine and science: the consciousness or mind or identity is a result of – and controlled by – the material structure of the brain and the exchanges between the little particles (which we now call molecules) throughout. It is the status, substance, and activities of these particles that make Willis's works so meaningful today, when our medical-technological culture predominantly views emotions, memory, and thought as physical and not spiritual processes, and that made them so potentially seditious to publish and read in Willis's own time.

Willis was ostensibly more concerned with describing the structures than the mechanism of the soul in the *Anatomy* (see figures 11 and 21). He confirmed that the rational soul, immortal and incorporeal, is unique to humans and stated that the sensitive, inferior soul common to man and beast is composed of the vital or 'flamy' soul in the blood and the sensitive or 'lucid' soul in the head, as well as a third, the genital soul, derived from the other two and hidden within the spermatic bodies waiting to be 'inkindled into another Vital Flame,'[6] that is, a new individual. The sensitive/corporeal soul is comprised of the 'very subtil little Bodies' that reside throughout the brain, nerves, fibres, and blood and interact in a circuit typified as both the movement of troops and as commerce to and from various regions in the brain and body. Willis presents the body as a nation inhabited by the subtle bodies (atoms or particles) of the animal spirits. The 'Brain' (cerebrum) is a metropolis or castle, the 'oblong Marrow' (medulla oblongata) a highway, the 'Cerebel' (cerebellum) a 'free and municipal city,' and the nervous stock the provinces. Willis characterizes the brain (cerebrum) as a 'Castle, divided into many towers' or the 'Metropolis.' The metropolis generates the animal spirits (through a mechanical process of distillation from the blood) and sends them to the 'Callous Body' (corpus callosum, the thick band of nerve fibres that connects the cerebral hemispheres). But interestingly, this fountain that originates ideas and actions is not solely governed by the monarch or rational soul. It is also governed by the gentle flow of commerce and communications of the animal spirits, the subtle little bodies populating and giving life to the human body politic. Crucially, the animal spirits moving throughout the brain and nervous system are active and vital of their own accord. This vitality is not conferred upon them by the brain or the rational soul and, unlike Harvey's early image of the heart as monarch, they are not dependent upon the brain for circulation. Willis objected to the notion that the brain beats like the heart, as some philosophers had supposed: 'We wholly deny that it hath in it self a perpetual Systole and Diastole,' he wrote of this 'vulgar Opinion.'[7] The key image here, perhaps not surprisingly given the turmoil Willis (who had enlisted in the king's forces at Oxford in 1645) had witnessed in the struggle for power between the monarch and his no longer passive populace, is of *orderly* circulation where every 'Atom' has a particular 'office.'

The terms *commerce* and *circulation* as Willis uses them here seem to refer both to the notion of the spirits 'communicating,' and to a commercial gathering. The 'Callous Body' is a 'spacious field; where, as in a free and open place [the animal spirits] ... do meet together, and remain as in a publick *Emporium* or Mart; from whence, as occasion serves, they are raised up, and drawn forth for the uses of every Faculty.' From this central marketplace they travel to the common passage of the medulla oblongata, 'a broad or high Road,' or 'as it were the Kings Highway,' from the brain and the cerebellum, 'to be derived from thence into all the nervous parts of the whole Body.' They are 'orderly disposed ... as it were by series and orders' here, and thence flow outward towards the nerves, 'when they exert the loco-motive Faculty, or ... inward towards their Fountains, when the acts of sense, or rather the apprehensionsof sensible things are performed.' There are many diverting places and lesser paths grown to this medullary trunk, 'into which, the Spirits destinated to some peculiar offices, go apart: lest all the Spirits travelling this way and that way in the same path, should meet one another and disturb one anothers offices.'

According to Willis's model, the cerebellum, responsible for involuntary motions, generally receives little governance by the conscious brain, and has certain privileges to govern the rest of the body on its own. Willis proceeds to describe the 'acts' of the animal spirits in the cerebellum, the 'manner of œconomy or government the Spirits inhabiting the Cerebel and made free,' busied both by motions within and without this dwelling place to fulfil the needs of other parts of the system.[8] The cerebellum, then, is 'a free and municipal City [to which are granted] certain Priviledges and a peculiar Jurisdiction.' Willis emphasizes that even though the nerves of this free city are 'of another Dominion,' sometimes they are 'compelled to obey the beck and government of the Brain' because 'the Nerves of either Government communicate variously among themselves.' Nevertheless, it may be called a 'free and municipal City ... because the Brain it self in many things is compelled to serve the Cerebel and its Government ... and is necessarily bound to it ... so indeed that both these parts, though Principals, perform mutual offices, and as it were in a circle, require and accomplish services one for another.' The animal spirits in the cerebellum, Willis suggests, function 'as it were in a certain artificial machine or Clock' – that is, they are 'orderly disposed ... without any driver, which may govern or moderate their motions.' And why shouldn't we imagine that certain motions might occur 'without consulting the government of the will or appetite within the Brain,' Willis comments a few pages earlier, for 'it seems inconvenient' that the spirits should be 'called out of the Brain' when they are 'continually driven into fluctuations with the winds of Passions and Cogitations.' Not only do the 'subtil little bodies' of this body-nation influence the monarch-soul, but they also have freedoms not controled by the monarch.

The animal spirits in the brain and the cerebellum 'transmit' impressions or communications in distinct rays, 'as through Perspective-glasses' via those animal spirits which are the '*Internuncii* or Messengers going between.'[9] Conflating the communications system in the micro-governmental body with an almost prescient image of electronic communications, Willis described the motions of the animal spirits as undulating waves or a diffusion of particles: they are a breath moving upon the waters of the nerves; or as 'so many lucid particles' of rays of light. In either case, it is a rapid communications system: 'light is scarcely carried swifter through a diaphanous Medium,' Willis explains, 'than the communication of the Spirits is made from one end of the nervous System to the other.' Finally, the nerves extending directly from the brain and the nervous fibres extending into the muscles and tissues complete this circuit of communications within the body.

Even if Willis did not discern a difference between sensory and motor nerves, he did typify the commerce of the animal spirits or their communications as circular: the 'inflowing' animal spirits in the head are 'purified or refined more and more by a continual circulation' in the 'Marts.' They might from there go into the forepart of the corpus callosum to be used in the imagination, or into the medulla oblongata for other nervous activities, and whatever spirits remain 'are remanded back again to the hinder region of the brain by a certain circulation.' Notably, when he describes the animal spirits flowing from the nerves into the nervous fibres – from the aethereal locale of will and thought to the grosser material locale of muscular action – Willis changes the offices of the animals spirits from a gentle flow of commerce and communication to a more active military force. The nervous fibres, the organs of sense and motion, are roadways into the muscles. They depend on the nerves (the roadways or rivers from the brain) for military 'forces and supplements of the animal Spirits'; the nerves carry only the instinct for performing a motion, while the nervous fibres extending into the muscles cause the motion itself.[10]

At the junction of nerves and muscles the animal spirits are characterized as troops or companies of soldiers set, 'as it were in a watchtower' in the nervous fibres leading into the muscles and tissues. These 'implanted' animal spirits are 'more stable and constant' and only move according to the need for motion or sense. In this position, they both send 'impressions' (suggestive of imprints, as in printed materials) of sensory objects to, and receive impressions or commands from, the brain and cerebellum. The spirits responsible for motion receive food from the blood, and in a chemical reaction they are enkindled through nourishment from the arteries, the 'marriage' of the female seed of arterial juices with the male seed of the nervous juices, or the augmenting of the animal spirits by the 'auxiliary forces' of the blood. Like an 'explosion of Gun-powder' they thus expand the muscles for active motions.[11]

Once again, Willis emphasizes the orderliness of motions along the body's roadways: moving from the roadways of the nerves to their stations in the muscles, the animal spirits 'break forth not in heaps, or in a thick troop but only contracted orderly, and as it were by bands or divisions: but they being carried beyond the extremities of the Nerves, and there possessing the Membranes, Muscles, and other sensible parts, dilate themselves as it were into a most ample field, and with a very diffuse Army they dwell in the Pores and passages of the Fibres planted every where about; where also being endowed from the blood with new food, they become more lively and more expeditious or ready for the designed offices.' The nerves are small here, Willis explains, lest 'by too great a supplement of the animal Spirits, and the too thick gathering of the fresh ones still into the nervous parts, the army of the Veterans, before instructed, should be confounded, and so the orders of all being disturbed.' That is, there is clearly a hierarchy and an order established upon not only the correct instructions conveyed from above, but also upon orderly and sanctioned gatherings of the body's citizens. Indeed, Willis comments, when the animal spirits 'rush upon the nervous System with tumult and impetuosity,' the result is 'a great unquietness and continual throwing about of the Members are wont to be excited, to which sometimes madness and fury succeed.'[12] In this scenario madness and fury is due to the disorderly assembly of the 'subtil little bodies' in the muscles (those parts of the state responsible for action and labour rather than government or communications) obscuring the orderly commands issued from the brain and cerebellum. Willis seems to be tempering the analogy, and in 1664 does not use the word 'war' at any point in this description of the brain and nerves. As we will see, Willis's later tract changes this metaphor to the animal bodies of the corporeal soul actually involved in a civil war with the rational soul in the brain.

Whether the 'muscles' being described are the actual muscles that effect an individual's motions and actions or the labouring bodies of the

nation as in Thomas Hobbes's version of the body politic as artificial
man, the conditions under which they act and move are metaphorically
potent. Willis's description of animal government shares certain assumptions with Hobbes's *Leviathan* (1651), written while Hobbes was attend-
ing the court of the exiled Charles I and published two years after the
king's public execution (thirteen years before *Anatomy of the Brain*).
Although a more sustained comparison of the government of the man-
machine described by these two authors might prove to be fruitful,
discussion is here limited to how both Hobbes and Willis concentrated
their metaphors of the unlawful or the unhealthy in terms of too-large
gatherings by those parts of the state that are responsible for action and
motion. Hobbes envisioned a system of ethics and politics to circum-
scribe the roles of citizens in the commonwealth as an 'artificial man'
ruled by a unitary supreme power: '*Soveraignty* is an Artificiall *Soul*,' he
explained at the outset, 'as giving life and motion to the whole body ...
Reward and *Punishment* (by which fastned to the seate of the Soveraignty,
every joynt and member is moved to performe his duty) are the *Nerves*,
that do the same in the Body Natural.'[13]

Hobbes's notion of a mutual covenant of all wills to make one man the
'author' conflicts with Willis's later version of a variety of different
communications from sources all over the body. As one editor of *Leviathan*, Richard Tuck, has commented, Hobbes's proposal to transfer
citizens' individual judgment to one man or assembly of men was to
provide the sovereign above all 'an *epistemic* power, to determine the
meanings of words in the public language ... that is, they could attach
what personal meanings they chose to words where they were not required to communicate with other people. But as soon as any act of
communication was undertaken, they had to follow the sovereign's
ascription of meaning ... The consequence of this theory was that Hobbes
handed the sovereign unlimited ideological authority, over morality and
religion as well as day-to-day politics, and it is this power which has most
troubled his readers.'[14] Control in Hobbes's body politic occurs through
limiting communications, and specifically, the communications by the
workers or muscles of the body.

In Hobbes's vision of commonwealth, healthy gatherings are at marts
or markets, and disease results from too many bodies gathered together
in an assembly rather than flowing in an orderly fashion through their
designed offices. Each of the mechanisms within the body has its own
purpose and function, governed by the sovereign/soul at the head,
distributing through the nerves and tendons, or public ministers, the

sovereign's will to effect motion of the limbs. The muscles, as groups of 'men joyned in one Interest, or one Businesse,' serve commercial interests such as provinces, colonies, towns, or merchants of foreign trade. Hobbes's is a hierarchical system that reflects a distinction between the higher-order nerves distributing and effecting the will and laws of the sovereign brain, and the lower-order muscles effecting action and commerce. In describing the systems resembling the 'Muscles of a Body naturall,' Hobbes explained that a 'conflux of People to markets, or shews, or any other harmlesse end ... are Lawful. But when the Intention is evill, or (if the number be considerable) unknown, they are Unlawfull.' Hobbes concluded that factions 'are unjust, as being contrary to the peace and safety of the people, and a taking of the sword out of the hand of the sovereign.' Although 'it may be lawfull,' he suggested, 'for a thousand men, to joyn in a Petition to be delivered to a Judge or Magistrate; yet if a thousand men come to present it, it is a tumultuous Assembly ... And this is all I shall say concerning *Systemes*, and Assemblyes of People, which may be compared (as I said,) to the Similar parts of mans Body; such as be Lawfull, to the Muscles; such as are Unlawful, to Wens, Biles, and Apostemes, engendred by the unnaturall conflux of evill humours.'[15] In Hobbes's judgment, large gatherings of the populace were lawful when they were commercial (muscular) in nature, but not when they were intellectual – the expression of individual judgment – such as in presenting written petitions.

This is, of course, an oversimplification of Hobbes's philosophy. Nevertheless, the two texts emphasize important cultural assumptions about the designated systems of body and society. For both Hobbes and Willis, healthy gatherings are at marts or the marketplace (for Willis, in the region of the brain), and disease or madness and fury result from too many bodies gathered together in a 'tumultuous assembly' rather than flowing in an orderly fashion through their 'designed offices.' For Hobbes, as for Willis, commercial activities such as markets or shows are lawful. For Hobbes, however, it is a single representative that communicates to the higher-order soul, the sovereignty that gives life and motion to the whole body. The commonwealth, according to Hobbes, is defined in terms of many wills giving over their authority to one. The agreement of men is 'by covenant only,' writes Hobbes. The only defence, then, from invasion by foreigners and the injuries of one to another is to confer everyone's wills, or multiple voices, 'unto one Will ... and every one to owne, and acknowledge himselfe to be Author of whatsoever he that so beareth their Person shall Act, or cause to be Acted.' (Note that this is a statement that resonates with contemporary invocations of viral or foreign bodies invading the cyborg networked and decentralized communications systems of today). Moreover, stated Hobbes, censorship is a power that should be 'annexed to the Soveraignty to be Judge of what Opinions and Doctrines are averse, and what conducing to Peace; and consequently, on what occasions, how farre, and what, men are to be trusted withal, in speaking to Multitudes of people; and who shall examine the Doctrines of all bookes before they be published. For the Actions of men proceed from their Opinions; and in the wel [*sic*] governing of Opinions, consisteth the wel [*sic*] governing of men's Actions, in order to their Peace, and Concord ... It belongeth therefore to him that hath the Soveraign power, to be Judge, or constitute all Judges of Opinions and Doctrines, as a thing necessary to Peace, therby [*sic*] to prevent Discord and Civill Warre.'[16]

For Willis, conversely, orderly communications throughout the body – in a series of pathways to and from the brain – are *necessary* to keep the government of the brain and cerebellum functioning in health. What is crucial in Willis's analogy is not censorship, restriction, or the singular authority of the monarch, but orderly conduct by the body's citizens. When too many troops are sent into the muscle fibres, the animal spirits may be made 'too sharp' because the instructions from above are 'confounded,' in turn causing the animal spirits to become 'tumultuous' and 'impetuous.' A 'great unquietness' can result from troops rushing into the muscles, from which madness and fury *may* ensue; but not, as in Hobbes's commonwealth, the inevitable abscesses of unlawful gatherings in the muscles. It may be significant that Willis, who had held his own 'unlawful gatherings' of worshippers in his home during Cromwell's rule, explains in the chapter immediately following his analogy of the body politic that 'Tumors, or Strumous Ulcers or the running Sores of the [King's] Evil'[17] do not contain or pour out only a nervous fluid, as was commonly assumed at the time. Such matter is largely food from the blood, he explained, which collects when some of the nervous juice, 'being made vicious,' prevents large amounts of the nutritious humour provided by the blood from being assimilated by the body. This 'evil Ferment of the nervous Juyce puts on strange form, and that diversly degenerous' – meaning degenerated or fallen from virtue. Having noted that nutrition for the body might be provided by both blood and nervous juice together, Willis postulates that a 'twofold Juyce' might be necessary for nourishing the body. Thus concludes *The Anatomy of the Brain*.

It is tempting to speculate that Willis's *Anatomy* is not merely analogy but also, published four years after the restoration of Charles II, a subtle comment on the larger body politic, especially given Willis's interesting choice of words at the end of the penultimate chapter, on 'the Nervous System in General, where ... a prospect of the whole Animal Government is exhibited.' Immediately following his characterization of troops going too quickly into the muscular tissues as described above, Willis concludes with what might be one of the most ambiguous series of sentences in the book: 'In the order and ordination of the animal Spirits, such as was now described, the *Hypostasis* or the Essence of the sensitive Soul consists, to wit, which is only a certain *Systasis* or shadowy subsistence of those Spirits, which like Atoms or subtil Particles, being chained and adhering mutually one to another, are figured together in a certain Species. Moreover, the faculties of the same Soul depend upon the various *Metathesis* and gesticulation of those Spirits within the aforesaid Organs of the Head and nervous System.'[18]

In the glossary at the end of his text, Willis defines *hypostasis* as a substance or settlement, such as in the bottom of 'a urin,' thus discreetly emphasizing the materiality of the sensitive soul; *systasis* he defines simply as 'Constitution.' *Metathesis*, a linguistic term, is defined by Willis intriguingly as if it were a literary rather than a chemical or medical term.[19] According to Willis *metathesis* is 'transposing, the put[t]ing of one thing for another.' Does constitution refer to composition of the soul only, or does it also allude to the set of laws dictating how the nation was governed? Was Willis commenting on the uncertainty of the effects of the constitution of 1660, and the ensuing struggles for sovereign control over government and populace? Willis's tract, with its monarch firmly in place, implies not exactly a democratization of communications; but, quite possibly, in a time when the Militia Act of 1661 had given the king unprecedented authority to maintain a standing army, when new legislation renewed strict limits on the press and on public assembly, it may be advocating greater freedom of communication and movement for orderly citizens, and fewer supplements of troops gathered in the muscles than Hobbes's version of the body politic had recommended.

For both Hobbes and Willis, the healthy constitution requires an orderly flow through designed offices and a controlled system of messages to and from the sovereign government. Hobbes's version of absolute sovereignty, however, conflicts with Willis's dual sovereignty. Hobbes's emphasis (in Section XXIX of *Leviathan*) on the things that weaken or threaten dissolution of a commonwealth characterizes a crucial difference that insists on the limitations to the circulation of the populace. Although some doctors hold that there are two or three souls in a man, this was a dangerous situation for the commonwealth body, Hobbes said, for it would necessitate every subject being loyal to more than one master: if these powers disagreed, there would be great danger of civil war and dissolution, a disease he likened to epilepsy. More than one soul in the civil government (i.e., the power of representation in the nutritive, motive, and rational faculties of in the body politic), Hobbes submitted, would endanger the commonwealth as well, engendering a disease like the monstrous condition of a man with another man's head and torso growing out of his side. For Willis, however, the two souls contributed to a certain stability and regulation in the healthy body: disease does not result from a multiplicity of communications in the animal government; nor does it arise from independent forms of government within the political body. Willis is cognizant of the importance of the little bodies' participation in designed offices, and of their orderly motions through proper channels. For Willis, the damage to the body is wrought instead by 'clamor,' and it is from clamour that the possibility of civil war (disease) in the political body arises.

The Extension of the Soul in *Two Discourses Concerning the Soul of Brutes*

In his *Anatomy of the Brain*, Willis had avoided delving into the matter and mechanisms of the soul, explaining only rather vaguely that the essence of the sensitive soul consists in the 'order and ordination' of the animal spirits. The faculties of the soul depend upon the 'gesticulations' of these 'Atoms or subtil Particles,' he had written, 'But the consideration of this Soul and its powers requires a peculiar Tract, which hereafter (God willing) we intend.'[20] Willis's *Two Discourses Concerning the Soul of Brutes* was first published nearly a decade after the *Anatomy*. It was a widely read and influential treatise – going through eight editions between 1672 and 1683 – printed at Oxford, London, Amsterdam, Lyon, and Cologne; it was translated into English in 1683.[21] *Two Discourses* was both a cautious and a revolutionary explication of the rational soul as actually sharing its power and authority with the motive and sensitive, or corporeal, soul. This is not to say that Willis advocated greater power and voice for all in the body politic. Indeed, he prefaced this work with an appeal to the now established hierarchy of authority. The front matter includes a note to the 'Most Learned and Worshipful / By me ever

Respected / The Vice-Chancellor, Doctors, and Masters' at Oxford University. Since they had not objected when he lectured about these matters, he asks them respectfully to defend, if not the work in print then their own judgment, 'if perchance the literate *Thrasoe's* of this Age, who are wholly ignorant in Philosophy, every where wandring about, attempt to overthrow me with their Clamors, which is their chief Eloquence, to oppose your Authority against them, by which, if they are not put to Silence, it will however an high Confidence and inviolable Security to / *Honored Sirs, / the Admirer of you all, / Tho. Willis.*'[22] In a work that acknowledges the significant powers of the corporeal soul – the little citizens inhabiting the body-microcosm, as opposed to sole authority in the monarch seated in the brain – Willis was understandably cautious.

In *Two Discourses* the analogy of the body politic is present but not sustained to such great lengths as in the earlier *Anatomy of the Brain*. Here, Willis wrestled with guardedly explicating the materiality and the powers of the lower soul. By ostensibly discussing the corporeal soul of animals (which is the lower soul in humans), he attempted to avoid obviously ascribing to his spiritual monarch in the brain any association with material or mechanical action. However, the powers and functions of the two souls governing the body required, in fact, a more complex interaction that at times undermined the authority and autonomy of the rational soul. Willis's Preface, devoted to contrasting the rational soul of men to the corporeal soul shared by both man and beast, makes careful distinction between the mechanistic motion, sense, and limited knowledge of all living bodies versus the rational and immortal aspect of humankind. Moreover, 'if any one shall affirm, that most subtle Substance, and wholly Etherial, which serves for the Vital Oeconomy or Government to be immaterial, for that it enters upon the sluggish Disposition of inanimate Bodies, let him remember to be indulgent to me, if by chance I call it material, for that it subsists very much below the Prerogatives of Reason.' His work, he hopes, will offend no orthodox views of man's immortal soul. He should not need to fear that his work will be 'censured for Pernicious or Heretical,' he wrote, but 'on the contrary, we hope it is altogether Orthodox, and appears agreeable to a good Life, and Pious Institution: from hence the Wars and Strivings between our two Appetites, or between the Flesh and Spirit, both Morally and Theologically inculcated to us, are also Physically understood.'[23]

Willis was not being overly cautious in his frequent reassurances of the existence of an immaterial soul governing the mechanical body. The inferior soul of man, Willis would claim here, was material and divisible, 'co-extended' with the body, and its faculties of motion and sense were to some extent independent of the brain and nervous system within the various parts of the body. This could be demonstrated by cutting worms, eels, or vipers into pieces because these continue to move and show sensitivity when pricked.[24] The material soul diffused through both the blood of the circulatory system and the nervous juices of the brain and its appendices, gave its own inherent motion, vitality, and even rationality to the body. Although Willis was a devout and religious man, such a system could certainly be interpreted as a threat to religious orthodoxy and possibly the divine right to regal authority as well. Moreover, the political analogy that he himself used could be read as a confirmation that royal powers should be limited. Indeed, over half a century later when in Holland Julien Offray de la Mettrie, influenced in part by Willis, published his thesis that matter and mind were a single entity (in *L'homme machine*, 1747) he was forced to seek refuge at the court of Frederick the Great of Prussia. The publisher, Elie Luzak, was condemned by religious authorities. *L'homme machine* inspired tremendous hostility and numerous written attacks and rebuttals (one of them by Luzak himself). Sale of the book was forbidden in France; it was banned in Leyden, and burned in the public square of The Hague.[25] 'Thought is only a capacity to feel,' La Mettrie would proclaim, and 'the rational soul is only the sensitive soul applied to the contemplation of ideas and to reasoning!'[26] This assertion that the active principle in humankind is an inherent quality of matter itself and not distinct as soul was a highly political claim: if the animal spirit was purely mechanical, located in and affected by the material body rather than by the spiritual steersman, then any claims to the body's government based upon a theological hierarchy would be in effect, obliterated.

Nonetheless, Willis argued that there were two distinct souls, and he confirmed a mechanistic understanding of the corporeal soul with its two 'imperial seats,' the first being in the blood – which supplies nutrients to the body (the vital soul) – and the second in the nervous juice responsible for involuntary functions (the animal soul). The rational soul was losing authority over the 'citizens' of the body. No wonder, then, that Willis's discourses, published more than fifty years before La Mettrie's scandalous work, emphasize repeatedly that the man-machine is the work of God, and moved by God: 'I profess the great God, as the only Work-man, so also as the first Mover, and auspiciously present, everywhere, was he not able to impress strength, Powers, and Faculties to Matter, fitted to the offices of a Sensitive Life?'[27] This is a very 'nervous' introduction, indeed, to a tract that goes on to demonstrate that certain forms of knowledge, intelligence, and character in both man and beast are due to material mechanisms of the body.

Although Willis restates generally the structures of the animal government and reiterates the functions delineated in the *Anatomy*, mechanical imagery assumes more prominence and emphasis in the *Two Discourses*. Here, the animal spirits 'shut up within Passages, as it were Pipes and other Machines, abound with both an objective Virtue' of sensory faculties, and an active one of 'loco-motive powers' and 'Spasmodic Affections.' Willis emphasizes that if the human is a machine driven by an energy analogous to that used by human industry, it is nevertheless a superior machine because of the excellence of workmanship that created it and imparted the particles of its soul with 'supernatural' qualities. These qualities, like fire, air, and light used in the mechanical things of human industry, are 'chiefly Energetical.' In elaborating, he stated that 'in like manner we may believe, that the Great Workman, to wit the Chief Creator, from the Beginning, did make the greatly active, and also the most subtil Souls of Living Creatures, out of their Particles, as the most active; to which he gave also a greater, and as it were a supernatural Virtue and Efficacy; from the Excellent Structure of the Organs, the most Exquisitely laboured, beyond the Workmanship and artificialness of any other Machine.'[28] Matter, Willis argues, is not 'meerly passive,' although the assumption is common that matter cannot be moved unless by some other thing. Atoms, he explains, the smallest constituents of matter and thus also of the corporeal soul, are active and self-moving – thus, a material soul could move and perceive.

Significantly, the human is more than an energetical engine: it is one that reflects images and impressions. 'The Soul of the Brute is strong in sense and motion as a Machine,' Willis writes, while the 'Rational Soul, as it were presiding, beholds the Images and Impressions represented by the sensitive Soul, as in a looking Glass, and according to the Conceptions and notions drawn from thence, exercises the Acts of Reason, Judgment, and Will.' It is clear that the animal spirits are the 'Authors of the Animal Function,' Willis explains, but exactly what the animal spirits are seems 'hard to unfold.' They cannot be compared to fluid spirits such as turpentine or wine, but rather they are like 'Rays of Light' sent from the flame of the blood: 'For as Light figures the Impressions of all visible things, and the Air of all audible things; So the Animal Spirits, receive the impressed Images of those.' Although the metaphor is taken from printing, Willis describes here no grossly physical stamp or imprint on the brain: it is a premise, based upon observed properties of optics, that light can convey images or impressions from a source to a receiver, that is, to use the word *impression* is clearly a printing analogy, but here the 'imprint' is made solely with something like rays of light. 'I say none ought to wonder,' he explains, using the analogy of the camera obscura to describe visual perception, 'who hath beheld the Objects of the whole Hemisphere, admitted thorow an hole into a dark Chamber, and there on a sudden upon Paper exactly drawn forth, as if done by the Pencil of an Artist: Why then, may not also the Spirits, even as the Rays of light, frame by a swift Configuration, the Images or Forms of things, and exhibit them without any Confusion or Obscuring of the Species?'[29] The part of the soul that is the sensory in the brain, then, has 'a most speedy Communication, with the whole, and also with the several parts' of the soul, which then according to the 'Impressions there received,' choose certain appetites or acts.

Willis's is a distinctly anti-Cartesian representation of the man-machine, and in effect this corporeal soul inhabiting a mechanical body, representing the world to the conscious self as through the mirror and reflected light of a camera obscura, extended in space, and animated by an 'energetical' force like the rays of light, shares unmistakable affinities with the metaphorical cyborg whose mind is so compatible with the computer ethernet that it can exist there without the body. The suggestion is made more compelling by examining Willis's elaboration of the nervous system as communications system receiving and communicating impressions or 'news.' In *Two Discourses*, Willis reiterates the image of soldiers carrying out the orders for the body's motions. However, in the *Anatomy* Willis had described the animal spirits as troops 'flowing' to the fibres interwoven with the muscles by the passages of the nerves, and then being set, as in a watchtower, in the muscles. A decade later in *Two Discourses*, he describes the spirits as distributed throughout the nervous stock from the brain's fountain, but stationed throughout the nervous system to send 'news' to and from the brain. The spirits do not, he writes in this later text, travel back and forth 'as it were from one end of the Course or Circuit to the other,' as is commonly assumed; it is, rather, the information that travels through the circuit:

> Every impulse or stroke, which is inflicted from without to any Member, or to the Sensitive Body, is communicated instantly to all the Parts within the Head. If that an Impression or force tends from the Brain outwards, thorow the Nerves into the moving Parts, Motion is produced; but if they being

made outwardly are directed inwards towards the Brain, Sense arises ... the
Soul is stretched forth, thorow the whole, with a certain Continuity, its
Particles, *viz.* the Spirits contiguous one with another are set like an Army in
Array ... move not from their station ... and whether they be set in Battel
Array, or on the Watch, they perform the Commands carried outward from
the Brain, themselves being almost immoveable, and effect Motion, and
deliver presently to the Brain the news of any sensible thing impressed,
whereby Sensation is made.[30]

The soul in 1672 assumes both a more obviously detailed materiality and
extension, 'stretched forth, thorow the whole, with a certain Continuity,'
and assumes a more prominent role as communications system within
the body, in the fact that it is the 'news' rather than the militia that
travels back and forth through the nervous system and maintains its
equilibrium.

The sensitive soul, Willis explains 'as to all its Powers and Exercises of
them, is truly within the Head, as well as in the nervous System, meerly
Organical, and so extended, and after a manner Corporeal.' The rational soul is 'purely Spiritual,' but, he acknowledges, to differentiate
between the material corporeal and immaterial rational souls becomes
particularly problematic when we consider that knowledge, decision-
making, and the will to act is discernible in some of the actions of
animals. If the soul of brutes is 'much inferior and Material,' by what
mechanism do beasts come by their knowledge? For we know, Willis
comments, that beasts might also 'choose Acts, which seem to flow from
Council, or a certain Deliberation.'[31] Knowing that animals do not have
an immaterial rational soul, he concludes, acts resulting from reasoned
thought are not always guided by the immortal rational soul-monarch:
willed acts must also originate from the material soul extended through
the body.

'In most Mechanical things,' Willis elaborates, 'or those made by
humane Art, the Workmanship Excels the matter': we do not admire
that 'rude and simple sound' when wind is blown into a pipe, but we are
amazed by the complexity and the harmony of musical organs, the effect
of which excels *both* 'the matter of the Instrument, and ... the hand of
the Musitian striking it.' A machine can be built to play a complex
harmony without the direction of a musician, but nevertheless, it can
only play what is prescribed:

Further, altho the Musical Organ very much requires the labour of him
playing on it, by whose direction, the spirit or wind being admitted, now

into these, anon into those, and into other Pipes, causes the manifold harmony, and almost infinite Varieties of Tunes; yet sometimes I have seen such an Instrument so prepared, that without any Musitian directing, the little doors being shut up, by a certain law and order, by the mere Course of a Water, almost the same harmony is made, and the same tunes, equal with those Composed by Art. And indeed Man, seems like to the former, in which the rational Soul, sustains the part of the Musitian playing on it, which governing and directing the animal spirits, disposes and orders at its pleasure, the Faculties of the Inferior Soul: But the Soul of the Brute, being scarce moderatrix of its self, or of its Faculties, Institutes, for Ends necessary for it self, many series of Actions, but those (as it were tunes of harmony produced by a water Organ, of another Kind) regularly prescribed by a certain Rule or Law.

The distinction between man, beast, and machine, once again, is that little governor (or musician directing) in the brain. Although the actions of brutes may correspond to the harmonic tunes produced automatically by a water organ, Willis concludes, they are 'almost always determined to do the same thing,' that is, pre-programmed, and without free will. Indeed, he notes, the natures or souls of the more imperfect animals are 'inscribed ... *They do not so much act, as are acted.*'[32] The government of human beings, where harmony cannot be programmed, however, is another matter.

Even while 'the Corporeal Soul is the immediate Subject of the Rational Soul, of which, as she is the Act, Perfection, Complement, and Form by her self, the Rational Soul also effects the Form, and Acts of the humane Body ... [and] easily performs the Government of the whole Man,' they 'sometimes are wont to dissent among themselves, yea sometimes are wont to dissent and move more than Civil Wars.' Concupiscence, the lust for earthly pleasure, is the primary cause of civil war. The material corporeal soul 'co-extended with the whole body' and occasionally extending beyond the body could 'extend its Sicknesses, not only to the Body, but to the Mind or rational Soul' and infect the latter with its 'failings and faults.' Willis characterizes this influence over and usurpation of the rational soul's power as psychological sickness extended through the body. From this comes a physical understanding of 'the Wars and Strivings between our two Appetites, or between the Flesh and Spirit,' he writes, 'for that, *I see and approve the better things, and follow the worser;* and this, *The Flesh lusts against the Spirit, and the Spirit against the Flesh.*' These failings generally come to pass in us, he explains, because 'the Corporeal Soul adhering to the Flesh, inclines Man to Sensual

Pleasures, whil'st in the meantime, the Rational Soul, being help'd by Ethical Rules, or Divine favours, invites it to good Manners, and the works of Piety.' What Willis is diplomatically suggesting is that the soul-monarch, however spiritually superior, is fallible, and influenced by material processes. Unlike Descartes' comparison in *Meditations on First Philosophy* of a dropsical man drinking and doing himself harm as being like a poorly made clock – that is, the motions are caused by an inferior bodily mechanism and would be the same even if there were no mind present – Willis's rational soul could actually be perverted by the interaction with the material corporeal soul. 'Knowing Power,' however, is twofold – of body and mind – of the intellect (the 'Handmaid of the Rational Soul'), from which proceeds the will, and of the imagination (the 'Procuress of the Corporeal Soul') to which the 'Sensitive Appetite' cleaves.[33] There is always, Willis explains, a potential war in the 'Empire of the Mind' where reason might succumb its 'proper force' to the corporeal spirit.

Is it accidental that Willis specifically reminds his readers that 'it follows not' that the rational soul 'is propagated *Ex traduce* or of its Kind,' that is, the rational soul does not beget another rational soul, 'for as much as oftentimes the Son, in respect of Wit, Temperament, Ingenuity, the Affections, and other Animal Faculties, is exactly like the Father,' these qualities and faculties actually come from the corporeal soul. Willis makes this comment immediately prior to his announcement that he would 'now consider the Disputes and Wranglings of [the souls], which in respect of their Powers, often happen.' Could he be commenting on the ongoing struggles for power between Charles II and parliament? Between growing factions within parliament itself? Or perhaps between political factions developing from speculations about his illegitimate son and possible heir to the throne James, the Protestant Duke of Monmouth, versus the lawful succession of Charles's brother the Catholic James II? While Willis earlier had stated that the rational soul governs the corporeal absolutely, in *Two Discourses* he acknowledges that the corporeal soul 'does not so easily obey the Rational in all things, not so in things to be desired, as in things to be known: for indeed, she being nearer to the Body and so bearing a more intimate Kindness or Affinity towards the Flesh, is tied wholly to look to its Profit and Conservation.' Being responsible for and more attentive to the needs of the body, however, the lower corporeal soul may be apt to indulge in pleasures and grow 'deaf to Reason'; moreover, Willis avers, 'the lower Soul, growing weary of the yoak of the Other, if occasion serves, frees itself from its Bonds, affecting a License or Dominion ... " – *Where Ensigns Ensigns meet, / And where with Arms, they one another threat.*" This Kind of Intestine Strife, does not truly cease, till this or that Champion becoming Superior, leads the other away clearly Captive.' The passage Willis quotes here appears as '*Infestis obvia signis signa, pares aquilas, & vota minantia votis*' in the 1672 Latin edition.[34] In the Latin, the passage is a noteworthy amendment to the Roman poet Lucan's *Bellum Civile* (*Civil War*). Lucan's work has been described as an anti-Aeneid – a disenchanted rewriting of Virgil's epic myth of heroic war and conflict, and the dawn of Rome. In Lucan's poem, Aeneas's descendent Julius Caesar oversees not the dawn but the demise of a nation:

> Bella per Emathios plus quam civilia campos
> iusque datum sceleri canimus, populumque potentem
> in sua victrici conversum viscera dextra
> cognatasque acies, et rupto foedere regni
> certatum totis concussi viribus orbis
> in commune nefas infestique obvia signis
> signa, pares aquilas et pila minantia pilis.

> Of wars across Emathian plains, worse than civil wars,
> and of legality conferred on crime we sing, and of a mighty people
> attacking its own guts with victorious sword-hand,
> of kin facing kin, and, once the pact of tyranny was broken,
> of conflict waged with all the forces of the shaken world
> for universal guilt, and of standards ranged in enmity against
> standards, of eagles matched and javelins threatening javelins.
>
> (trans. S. Braund)

Lucan's poem tells of the Roman revolution and civil war between Caesar and his son-in-law Pompey, whose death signals the collapse of the Roman Republic.[35] Significantly, Willis substitutes *votum*, a vow or a promise to God, for *pilum*, the Roman legionnaire's javelin-like weapon in the last line of Lucan's poem.

It seems likely he was here proclaiming his own disenchantment with civil war and the demise of the republic. However, even with the restoration of Charles, the commonwealth that Willis had imagined in *The Anatomy* had been beset by conflict: Charles's Act of Uniformity, which would have given nonconformists the right to worship in their own way, had been defeated by parliament in 1663. The four acts known as the

Clarendon Code passed by parliament during the ministry of Edward Hyde, earl of Clarendon, placed punishing restrictions on the ability of the nonconformists to hold municipal offices, to hold gatherings – even private ones – of more than five people, or to visit or live in towns or cities where they had preached. In 1665–6 the government had suffered the crises of diplomatic isolation, economic strain, and a growing public dissatisfaction all as a result of the Anglo-Dutch war; in 1665 the Great Plague and the Great Fire of London had devastated the city. As Tim Harris has noted, Samuel Pepys recorded in 1667 the scandal that the Archbishop of Canterbury (Willis's dedicatee) 'doth keep a wench and that he is as very a wencher as can be.' Furthermore, the state of the body politic was by no means hale and robust: the proverbial maxim in London by this time was that 'the Bishops get all, the Courtiers spend all, the Citizens pay for all, the King neglects all, and the Divills take all.' Squabbles and machinations had plagued parliament, which turned out to be no system of peaceful regulation: In the Easter holidays in 1668 rioters had attacked London's brothels to protest religious persecution and the court's sexual license. The supposed sexual indulgences of Archbishop Sheldon figured in the rhetoric accompanying these riots.[36]

To return, then, to Willis's 1672 discourse on the souls of brutes, Willis follows his misquotation of Lucan with the explanation that the rational soul might use its proper force, institutes, and sacred ethics in trying to 'draw the Faculties of the Corporeal Soul to its Party,' but the corporeal soul adheres to the flesh and may in fact corrupt the rational. Thus can the corporeal soul 'seduce us in the Mind or Chief Soul ... to roll in the Mud of Sensual Pleasures: So that Man becomes like the Beast, or rather worst.' Reason becoming brutal, Willis explains, 'leads to all manner of Excess' in the 'Empire of the Mind' whereupon the rational soul, 'returning at length, sometimes on her own accord, or awakened by some occasion, and knowing of its fall, arises up against the Sensitive Soul, as against an Enemy or Traitor, casting her out of her Throne, commands her to Servitude; yea, sometimes by reasons of some wickedness committed, it compels it to torment it self, and its Lover the Flesh, and so expiate as much as it may, its faults, by inflicting on it proper Punishments.' Willis specifically draws attention to the strife of political and religious doctrine in his body politic: When one of the souls 'opposes the other, and either strives for the obtaining of Proselytes, it happens that Man is hurried into contrary Endeavours, and is acted little less than like a Daemoniack possess'd with a Legion,' he concludes. Thus ends Willis's digression '(which we have touch'd only by the by, as besides our pur-

pose)' on the body's government by the rational soul.[37] *Two Discourses*
goes on more prosaically to conduct a thorough comparative anatomy
and a discussion of some of the exemplary cases in his medical practice.
For Willis, the man-machine exhibits several interrelated principles: the
human engine is, for him, a real anatomical structure with discernible
material properties; the human engine is an analogy or perhaps even a
microcosm of the systems of communications and control in the body
politic; and the human engine demonstrates that when crass materialism
corrupts spiritual loyalty to divine principles, the man-machine is no
better than a beast.

The analogy in *Two Discourses*, then, revises that found in the *Anatomy*
– which was a blueprint for a sound and orderly political œconomy
written in a more hopeful period immediately following the Restoration.
It is in this later text a cautionary rebuke against sensual pleasures and
sins of the flesh. Written in this period of dissent and the notoriously
sensualist and gaudy excesses of Charles and his royal court, the analogy
is noteworthy in that the rational soul as monarch is not only feminized,
but also in succumbing to the 'procuress' of sensual appetite it invites
war: could this analogy mask another subtle commentary by the pious
Willis, this time on the king whose much-satirized sensual pleasures and
many mistresses were supposed to have undermined the affairs of the
state, and perhaps too on the other ambitious and less than pious
members of government? The analogy between body and body politic,
however, is not extensively developed in this tract and I would hesitate to
read too much into the sensual corporeal soul's perversion of the rational monarch governing the body, except to emphasize once again the
obvious history in Willis's body politic: mechanical communications
between governor and governed might contribute to a healthy equilibrium, but the risk of materialism and mechanism, undoubtedly, is that
they consistently threaten a spiritually based order.

Literary Communications: Materialism and the
Mechanical Operation of the Spirit

If for Willis a spiritual hierarchy of king, commerce, and army communicating through controlled and moderate exchanges in a centralized
mart was an ideal, for an increasingly secular eighteenth-century culture
of chaotic economic expansion, trade, profit, and rising national wealth,
new hierarchies would define government and commerce in both the
body and the body politic. The rise of a chaotic and uncontrolled print

culture of newspapers, pamphlets, booklets, and broadsheets in a trade influenced more by the will of the populace than by the command of the monarch and aristocracy parallelled increased portrayals of the sensitive, nervous body of higher-order citizens. Throughout the eighteenth century, claims to moral authority relied upon differentiating lower-order mechanic intellect from higher-order nervous intellect, that is, dull, gross corporeality from refined sensitivity.

Furthermore, to imagine intellectual labour as a commercial, manual, and mechanical labour constituted an alarming diminishment in the status and profits of the educated gentleman. The years between 1640 and 1660 in England, with the civil wars and execution of the king, the formation of a republican commonwealth and its replacement by Cromwell's Protectorate, had witnessed the breakdown of the system of censorship by government and church authorities and a greater output of printed material than had ever been seen in that country. A confusion of ideas and opinions both conventional and unorthodox appeared in an outpouring of pamphlets, periodicals, newspapers, and books for political, religious, and commercial ends. For the first time, Parliament gained access to the press, which resulted in what Sheila Lambert has called 'the first mass propaganda machine of modern times'[38] while Quakers, Ranters, Levellers, and others published radical works with unprecedented freedom.[39]

Four years after the restoration of Charles II, the same year that Willis published his representation of bodily government in the *Anatomy*, Richard Atkyns published his own version of the body politic in *The Original and Growth of Printing ... Wherein Is also Demonstrated, that Printing Appertaineth to the Prerogative Royale; and Is a Flower of the Crown of England* (1664). In this pamphlet Atkyns deplored the grant of a royal charter to the Stationers' Company in 1557. The subsequent lack of controls over publication, he claimed, had weakened the power of the monarch in the 'Head' of the body politic by extending rights of communication to the 'Body,' and had also debased the 'famous ART' of printing by turning it 'into a mechanick Trade for a livelihood.' Here print becomes a sort of nervous juice of communications: 'If the Tongue, that is but a little Member, can set the Course of Nature on Fire; how much more the Quill, which is of a flying Nature in it self; and so Spiritual, that it is in all Places at the same time; and so Powerful, when it is cunningly handled, that it is the Peoples Deity,' Atkyns argues in the 'Epistle to the King.' Furthermore, 'that this Power which is intire and inherent in Your Majesties Person, and inseparable from Your Crown, should be divided, and divolve upon Your Officers (though never so great and good) may be of dangerous Consequence: You are the Head of the Church, and Supream of the Law; shall the Body govern the Head?' The fine (spiritual) art of printing, according to Atkyns, was 'depraved' when printers formed a body – 'Concorporated' or 'Incorporated' – with the tradesmen who were the bookbinders, booksellers, and type founders: 'the Body forgot the Head, and by degrees,' he remonstrates, 'they kickt against the Power that gave them Life.'[40] Atkyns's hierarchy is explicit: mere 'mechanicks' had usurped what should be a higher spiritual 'art.' Atkyns's pamphlet, it should be noted, was probably less a sincere accolade to the king than an attempt to vindicate his own position in a legal conflict with the London Stationers' Company over a number of profitable patents. Regardless of his own aims, however, Atkyns's metaphors are a telling conflation of material (commercial) motives and material (mechanical) spirit in early modern print culture.

Despite the establishment of the Licensing Act of 1662 to restore the state's control over printing that had lapsed during the Civil War, and its renewal in 1685, the act was finally allowed to lapse in 1694 as too difficult to enforce. The number of printers, which had been restricted to twenty in London, swelled, and with the requirement to register books with the Stationers' Company removed, publication expanded. Authorship had become a trade wherein learning and patronage counted for little while material production and consumption – booksellers, and the book-buying public – dictated subject matter. Censorship of print materials prior to publication was no longer an impediment to communications. Printing was a free market, expanding at an unprecedented rate. Around 6,000 titles had appeared in England during the 1620s; there were 21,000 in the 1710s, and by the 1790s the number was more than 56,000.[41] How did the materialist flow of commerce mesh with an idealized commerce of knowledge and genius? With the escalating commercialization and professionalization of print culture in the late seventeenth century and throughout the eighteenth, the new literary marketplace was simultaneously a force of democratization and reason, and a chaotic traffic in ungoverned ideas that both destabilized and threatened traditional letters, authority, religion, and civility.[42]

As might be expected, the eighteenth century witnessed a rapidly expanding multitude of heterodox publications in England. The literary chaos also, as Martin Battestin has noted, provoked a defence of spiritual orthodoxy by fiction writers from Jonathan Swift at the beginning of the century to Laurence Sterne at its end who, 'recognizing the danger posed by Hobbes and the deists,' attacked such materialist and mechanist ideas 'in force,' warning against these new publications as 'dangerous not merely to the moral and spiritual health of their readers, but to the social order itself.'[43] The new print culture represented a danger to traditionalist critics because it lacked centralized control. William Warner's analogy for eighteenth-century print culture is remarkably apposite: 'The print market in which *Pamela* appears,' he suggests in his important study entitled *Licensing Entertainment: The Elevation of Novel Reading in Britain, 1684–1750,* 'may best be described as an "open" system' (meaning the open source development and dissemination of computer programs whereby the 'text' is released to the network, and modified and redistributed by any participant in the community). Furthermore, 'I do not mean that it is random or chaotic, nor that it is free of constraints. But neither is it a self-regulating totality that sustains some essential character through the sort of homeostasis that is characteristic of, for example, many biological systems. The print market is a system of production and consumption in which no one can control or guarantee the meanings that sweep through its texts. It is open to seismic shifts and dislocations. Lacking centralized censorship or certification, the market is influenced by any who can get their writing printed. Here there are no commonly recognized standards, and remarkably few limits as to what may be said or written.'[44] One might quibble about whether the print culture exhibited homeostasis: the very fact that it continued to thrive and survive, despite a variety of upheavals and flux, suggests a tendency to reach a sort of tenuous though constantly disturbed equilibrium. Nevertheless, Warner's analogy both underscores the important and lasting metaphoric relationship of bodily control systems and the government of human communications, and it reminds us of how important this shift in media culture was during the eighteenth century.

The relationship of the ungoverned man-machine to ungoverned print culture is somewhat tenuous, but certain authors of the period made explicit the relationship between the physiology of the mind and the democratization of human expression. As we have seen with Richard Atkyns's complaint, the 'mechanick' world-view represented one of the most disturbing aspects of ungoverned communications in a materialistic society, for not only would uncontrolled communications for profit threaten the stability of the body politic, but also the established social hierarchy that had been based firmly upon church and monarchy. Indeed, 'we think, and we are even honest citizens, only in the same way as we are lively or brave; it all depends on the way our machine is constructed,' La Mettrie would write at mid-eighteenth century in the more inflammatory publishing milieu of French authors: 'Words, language, laws, science and arts' are merely a mechanism of education stamping ideas into our brains. 'Man was trained like an animal,' La Mettrie claimed, 'he became an author in the same way as he became a porter.' The old Hobbesian hierarchy of the mind over the muscles of the state was effectively destabilized by a man-machine lacking the centralized spiritual sovereign that Hobbes and Descartes had celebrated. 'All the windy learning which inflates the balloon-like brains of our haughty pedants,' wrote La Mettrie, 'is therefore nothing but a mass of words and figures, which form all the traces in the head by means of which we discern and recall objects.' The relative equality of authors and porters is a matter of mechanical motion with significant sociopolitical implications. 'Given the slightest principle of movement,' La Mettrie conjectures, 'animate bodies will have everything they need to move, feel, think, repent, and, in a word, behave in the physical sphere and in the moral sphere which depends upon it.' His proof is a list of experiments that suggest the body or parts of the body separated from the brain and nervous system can move or be reanimated on their own, or polyps cut into tiny pieces may even regenerate spontaneously. 'I have given many more facts than are needed,' he concludes, 'to prove beyond all doubt that each tiny fibre or part of organized bodies moves according to its own principle, whose action does not depend, like voluntary movements, on the nerves.'[45] For La Mettrie, if there is no spiritual steersman in the brain controlling the body's movements, then knowledge and authority were no longer the perquisite of traditional powers.

If knowledge could be a physical imprint on the brain, judgment, reason, memory, all those aspects that make us human, would merely be the result of this physical conformation. Thus, while the man-machine image implied a certain equality of potential that negated an earlier hierarchy of nobility and status based on birthright, it also now came to imply a hierarchy based on education and structural superiority. In England and on the continent, a delicacy of the nervous structure would reinforce an older social order of the commonwealth's muscles and minds.

Recall Swift's targets of satire early in the century, 'the numerous and gross Corruptions in Religion and Learning,' in *A Tale of a Tub Written for the Universal Improvement of Mankind,* wherein Swift directed his criticism at both the arrogant faith in the new science to explain human identity, and at the implication that higher-order knowledge or spirituality could be a mere mechanical process and a purchased prescription. Scoffing at the presumptions of ill-educated or pointless and trivial authority, he created the recipe for learning by imbibing 'an universal System in a small portable Volume, of all Things that are to be Known, or Believed, or Imagined, or Practised in Life': Take a collection of learned books, distil them seventeen times, and let all that is volatile evaporate; then snuff the elixir up through the nose whereupon 'it will dilate itself in the brain (where there is any) in fourteen minutes, and you will instantly perceive in your head' the entire sum of distilled knowledge. For Swift too, the 'mechanick' was bound up with the lower order materialist functions of labouring bodies. Recall, for example, his Grubstreet Hack commenting that while the (now extinct) critic had once been a restorer of 'Antient Learning' and a noble hero, the 'true modern' critic 'is a sort of mechanic, set up with a stock and tools for his trade, at as little expense as a tailor; and ... there is much analogy between the utensils and abilities of both.' The modern critic is entirely contrary to true intellect: 'before one can commence a true critic,' Swift's Hack suggests, 'it will cost a man all the good qualities of his mind; which, perhaps, for a less purchase, would be thought but an indifferent bargain.' The conflation of mechanical laws of nature to describe the mind, and the mechanical labourer as opposed to the learned intellect, reflected the long-term concerns of the educated elite regarding control of communications within the body politic. He may have been satirizing the materialist tradition inherent in the works of Hobbes, but Swift certainly allied with Hobbes in terms of confirming the need for a hierarchy of mind and muscle within the body politic.

The 'certain great prince' intent on raising a mighty army and invincible fleet, for example, motivated by the effect of pent-up sexual spirits upon the brain, evokes Hobbes's monstrous mechanical man. The same spirits, Swift's narrator concludes, 'which, in their superior progress, would conquer a kingdom, descending upon the *anus* conclude in a *fistula*.' Swift's satires functioned not only to recomplicate those reductionist systems that defined and reduced the complexities of spirit, character, and morality to mechanism. His exaggerated caricature of the effects of a mechanistic spirit on the body politic also emphasized his ire at the 'modern' cultural changes espoused and enacted by the ill-educated fools threatening his own brand of authority as an Anglican clergyman and as a classically educated reader and writer. Swift's *A Discourse Concerning the Mechanical Operation of the Spirit*, appended to *A Tale of a Tub*, was perhaps his most scathing remark upon the mechanical construction of human spirit: it is at once a harsh invective against mechanistic explanations for the soul, and a parody of what Swift saw as a false religion purported by the 'mechanick,' unlicensed 'Enthusiastick' preachers who claimed direct and unmediated illumination by the Holy Spirit. The material/spiritual play in this piece is an elaborate defence of the status of the true minds, spirits, and voices of the commonwealth over the laughable baseness of its uneducated 'manual' tradesmen. By digesting 'Theological Polysyllables and mysterious Texts from holy Writ, applied and digested by those Methods, and Mechanical Operations already related,' he writes, one may gain a 'competent Share of *Inward Light*' – which then transforms to material liquid shared with the congregation via nose blowing, hawking, spitting, belching, or snuffling. In his most barbed coupling of true spiritual illumination with mechanical physicality Swift creates a double entendre that conflates the human spirit (which creates the moral being) with semen (which creates the material being). This form of worship provides three ways, the narrator opines, of 'ejaculating the Soul, or transporting it beyond the Sphere of Matter': 'It is, therefore, upon this *Mechanical Operation of the Spirit* that I mean to treat, as it is at present performed by our *British Workmen*. I shall deliver to the reader the result of many judicious observations upon the matter, tracing as near as I can the whole course and method of this *trade*.'[46] Swift's satire is directed against what he saw as the deluded and self-obsessed blindness of fanatics, whose notions of spiritual inspiration he reduced to the crassest material and physiological terms. It is also, however, a haughty ridicule of the lower orders, the tradesmen, the workmen, the mechanics who for Swift had no authority to communicate their own form of spirituality. Characterizing the human as mechanical was also associated with a natural law that potentially equalized the capacity of all humanity to think, reason, make decisions, and communicate ideas. At the centre of the human-machine image is the problem not only of how we come by, and to whom we owe, the intelligence and knowledge that governs our actions, but also the relative value of such knowledge.

Swift's works deride an increasingly secular and commercialized culture attentive to material and economic rather than spiritual gain. For Swift, the modern symbolizes an endorsement of intellectual and spiritual laziness as well as crass materialism. The image of mechanism and materialism in the modern thinker was adopted by Alexander Pope years later in his *Peri Bathous: Or, Martinus Scriblerus, his Treatise of the Art of Sinking in Poetry* (1728). Pope's ironic recipe for making an epic poem –

the 'greatest work human nature is capable of' – purports to contribute additional advice to the already existing 'mechanical rules for compositions of this sort.' The pedantic Martinus proclaims: 'I shall here endeavour (for the benefit of my countrymen) to make it manifest, that epic poems may be made *without a genius,* nay without learning or much reading.' Pope's obvious target is the crass materialism in print culture, where mere mechanics may take the place of genius and ancient learning: the section concludes with the caution that, while similes and metaphors 'may be found all over the Creation; the most ignorant may *gather* them, but the danger is in *applying* them. For this advise with your *bookseller.*'[47] Swift made his living as chaplain, and eventually as dean of St Patrick's Cathedral in Dublin; Pope made his living from sales of his poetry. These men were not averse to financial gain from their own intellect. What was at stake was tradition, authority, and morality. The gradual acceptance of mechanism and materialism in the realms of both physiology and the literary arts thus engendered a new hierarchy of sublimity and genius that challenged mere mechanism.

With understanding, reason, and knowledge no longer the prerogative solely of a privileged gentility, the hierarchy of mind and muscle in the body politic that was still paradigmatic in the works of Harvey, Hobbes, and Willis, was in some ways being levelled. The rise of a merchant class had created a 'mechanical' identity out of the very tools of a man's trade, whether as publican, apothecary, or writer (see figures 22, 23, and 24), while women's 'labour' was still identified as sexual, reproductive, and domestic (see figure 30). The victualler or publican (see figure 23) 'erected out of his own implements without the assistance of Nature,' however, is a somewhat crude and laughable figure, with his head a 'Piss Pot Engrav'd' and his sole reading material, it would seem, *The London Gazett Extraordinary,* tucked under his chin as a napkin. The implications of a mechanically formed reason and knowledge, as we have seen, along with the 'mechanick' production of texts was that trash was dangerously vying with 'true' art. Linda Zionkowski has drawn attention to eighteenth-century disputes over 'mechanic' and commodified literature versus that of the 'legitimate' minds of the '"guardians of the commonwealth" [who] could participate in shaping social, political, and aesthetic beliefs.' Zionkowski describes the resentful reactions of Goldsmith, Fielding, and Johnson to the challenge posed to the legitimate minds of the state by the merely mechanical labour of lower classes: 'Goldsmith blames such anarchy,' she notes, 'on "that fatal revolution whereby writing is converted to a mechanic trade; and booksellers, instead of the great, become the patrons and paymasters of men of genius.'" Similarly, she observes, Fielding as Sir Alexander Drawcansir in the *Covent-Garden Journal* 'erects a binary opposition in these passages. On the one side stand authority, hierarchy, and writing as an exercise of the intellect; on the other, anarchy, equality, and writing as mechanical work.'[48]

The chaotic circulation of texts in the marketplace demanded a new image of a man-machine with the innate quality of fine taste in combination with a superior education: The new man-machine for the self-appointed purveyors of fine taste was an ideal constitutional balance of understanding and genius. How to distinguish oneself in a literary marketplace of mechanical writers and other such dunces? How to establish the worth of poetry amidst the rising importance of the new philosophy as the means to understanding humankind? In literary terms, the human as machine reinforced a hierarchy of mind and body. The man-machine may have been reviled and ridiculed by the likes of Swift, Pope, and Arbuthnot, but it also functioned to reinforce a desired hierarchy of function and purpose. The term *mechanical*, as we have already seen, was complicated by its associations with lower status. In this context the term presents an oppositional hierarchy between dullness – physical mechanic labour – and genius, the refined spirit of humanity. At the macro-level of social and political organization, these problematic possibilities were reflected metaphorically in the commonwealth body, where a wholly mechanistic and materialistic view meant that the higher order 'minds' of society also risked being reduced to the same status as society's 'muscle.'

As we have seen, the 'mechanick' and physical is associated not only with labouring tradesmen, but also with the new materialist philosophy that seemed to level humanity to mundane mechanism. To some extent the new literary culture was also the beginning of a lengthy tradition of the critique of science by those affiliated with the literary arts of poetry and fiction. Alexander Pope's adaptation of the old metaphor of the pilot in the brain steering the human body clearly articulates a new hierarchy of knowledge where the value of studying the 'true Nature and Measures of Right and Wrong' far exceeds the goal to 'settle the Distance of the Planets, and compute the Times of their Circomvolutions,' as Pope characterized it in *The Spectator* in 1712. 'I always thought the most useful Object of human Reason,' he wrote, is 'human Nature.' Furthermore, 'other Parts of Philosophy may perhaps make us wiser, but this not only answers that End, but makes us better too.' What moves men,

Pope argues, is not mere reason but also the passions, which 'are to the Mind as the Winds to a Ship,' that is, 'they only can move it, and they too often destroy it ... In the same manner is the Mind assisted or endangered by the Passions; Reason must then take the Place of Pilot, and can never fail of securing her Charge if she be not wanting to her self: The Strength of the Passions will never be accepted as an excuse for complying with them, they were designed for Subjection, and if a Man suffers them to get the upper Hand, he then betrays the Liberty of his own Soul.'

We see here some noteworthy changes from Willis's much earlier representation of reason as governing monarch. Whereas Willis had described the corporeal soul that tempts and wars with the superior rational soul as a procuress, Pope suggests that reason as the pilot or governor steering the human vessel should not dominate the passions. While for Pope reason is a pure fountain and the passions 'designed for Subjection,' what elevates man – indeed, what give him his 'Humanity' are the passions – 'the very Principles of Action':

> The Understanding being of its self too slow and lazy to exert it self into Action, it's necessary it should be put in Motion by the gentle Gales of the Passions, which may preserve it from stagnating and Corruption; for they are as necessary to the Health of the Mind, as the Circulation of the animal Spirits is to the Health of the Body; they keep it in Life, and Strength, and Vigour; nor is it possible for the Mind to perform its Offices without their Assistance ...
>
> We may generally observe a pretty nice Proportion between the Strength of Reason and Passion; the greatest Genius's have commonly the strongest Affections, as on the other hand, the weaker Understandings have generally the weaker Passions; and 'tis fit the Fury of the Coursers should not be too great for the Strength of the Charioteer.

Not trade, not the even and steady flow of commerce, as for Willis and Hobbes, but passion prevents stagnation and corruption in the human œconomy. Pope uses the rationale of a carefully (but not excessively) governed equilibrium: we must 'be very cautious,' he explains, 'lest while we think to regulate the Passions, we should quite extinguish them, which is putting out the Light of the Soul ... we must endeavour to manage them so as to retain their Vigour, yet keep them under strict Command; we must govern them rather like free Subjects than Slaves, lest while we intend to make them obedient, they become abject, and unfit for those great Purposes to which they were designed.'[49] Thus,

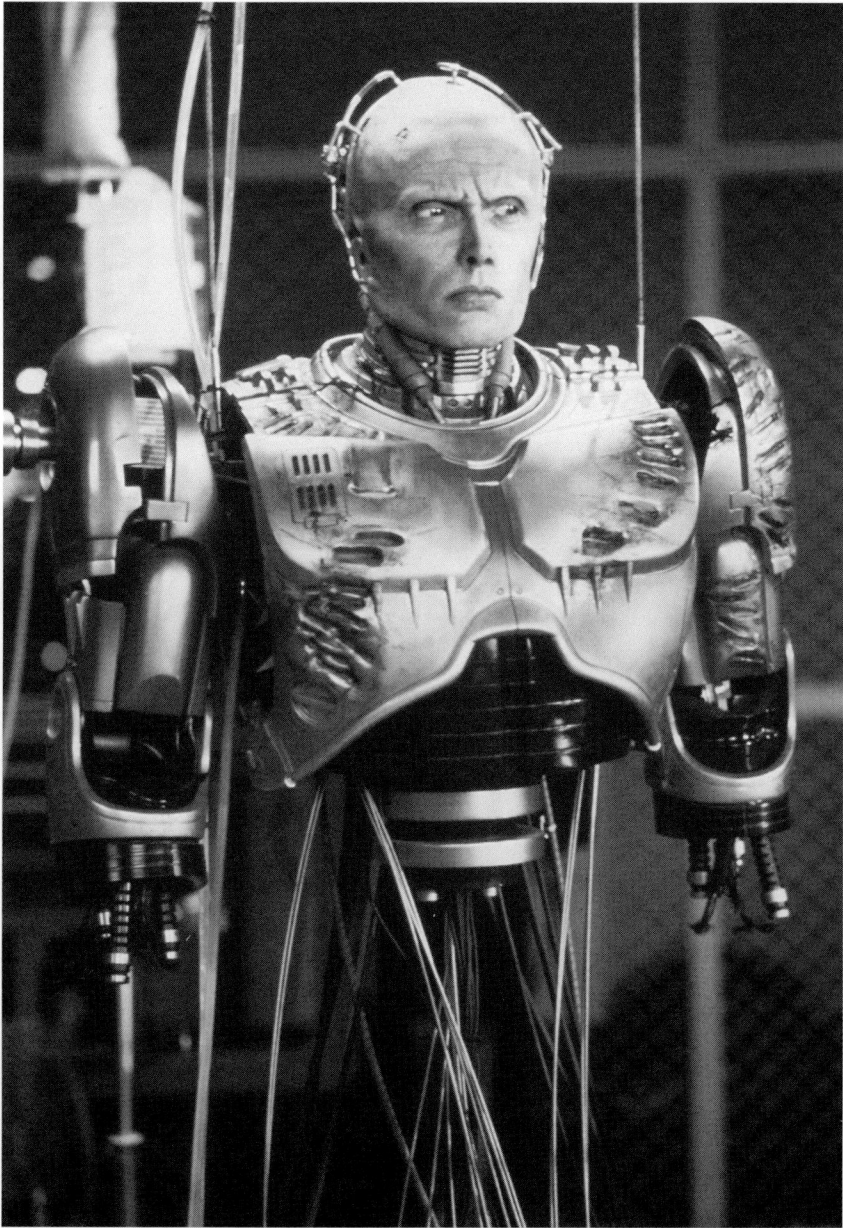

1. Peter Weller as Robocop, in *Robocop 2* (1990).

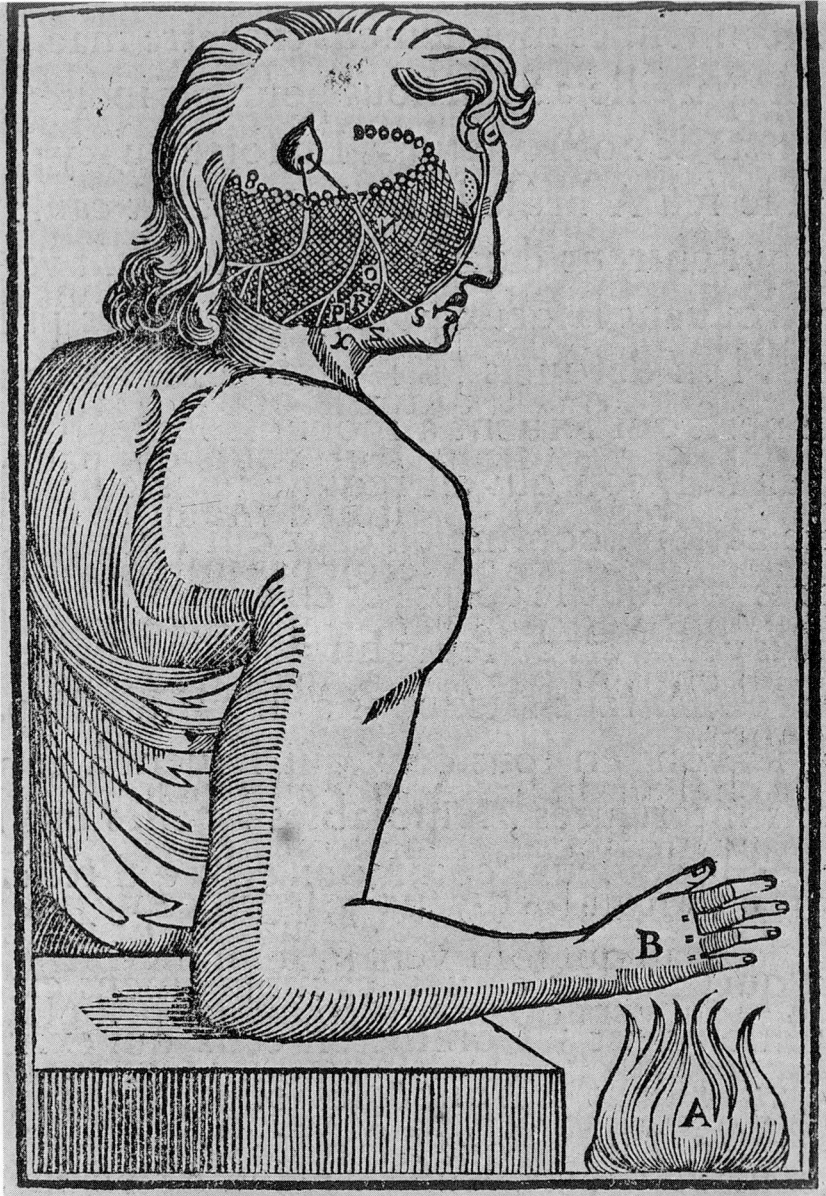

2. 'Sensation.' René Descartes, *L'homme ... et un traitté de la formation du foetus du mesme autheur* (Paris: Charles Angot, 1664).

3. The digitally networked body/mind in Mamoru Oshii's *Ghost in the Shell* (1996); in this version of the future, the human consciousness or 'ghost' can be transferred from one body to another, to a wholly artificial cybernetic body, or it can exist as pure data in the network.

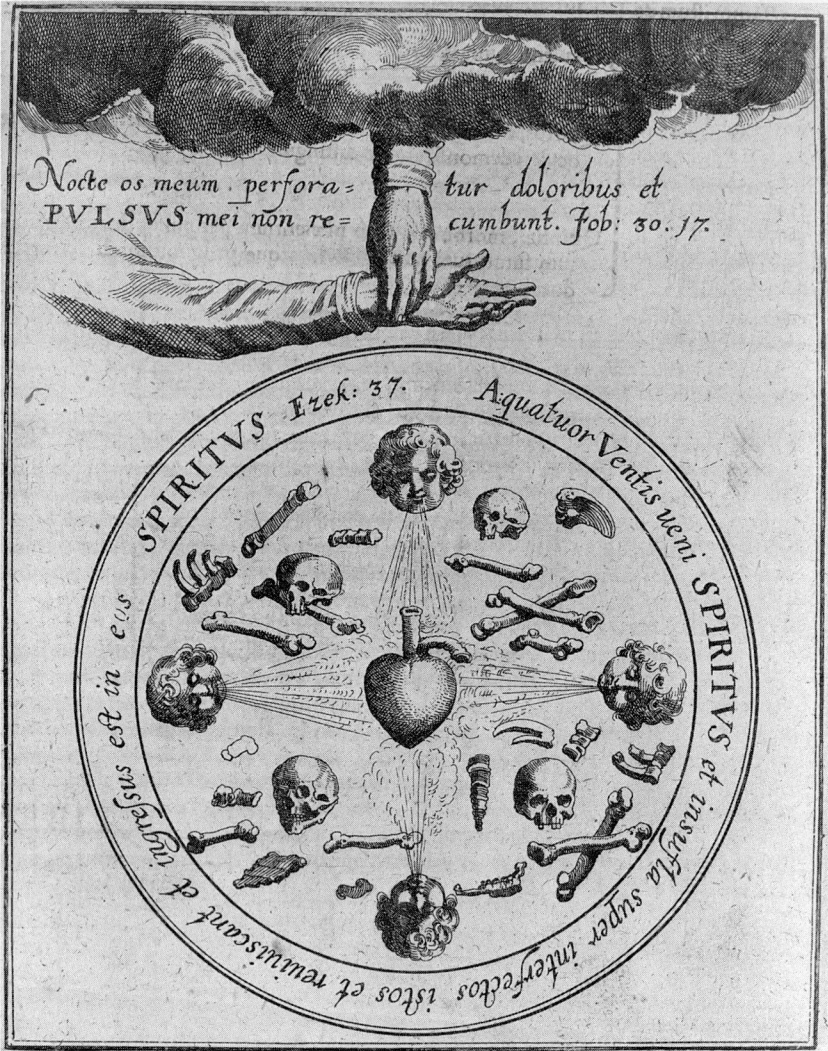

4. Robert Fludd, 'Nocte os meum perforatur doloribus et pulsus mei non recumbunt. Job: 30.17,' in *Medicine catholica; seu Mysticum artis medicandi sacrarium* (Frankfurt, 1631). The hand of God takes the pulse of an arm, while the four winds revive and infuse the heart with spirit.

5. J.J. Scheuchzer, *Kupfer-Bibel* (Augsburg und Ulm: C.U. Wagner, 1733), volume 3. The heart, illustrated as a pumping machine.

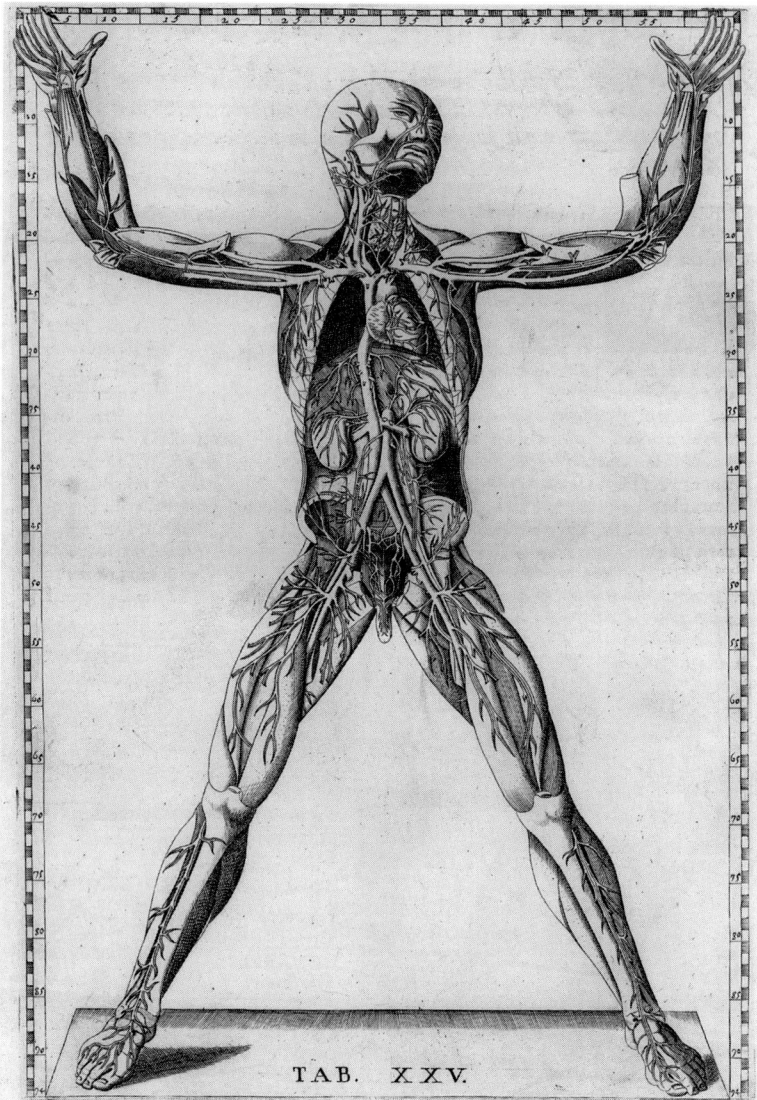

TAB. XXV.

8. Mapping the body's veins and arteries. Eustachius. Although the engravings for his anatomical text were completed by 1552, they were not printed until rediscovered in the eighteenth century, purchased by Clement XI, and printed in Rome in 1714. Plate 25 in *Tabulae anatomicae clarissimi viri Bartholomaei Eustachii quas est tenebris tandem vindicatas.*

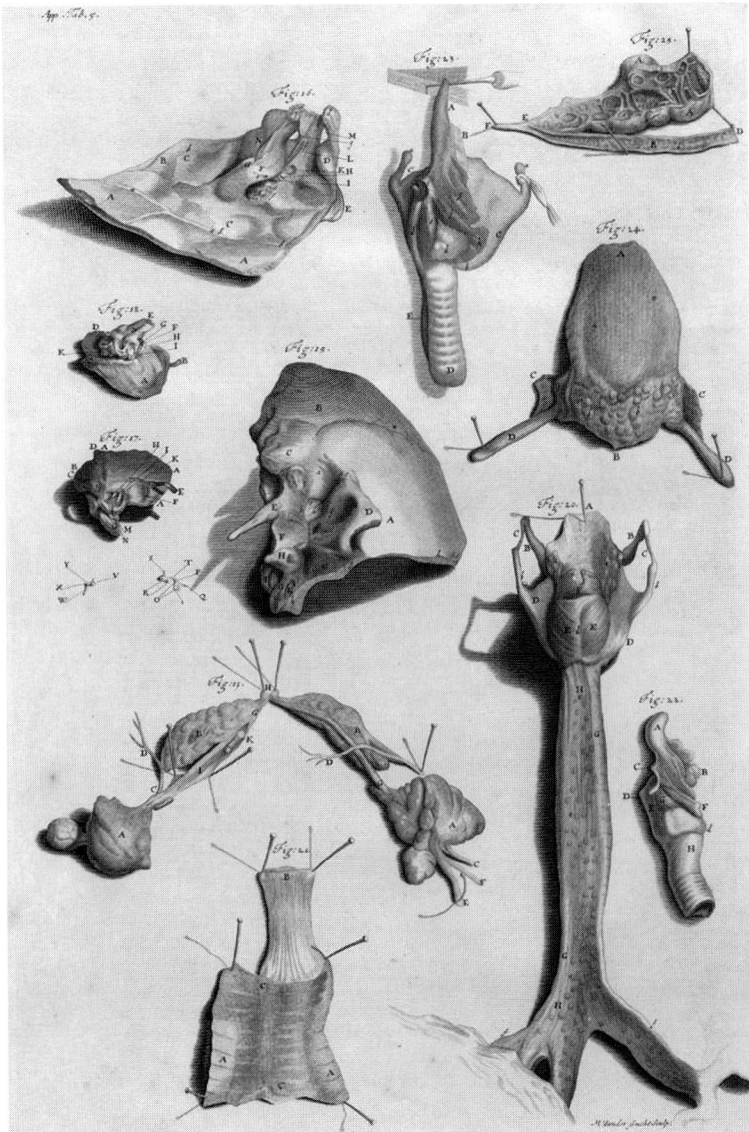

9. The body as components. 'Temporal bone, ear ossicles, salivary glands, trachea and laryngeal cartilages, tongue, seminal vesicles and vas deferens,' engraving/etching by Michael van der Gucht, in William Cowper's *Anatomy of Humane Bodies* (Oxford, 1698).

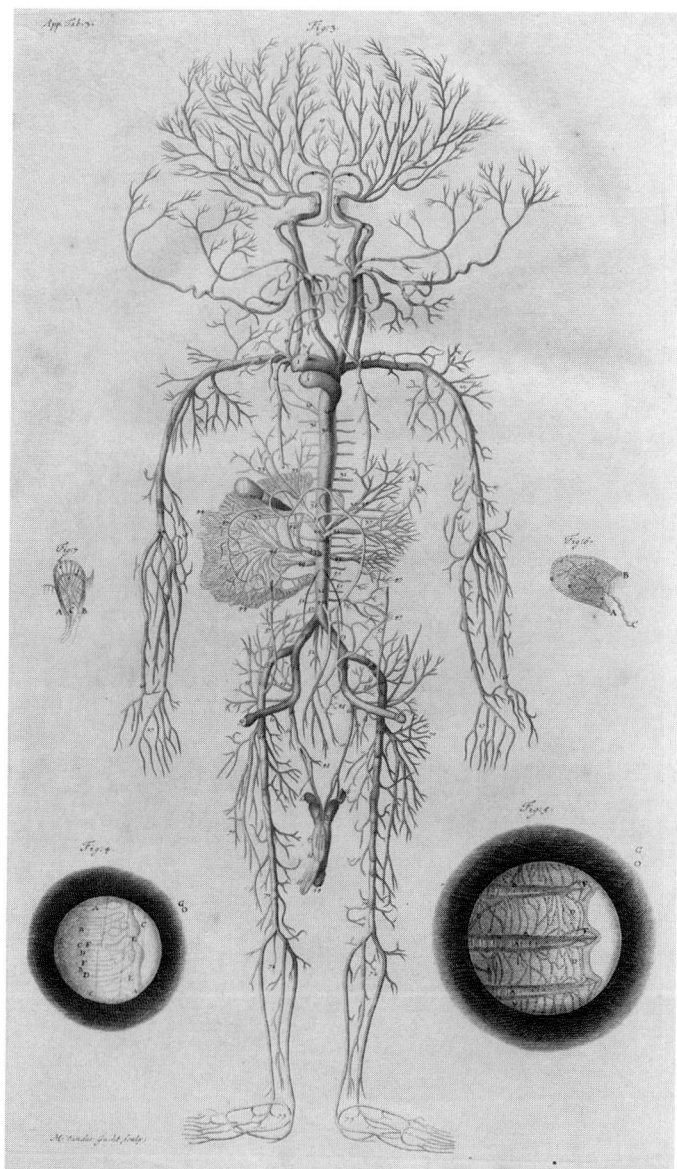

10. The body as abstracted component systems. 'Arteries of the Fetus,' engraving/etching by Michael van der Gucht, in William Cowper's *Anatomy of Humane Bodies* (Oxford, 1698).

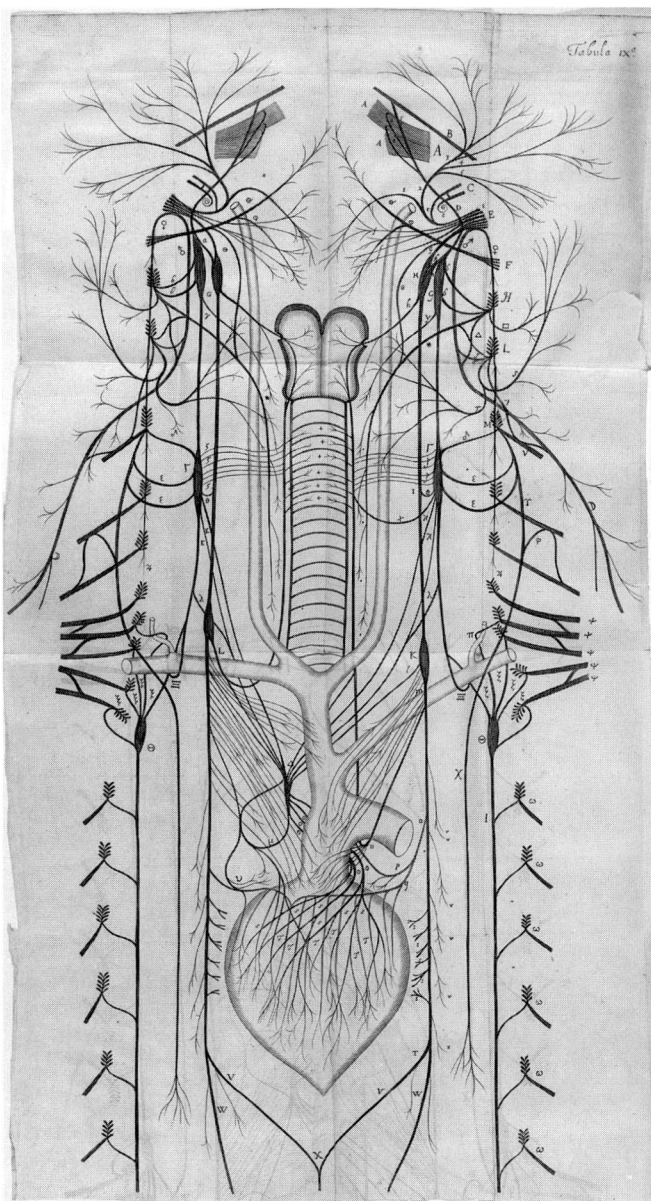

11. Thomas Willis, *Cerebri anatome* (1664). The nervous system as an abstract map.

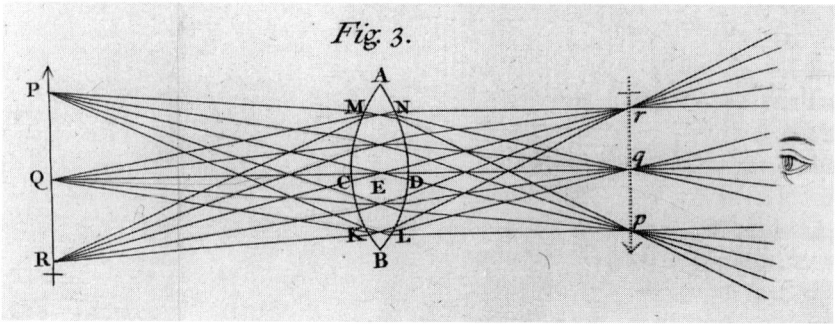

Fig. 3.

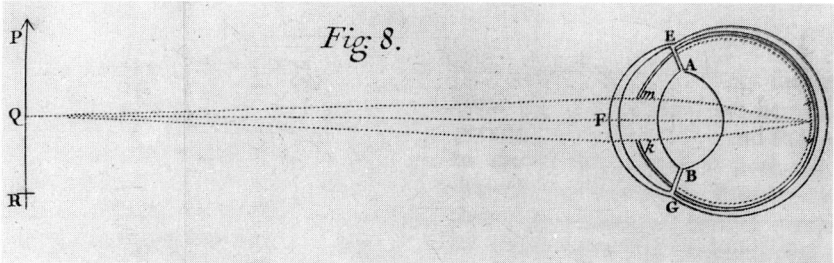

Fig. 8.

12. The lens as a mechanism of the reflection of light. Isaac Newton, *Opticks* (London, 1704).

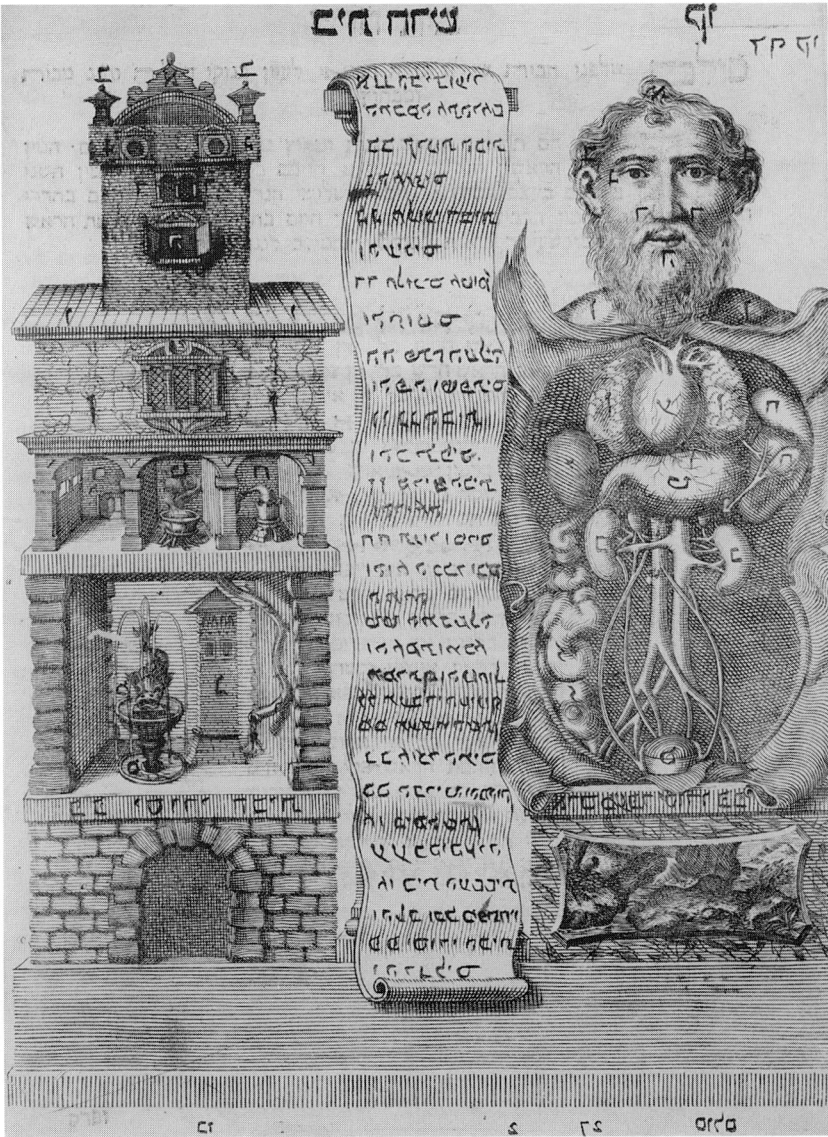

13. The human body as a house or temple with the organs as separate mechanisms. The lungs are the ventilated storey, the stomach a cauldron and heat
source, the kidneys a water reservoir, and the bladder a fountain. Tobias Cohen,
Ma'aseh Toviyah (Venezia, 1708).

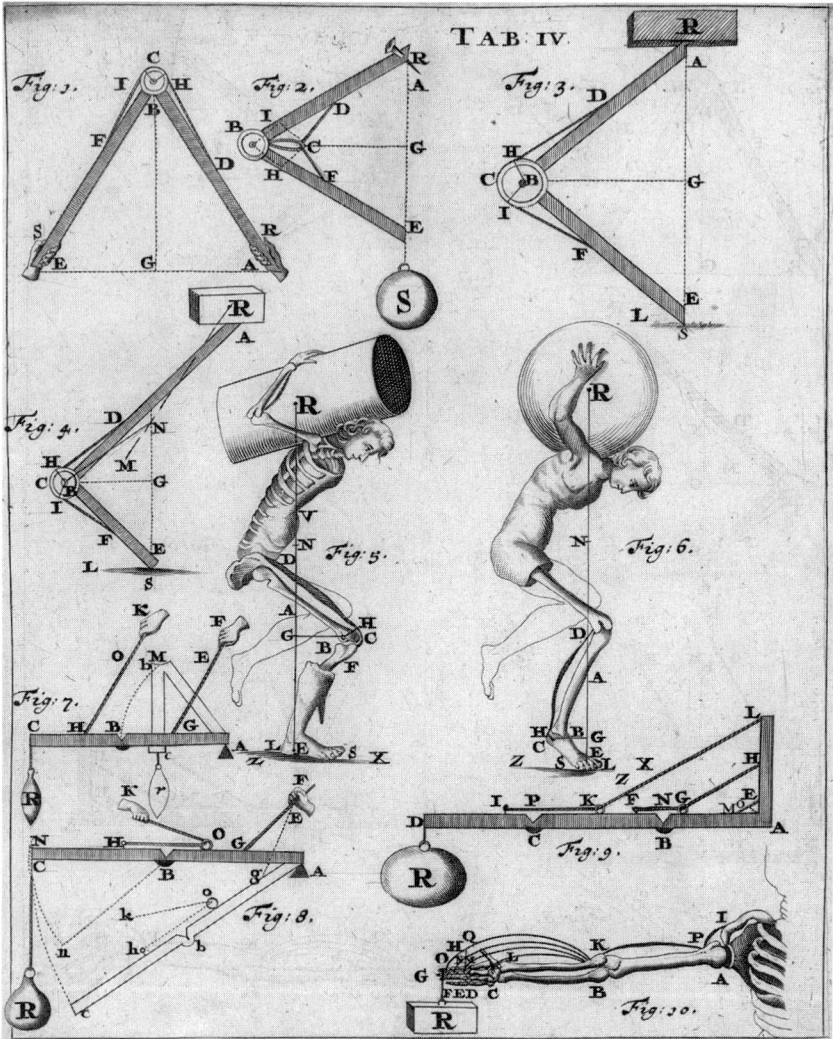

14. Physiology as a mechanism. Giovanni Alphonso Borelli, *De motu animalium*
(1710).

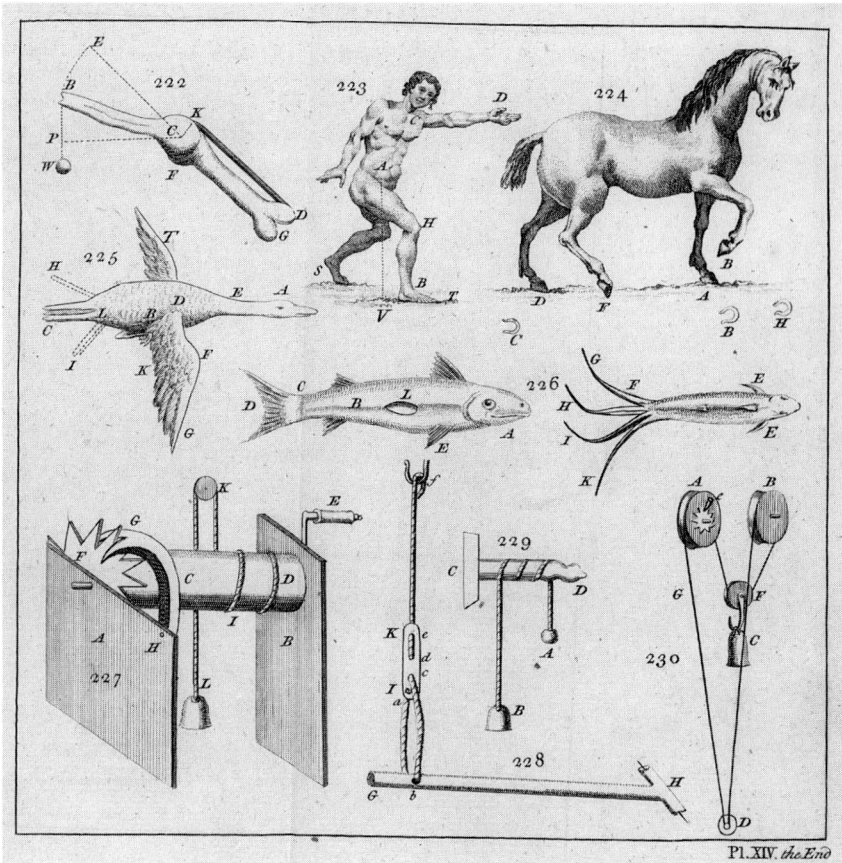

15. Machines governed by mechanical forces. Plate 14 in William Emerson, *The Principles of Mechanics* (1758).

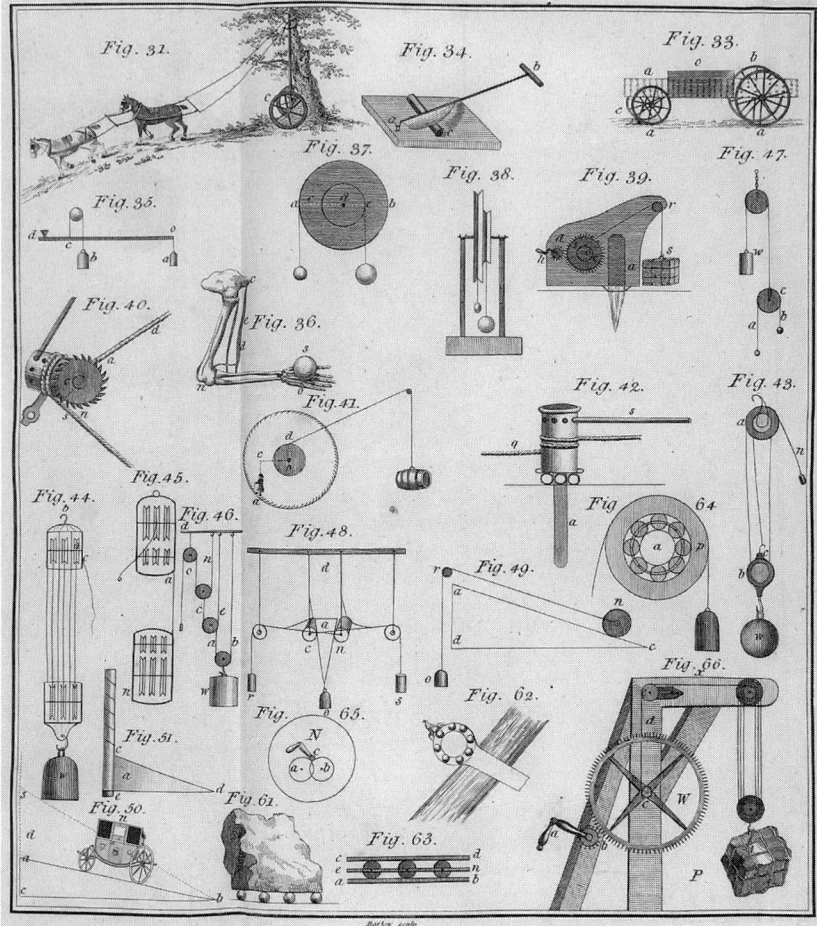

16. The arm as a mechanical lever. Adam Walker, *A System of Familiar Philosophy* (1799). 'The lever of the *third kind* has its acting part shorter than its resisting part, requiring more exertion than would be necessary to lift the body by main strength … Fig. 35 will explain it better; *a* is a weight to be lifted by the weight *b* pulling at *c*, over a pulley; *d* is the prop, or fulcrum. Now as the acting part of the lever, *d c*, is but one-fifth of the resisting part, *d o*; therefore *b* must be five times as heavy as *a*, or it will not balance it. Of this sort of lever are the limbs of all animals. My arms, legs, fingers, are such. The legs of quadrupeds, the wings of birds, and the fins of fishes' (p. 86).

17. René Descartes, *Tractatus de homine, et de formatione foetus ...* (1664). 'Through these pores the animal spirits in the cavities of the brain immediately begin to make their way into the nerves and so to the muscles which serve to cause movements in the machine ... if fire A is close to foot B, the tiny parts of this fire ... have the power also to move the area of skin which they touch. In this way they pull the tiny fibre *cc* which you see attached to it, and simultaneously open the entrance to the pore *de*, located opposite the point where this fibre terminates – just as when you pull one end of a string, you cause a bell hanging at the other end to ring at the same time' (*The Philosophical Writings of Descartes* vol. 1, trans. John Cottingham et al., 101).

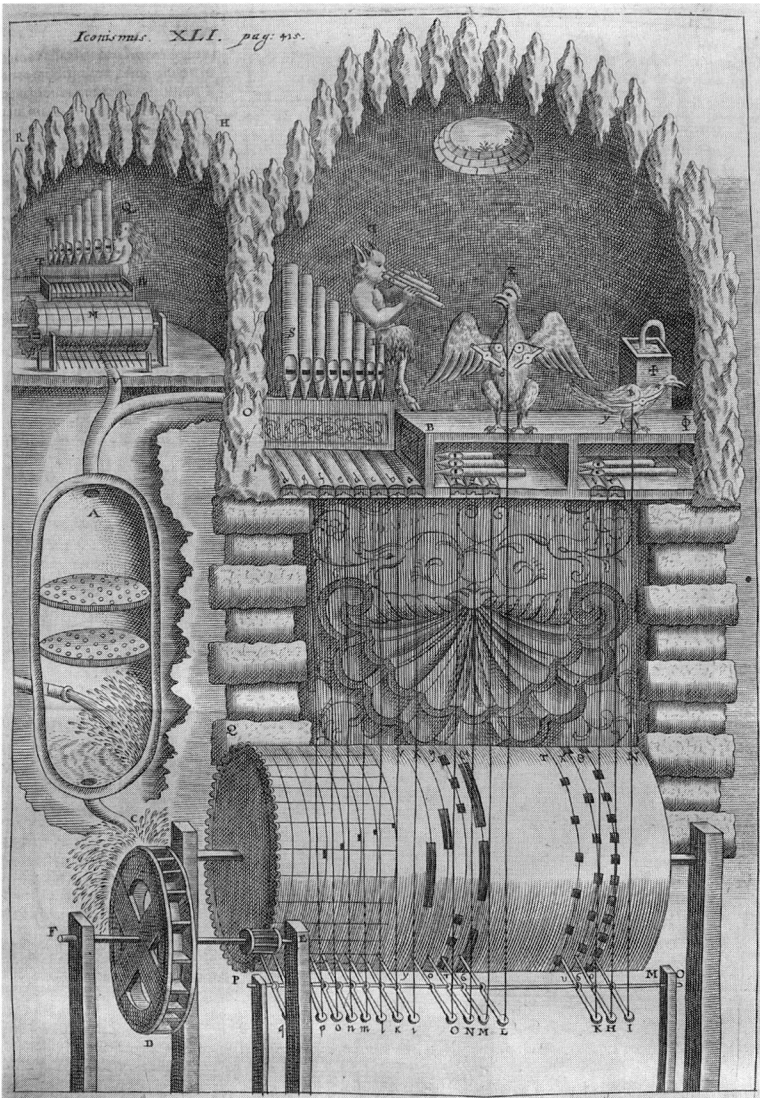

18. Kaspar Schott (P. Gasparis Schotti), Water Organ, Piping Pan, and Rooster Automata in a Grotto (1656) from *Mechanica Hydraulico-Pnevmatica*. The music box cylinder of the organ mechanism is a binary on–off switch that would later inspire the Jacquard loom and subsequently the designs by Charles Babbage for the computer he called an Analytical Engine.

19. 'M. Walker's improved Steam Engine.' Adam Walker, *A System of Familiar Philosophy* (1799). 'These kind of machines approach nearer to the animation and powers of animal mechanism, than any yet invented by man.'

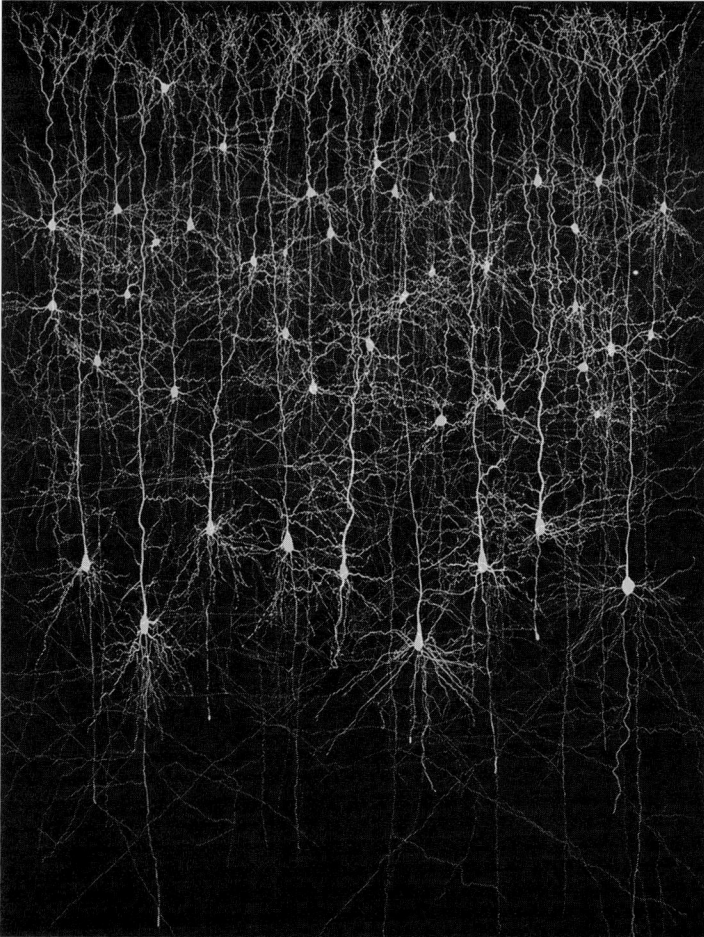

20. 'Neurons in a column,' an image from IBM Research's press release (June 6, 2005): 'IBM and EPFL Join Forces to Uncover the Secrets of Cognitive Intelligence: IBM's Blue Gene supercomputer tackles major scientific challenge-scientists to create a complex digital 3D model of the brain.' The model is based on the premise that human 'circuitry' and machine circuitry are to a great degree functionally equivalent. This image is described as 'a view of the neocortical microcircuit emphasising the downward axons of the pyramidal neurons which pass information out of the neocortex and which make up 80% of the neurons in the neocortex.' (http://domino.research.ibm.com/comm/pr.nsf/pages/news.20050606_CognitiveIntelligence.html)

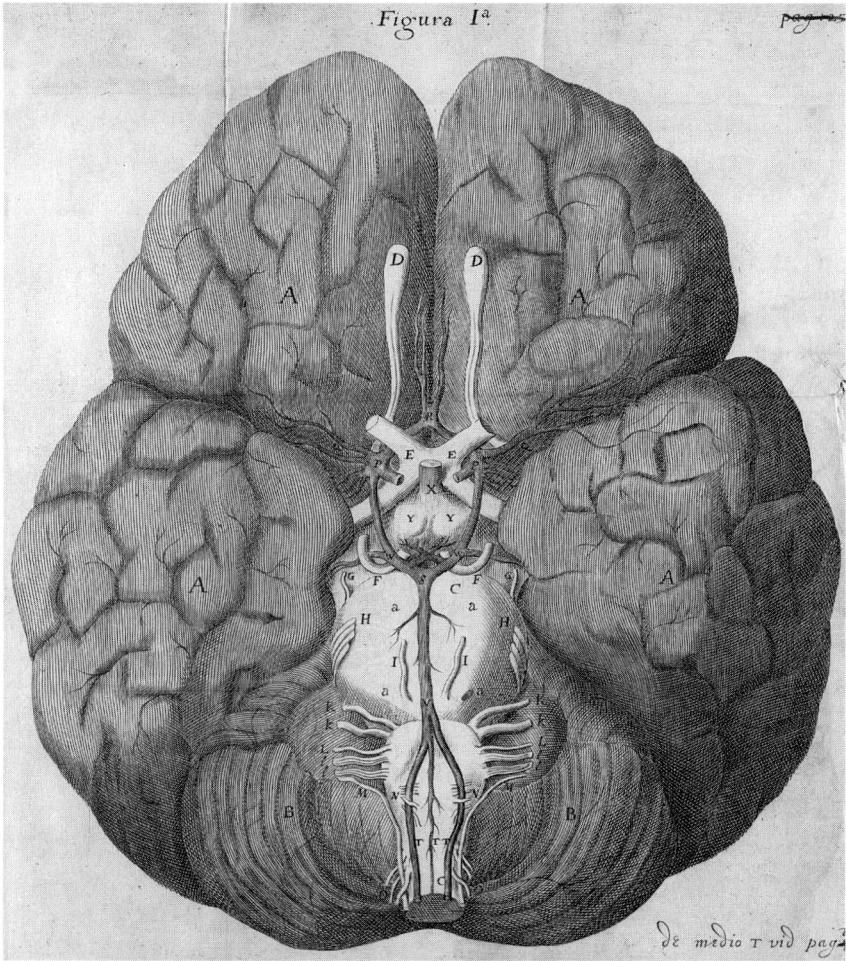

21. 'The humane Braine taken out of the Skull,' by Christopher Wren in Thomas Willis, *Cerebri anatome* (1664). 'AAAA. The anterior and posterior lobes of the brain' (cerebrum); 'BB. The cerebel or hinder part of the brain' (cerebellum). 'CC. The long marrow or pith' (medulla oblongata).

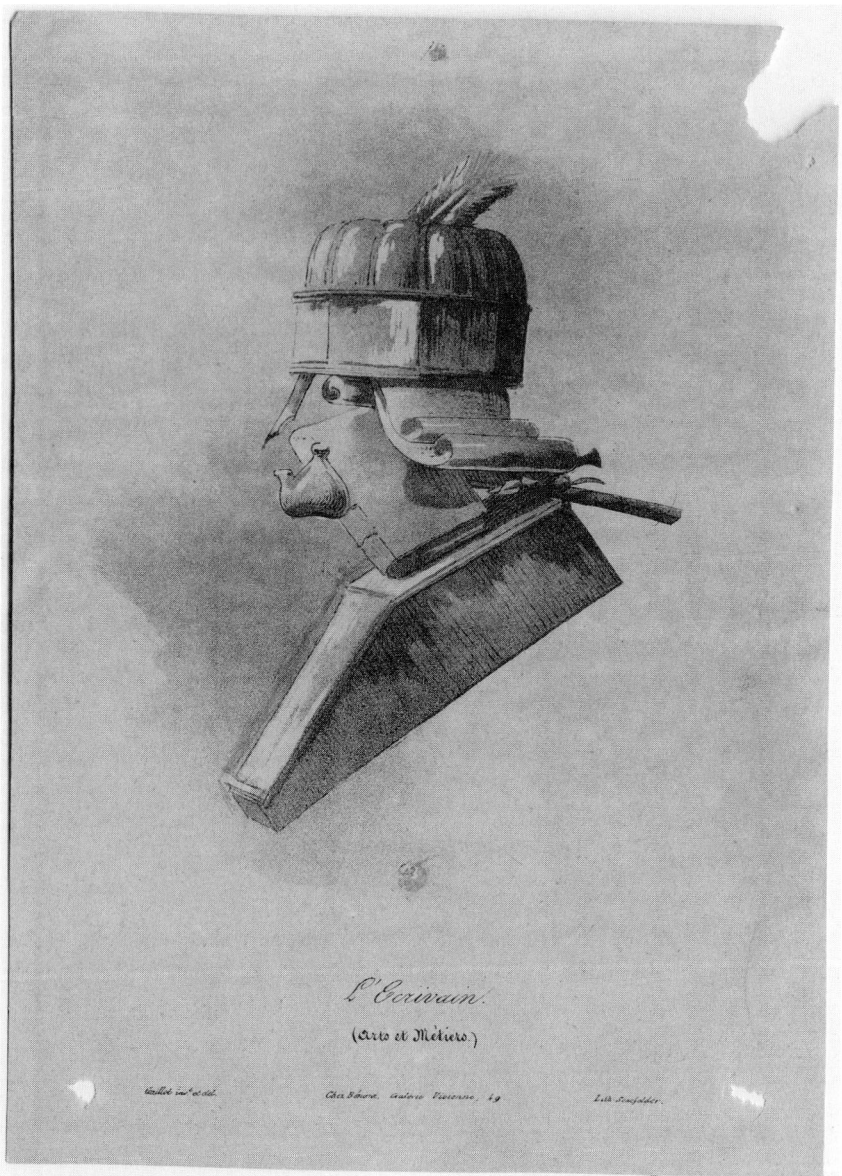

22. 'L'Ecrivain (Arts et Metiers),' by Bernard Gaillot (1780–1847). Lithograph.

23. 'A Victualler or Publican. Erected out of his own implements,' probably by George Bickahm, in *Bowles and Carver's Caricatures* (1740).

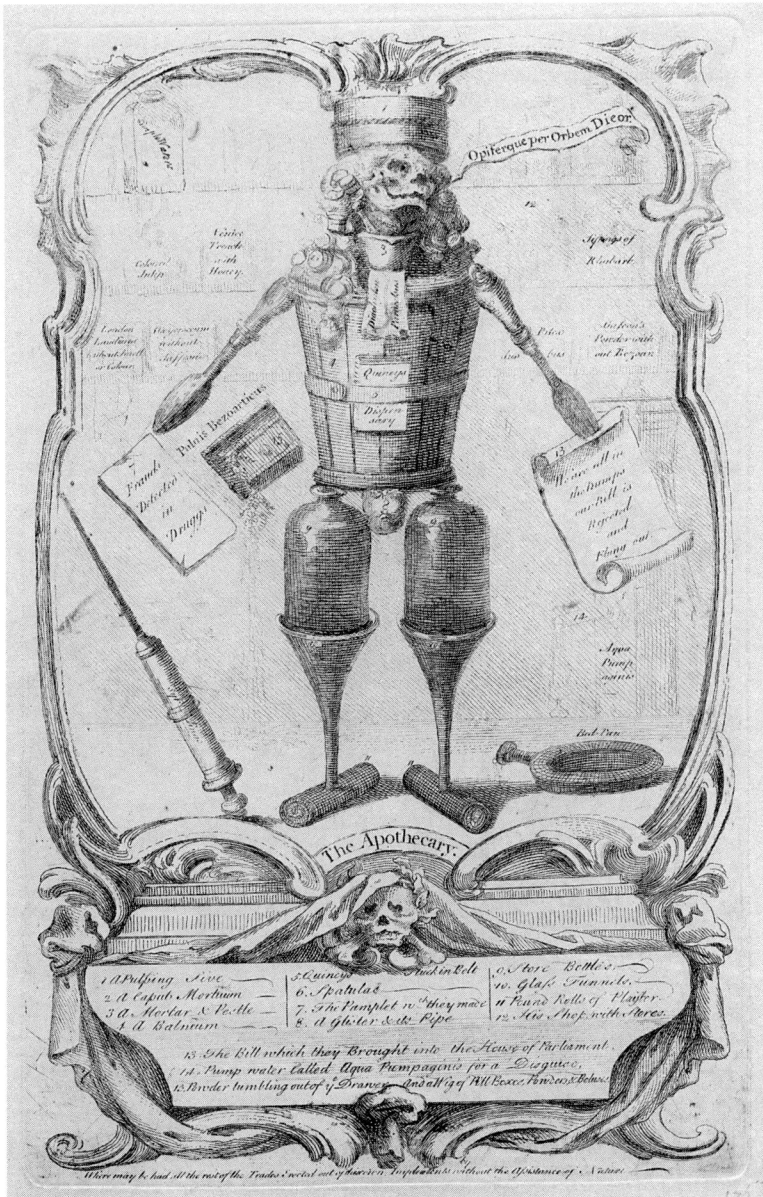

24. 'The Apothecary,' probably by George Bickham, in *Bowles and Carver's Caricatures* (1740).

Pope in one carefully chosen analogy proclaims the superiority of passionate human nature over mere calculating science, the association of genius with higher-order purpose and the light of the soul itself, and the natural order and health conferred upon the body by these gentle and well-governed passions. The issue of *The Spectator* immediately following Pope's letter is by Joseph Addison, on '*Fine Taste,* as the utmost Perfection of an accomplished Man.' The faculty of taste is 'to some extent born with us,' Addison claims, describing how an eminent mathematician had told him that the only pleasure he took in reading Virgil was in examining the voyage using a map. Modern compilers of history would similarly be interested in little more than bare matters of fact, he comments. Fine taste may, however, be improved by 'conversation with Men of a Polite Genius.' Moreover, the man with 'a finished Taste of good Writing,' suggested Addison, is well versed in the works of the best ancient and modern critics unlike the 'Man of very little Taste' discoursing upon 'Mechanical Rules.'[50] The point was made, and often, by others: whether it was called genius, imagination, sensibility, or fine taste, the elevation of the mind through art and poetry was directly opposed to a purely rational mechanism in thought.

Swift's rational and emotionless Laputans, for example, float on an island far removed from the earth, the grounding to which their wives occasionally escape in search of love or passion. The Laputan word machine, 'for improving speculative Knowledge by practical and mechanical Operations,' is a contrivance by which 'the most ignorant Person, at a reasonable Charge, and with a little bodily Labour, might write books in Philosophy, Poetry, Politicks, Laws, Mathematicks, and Theology, without the least Assistance from Genius or Study.' The contraption caricatures not only contemporary calculating devices that artificially reproduced human thought such as mechanical calculators inspired by Blaise Pascal's arithmetic machine, but also the overly rational reduction of all things mechanical to mathematics – the inference that language, words, thoughts, all those products of genius and education, are reducible to mechanical artifice. The only sciences the Laputans know are mathematics and heavily theorized but poorly expressed music: 'Imagination, fancy, and invention,' Gulliver explains, 'they are wholly strangers to.'[51]

The joke of mechanical intelligence is used similarly in *The Memoirs of Martinus Scriblerus* co-authored by John Arbuthnot and Alexander Pope. Here, the laughable Scriblerus, 'learned Inquisitor into Nature,' is sent a letter of greeting from the Society of Free Thinkers. The Society has heard 'with unspeakable joy' of his 'inquisitive Genius.' It is a great pity, however, they write, that this genius is not being put to better use 'than in looking after that Theological Non-entity commonly call'd the *Soul.*' Such a chimera does not exist, they explain, and offer to aid his studies with 'an easy *mechanical Explication* of *Perception* or *Thinking.*' Like a mechanical spit, which only has a meat-roasting 'quality' because of the composition of its several parts, 'the fly, the wheels, the chain, the weight, the cords, &c.,' so in an animal the quality of self-consciousness is 'with its several modes of sensation, intellection, volition, &c ... is the result from the mechanical composition of the whole Animal.' The power of 'thinking, self-moving, and governing the whole Machine, is communicated from every Particle to its immediate Successor; who, as soon as he is gone, immediately takes upon him the Government, which still preserves the Unity of the whole System.' It is well known to anatomists, the society pontificates, that 'the Brain is a *Congeries* of Glands, that separate the finer parts of the blood, call'd Animal Spirits; that a Gland is nothing but a Canal of a great length, variously intorted and wound up together. From the Arietation and Motion of the Spirits in those Canals, proceed all the different sorts of Thought.' Simple ideas arise from the motion of the spirits in a single channel, while several channels emptying into one results in a proposition, which might empty into the third to form a syllogism or ratiocination. There is no choice, no self-determination, but merely 'the bad configuration of those Glands' to explain 'why some people think so wrong and perversely.' The society, they conclude, has accordingly employed 'a great Virtuoso at Nuremberg' to make 'a sort of an Hydraulic Engine, in which a chemical liquor resembling Blood, is driven through elastic chanels resembling arteries and veins, by the force of an Embolus like the heart, and wrought by a pneumatic Machine of the nature of the lungs, with ropes and pullies, like our nerves, tendons and muscles: And we are persuaded that this our artificial Man will not only walk, and speak, and perform most of the outward actions of the animal life, but (being wound up once a week) will perhaps reason as well as most of your Country Parsons.'[52] The complexity of this image derives from competing agendas: a sophisticated joke satirizing the most up-to-date philosophy of the day, it reduced mechanical philosophies to a ridiculous hubris, contrasted the inquisitive genius in search of the soul with the arrogant free-thinking philosopher, and also made a pointed comparison to the mechanical reasoning of notoriously ill-read country clergymen. The hierarchy, again, is clearly one where genius comes out ahead of both the learned but merely logical thinkers and the poorly educated 'mechanical' thinkers.

These satires emphasize various cultural anxieties for the student of the ancients in a modern commercial society informed and rationalized by natural philosophy: materialism threatened the soul, both in the physical body and in the body politic. The pursuit of wealth, knowledge, and individual expression undermined the monopoly of authority (and wealth) of an older version of educated gentlemen. With a seeming dual synergy, the secularization of communications circulating in the body corresponded to that of the body politic.

Willis's mechanical nervous system formed the basis for the eighteenth-century culture of sensibility which, as numerous writers since have demonstrated, established a new social hierarchy equating an innate nervous sensitivity with cultural refinement, morality, and intellectual aptitude.[53] Drawing on the works of both Willis and Newton, the Scottish physician George Cheyne characterized the human body as mechanical instrument. Non-philosophers need only suppose, he writes in his best-selling *The English Malady* (1733), 'that the Human Body is a Machin of an infinite Number and variety of different Channels and Pipes, filled with various and different Liquors and Fluids,' and as a musical organ ('a finely fram'd and well-tun'd Organ-Case') with nerves like keys, which convey sound to a 'sentient Principle, or *Musician'* in the brain. If Willis's favoured analogy in a time of political turmoil was of the rational mind at war with the sensual corporeal soul, however, Cheyne's – in a time when 'our Wealth has increas'd, and our Navigation has been extended, [and] we have ransack'd all Parts of the *Globe* to bring together its whole Stock of Materials for *Riot, Luxury,* and to provoke *Excess'* – was of the body's quivering fibres frequently brought out of their 'Harmony' by the body's excesses and sensual pleasures. While the book decries the rich, lazy, luxurious, and inactive – those who 'fare daintily and live voluptuously' – whose nervous diseases were due to their own unlimited appetites, Cheyne's musician as intelligent principle is characterized by a distinct hierarchy of intelligence and sensibility: Nervous diseases do not occur in those living in 'barren and uncultivated' countries or to those who are 'rude and destitute of the *Arts* of Ingenuity and Invention.' However much Cheyne claimed the intelligent principle was a musician residing somewhere in the brain, he also portrayed intelligence as having in effect seeped out into the nerves that 'communicate' (Cheyne's term) action, motion, and sense. Sensibility was defined by the material of the nerves themselves, and was imagined in terms of society's brains versus society's muscles. The height of refinement was symbolized by the height of nervous 'sensitivity' – characterized by intelligence, literacy, and the expression of imaginative creativity. Those who are born with 'weak nerves' are of a higher moral and intellectual order than the merely rich and luxurious, Cheyne explains. The highest order of sensibility, the sign of upper-class status, was associated with the highest order of thinking and writing, as opposed to the 'dulness' of the mere insensible labourer who because of his mechanical configuration is neither intelligent nor likely to suffer from such diseases of the nerves as might befall the more refined: 'It is a common Observation, (and, I think, as great Probability on its Side)' he opines, 'that *Fools, weak* or *stupid* Persons, *heavy* and *dull Souls,* are seldom much troubled with Vapours or Lowness of Spirits.' And, while nervous disease might be 'original' or 'acquir'd,' those who are born with 'weak nerves' are of a higher moral and intellectual order. They are less likely, in fact, to succumb to the excesses of the merely rich and luxurious: 'I shall only here observe two things in regard to [those who are born with weak nerves]. The *first* is, that they are never to expect the same Force, Strength, Vigour and Activity, nor to be made capable of running into the same Indiscretions or Excess of sensual Pleasures (without suffering presently, or on the Spot) with those of strong Fibres and robust Constitutions. No Art hitherto known, can make an Eagle of a Wren, (for tho' a Wren, by Art and Management, may be made, as it were, a Nightingale, yet never a Carrion Crow or Kite).' Thus, 'the common Proverb is just and true,' Cheyne concludes, 'that a *Venice Glass* will last as long, if well look'd after, and even shine more bright, than a more gross and course one.'[54]

The physician S.A. Tissot would similarly write in *An Essay on Diseases Incident to Literary and Sedentary Persons. With Proper Rules for Preventing Their Fatal Consequences* (1769) that those of superior minds were more likely to suffer from nervous diseases than were fools: 'It is an old complaint, that study, though essentially necessary to the mind, is hurtful to the body,' he writes. Those who are of weak constitutions 'as most studious men are, should take greater care than others, that what is impaired by application to their studies may be repair'd by attention to their constitutions.' For example, 'it is observable that fools always eat and drink a great deal, and yet digest perfectly well, even though they lead a sedentary life, and do not surpass others either in the bulk or strength of their bodies: whilst men of genius and abilities, though they have strong muscles, and take exercise sufficient, are obnoxious to crudities in the stomach and slowness of concoction.'[55] Or, as Mark Akenside would write in 'The Pleasures of the Imagination,'

some within a finer mould
She wrought and temper'd with a purer flame.
To these the sire omnipotent unfolds
The world's harmonious volume, there to read
The transcript of himself.[56]

With the general acceptance of human identity and consciousness as a mechanical and embodied process, the markers of cultural authority had also changed. The new image of the man-machine required a new balance. If it was undeniable that humans had a mechanically constructed identity, then the very construction of nerves would become indicators of a humanistic hierarchy of sensibility: the lowest order of human machine would be a dull labourer or 'mechanick' thinker with a strong body and a heavy-duty nervous system; the overly rational 'empirick' philosopher would fare little better, while the highest order of creative genius was characterized as weak-nerved, susceptible both to the slightest irritations and perceptions. The 'minds' of the social body retained their earlier metaphoric elevation over its 'muscles.' In the early part of the eighteenth century, gentlemen writers reinforcing the new hierarchy reduced mechanick knowledge, mechanick writings, or mechanical operation of the spirit to foolishness, dullness, or insensitivity in contrast to (their own) genius and passion. The proliferation of print culture, authors, and readers created a parallel and complimentary hierarchy, coming out of the same gradual conceptual and cultural shifting: Texts became material commodities to be bought and sold, and to some degree this commerce created a version of the text comparable to the 'mechanick' in its baser materialist sense. The new print culture, however, both relied on and rejected materialism just as the new aesthetic both relied on and rejected materialist interpretations of human identity. As the man-machine, then, became the sensible machine, the all-important pilot as soul became less an indicator of mere knowledge or rationality (animals and lower order humans, after all, shared these characteristics) but of reason combined with sensitivity, passion, and imagination. The rise of a literature of sensibility was a response to the implications of materialism and mechanism. The rise of the 'sublime' did not replace mechanism, but reincorporated it and redeemed it.

As Laurence Sterne's sentimental Yorick, wondering why so many kind-hearted men 'fall out so cruelly as we do by the way' muses, 'when a man is at peace with man, how much lighter than a feather is the heaviest of metals in his hand! he pulls out his purse, and holding it airily and uncompress'd, looks round him, as if he sought for an object to share it with – In doing this, I felt every vessel in my frame dilate – the arteries beat all cheerily together, and every power which sustained life, performed it with so little friction, that 'twould have confounded the most *physical précieuse* in France: with all her materialism, she could scarce have called me a machine –' Yorick determines to share his wealth with someone less fortunate than he – until, that is, moments later when a monk enters the room begging for his convent. Yorick refuses him, then feels immediate regret when the monk leaves. Who knows what causes our virtues to change so like the ebb and flow of tides? he muses. Later he considers how to undo the bad impressions the poor monk's story might make (in case he has told it to the beautiful young woman Yorick has just met). When he next sees the monk, he offers him a snuff-box and a stream of sentiment. The monk blushes at Yorick's kind words; Yorick, holding the hand of the young woman, blushes too 'but from what movements,' he relates coyly, 'I leave to the few who feel to analyse.' Sterne portrays in Yorick a quivering machine, his actions steered by – what? the mechanism of sexual response? by the monetary transaction? or by sentiment, virtue, and benevolence? If we might imagine that Sterne is recalling Hobbes's mechanical man, who is generous only for reasons that are self-serving ('for no man giveth but with intention of good to himself'), we find the chapter ending with Yorick plucking nettles from the monk's grave and bursting into tears: 'I am as weak as a woman,' he weeps, 'and I beg the world not to smile, but pity me.'[57]

As Martin Battestin has so convincingly argued, 'in the evolution of Sterne's two great works of fiction over the last ten years of his life, we may trace the story of his attempt to come to terms with the disturbing implications of the prevalent philosophy of the Enlightenment: mechanistic materialism – a wholly secular philosophy representing men and women as mere automata, enclosed in self, without soul or will, and doomed at last to vanish, like Tristram's Parson Yorick, into total darkness ... [Sterne] was the first to struggle openly in his fiction with the mechanistic doctrines of La Mettrie and his followers.'[58] Battestin concludes that Sterne at last does reconcile the philosophies of Locke, Hume, and the French materialists and his own religious doctrines. The above passage from Sterne's last sentimental novel, nevertheless, is emblematic of the complex interrelationship of a newly commercial print culture and the concomitant rise of the culture of sensibility. A generation after the critiques by Swift, Pope, and Arbuthnot's, Sterne's preoccupations were not with Grub Street, but with making his own fame as a paid author: his novels of sensibility *sold*. And even if Tristram's family may be a machine consisting of a few wheels set in motion by many different springs, even if his book may be a machine, and man himself may be characterized as a machine, the individual human-machine is complex and sensitive – not merely a rational being, but also comprised of both judgment and wit:

> Now, my dear Anti-Shandeans, and thrice able critics, and fellow-labourers, (for to you I write this Preface) – and to you, most subtle statesmen and discreet doctors (do – pull off your beards) renowned for gravity and wisdom ... – not forgetting all others as well sleeping as waking, – ecclesiastical as civil, whom for brevity, but out of no resentment to you, I lump all together. – Believe me, right worthy,
>
> My most zealous wish and fervent prayer in your behalf, and in my own too, in case the thing is not done already for us, – is, that the great gifts and endowments both of wit and judgment, with every thing which usually goes along with them, – such as memory, fancy, genius, eloquence, quick parts, and what not, may this precious moment without stint or measure, let or hinderance, be poured down warm as each of us could bear it, – scum and sediment an' all; (for I would not have a drop lost) into the several receptacles, cells, cellules, domiciles, dormitories, refectories, and spare places of our brains, – in such sort, that they might continue to be injected and tunn'd into, according to the true intent and meaning of my wish, until every vessel of them, both great and small, be so replenished, saturated and fill'd up therewith, that no more, would it save a man's life, could possibly be got either in or out.[59]

The reception of Sterne's *The Life and Opinions of Tristram Shandy* had assured his livelihood and fame far more successfully than had his attempts at preferment in the political or religious realms. Print culture was developing a new literary order that both relied upon and rejected materialism, that both incorporated and complicated dull reason and knowledge with passion and genius, and that laid the foundation for a particular form of literary expression rejecting mere mechanism while celebrating – even revering – the mechanisms of imagination and sentiment.

The Man-Machine and Intellectual Electricity

The notion that the electronic text functions as prosthesis or extension of the mind has long served for metaphorical-cum-actual predictions of social transformation and networked identities, but as Chris Hables Gray and Mark Driscoll put it, the claims that new cybertechnologies would 'make the body obsolete, destroy subjectivity, create new worlds and universes, change the economic and political future of humanity, and even lead to a posthuman order'[60] have remained unfulfilled. Nevertheless, 'cyber' or electronic media continue to be studied in terms of their transformative effects on individual identity. J. David Bolter and Richard Grusin have observed in their work on remediation that 'we employ media as vehicles for defining both personal and cultural identity. As these media [film, television, virtual reality] become simultaneously technical analogs and social expressions of our identity, we become simultaneously both the subject and object of contemporary media.' As they comment, 'This is not an entirely new phenomenon.' As I have proposed, we need to examine more carefully the historical connections of mind, body, and technology not only as 'The Dissolution of the Cartesian Ego,'[61] as so many authors have in terms of visualization, graphic representation, and mind-body subjectivity; rather, we need to recognize the history of tropes that conflate material embodiment and material expression to such an extent that our media appear to be equivalent to (rather than manifestations of) our identity.

If the extension of the soul/mind throughout the body has become more acceptable to us now than it was during Willis's time, its extension beyond and subsequently imagined permeability with material technology is more troublesome. Perhaps it is no surprise that today the image of the cybernetic organism, which may not be governed by any spiritual steersman but by artificial mechanisms in an ungoverned network extending beyond the human body, has resulted in such extreme views of the imagined health or disease of the body politic. The exchangeability of our metaphors for the functions of the human mind and the electric/electronic device seems to suggest that the human soul or identity may escape the body network entirely into the ungoverned technological communications network: like Willis's materialist metaphors for the human mind as mechanism, our contemporary metaphors for mind and machine inspire both apprehension and anticipation of the future for the body politic.

Our 'new electric technology that extends our senses and nerves in a global embrace' could ameliorate the 'division and separation' that language, as a 'Tower of Babel,' has wrought, wrote Marshall McLuhan in an enthusiastic fit of theological polysyllables: 'The computer, in short, promises by technology a Pentecostal condition of universal understanding and unity. The next logical step would seem to be ... to bypass languages in favor of a general cosmic consciousness.'[62] McLuhan's ideas of the man-machine nexus and new global consciousness experienced a resurgence in the 1980s and 1990s, with the realization of the Internet as a communications system not only for the U.S. government's military use, but also for citizens in various countries around the world. It was in many instances a darker vision, particularly voiced by scholars and critics who were threatened by communications systems or by technology usurping the traditional authority of the printed word.[63] The notion of a sovereign voice of morality and will in the 'Cartesian brain' would seem to be inseparably and problematically linked to both the technologies of communication outside of the body, and the communications within the body's nervous system in our critical theory and fiction. However, that 'steersman' or pilot in the brain was banished by La Mettrie in 1747 and subsequently, as we have seen, in the increasingly secular studies of human physiology.

In *Histoire naturelle de l'âme* (1745), La Mettrie had argued against Descartes' version of the man-machine as a clock or Jacques de Vaucanson's flutist automaton. Benedict de Spinoza was wrong, he asserted, to claim that man resembles a 'Vessel without a pilot in the middle of the sea, which by its construction has the capacity to sail, but is determined by the winds and the currents' or a 'genuine automaton, a machine subjected to the most constant necessity, pulled by an impetuous fatalism like a vessel by currents of water.'[64] By the time he wrote *L'homme machine*, however, La Mettrie had utterly and dangerously reversed this opinion. 'Let us then conclude boldly,' he writes, 'that man is a machine and that there is in the whole universe only one diversely modified substance.'[65] Diderot too at the end of the eighteenth century contemplated the possibility that individual identity might be solely material. His *Rêve D'Alembert* explored the brain and nervous system metaphorically as a sensitive 'network' comprised of a spider's long legs extending through the body and sending 'messages' to the spider in the brain. 'The difference between the instrument called philosopher and the instrument called clavichord,' he postulates, is that the 'philosopher-instrument is sensitive, being at one and the same time player and instrument.'[66] The body steers itself: our keys may strike themselves,

Diderot suggests, or things in nature around us might strike them, but there need not be a separate musician. (As Sergio Moravia observes, Diderot's work in turn had been influenced by Théophile Bordeu, who envisioned the body not as a centralized monarchy but as a 'federation' of organs.[67]) Diderot did not publish this play, which comprised an offence against morality. Not surprisingly Julie L'Espinasse and Jean le Rond D'Alembert, upon whom Diderot had based two of his play's characters, insisted that he destroy what they thought to be the only copy of the text. The interplay between the human consciousness as a decentralized (and material) network, and the systems imagined to control or govern it, was a meaningful and much-contested issue: the censure of La Mettrie's written works, or Diderot's imprisonment for his *Lettre sur les aveugles* (1749), which supported Locke's theory of knowledge, or any other of many such examples during the period, forcefully underscores the complex relationship of communications, control, and the materialist man-machine trope. During this period, various cultural meanings and symbols accrued from the mechanistic model of man in the new medical sciences, which in parallel fashion would be used to reinforce a hierarchy of the governing intellect or soul versus the material muscles of the nation. The new medical and philosophical metaphors for the body played against, contributed to, and derived from cultural, political, and aesthetic functions of the new understanding of the body politic.

Nor did the theories of an interaction of human mechanics, communications, and individual and social bodies stop when dull mechanism was enlivened by electricity. While for Bordeu and Diderot the process that explained everything might be sensitivity, the discovery and theorization of the 'electric fluid' was also providing a means of explaining the material form of government for the human body. In England Richard Lovett, for example, claimed in his *Philosophical Essays* (1766) that 'nature's handmaid and chief agent; or ... what ... GOVERNS NATURE' was an electrical fluid – Newton's subtle medium or æther to be precise. Lovett explained that the truth had been there all along in Newton's *Principia* which described 'a certain most subtile Spirit, ... by the force and action of which ... all sensation is excited, and the members of animal bodies move at the command of the Will.' Lovett explained that matter is indeed imbued with motion: 'Before a subtile Medium or Æther was discovered, Motion was given up, along with other immechanical Causes, to be performed by the immediate influence of the first Cause. – Before such Discovery, we seemed very naturally to conclude that all Matter was inert, whereas we now find we were greatly mistaken, and that there are active Particles of Matter existing as well as passive.'[68] That there is motion in the smallest particles of matter, what La Mettrie called a 'natural oscillation' in all the elements of the body's fibres, is not a startling conclusion today when atomic configurations are taught in high school; but the questions of the mechanisms for governing those mechanical motions reverberate in images of both man-machine and cyborg.

In his *Zoonomia; or, the Laws of Organic Life* (first published in 1794), the physician Erasmus Darwin suggested that spirit was a substance of the same class as electricity. 'I beg to be understood,' he wrote as if in pre-emptive appeasement, 'that I do not wish to dispute about words and am ready to allow, that the powers of gravity, specific attraction, electricity, magnetism, and even the spirit of animation, may consist of matter of a finer kind; and to believe, with St Paul and Malebranch, that the ultimate cause only of all motion is immaterial, that is God.' Darwin espoused a materialist though not strictly mechanical version of the body. Earlier 'idly ingenious' philosophers had made the mistake, he explained in the opening paragraph of his Preface, of 'attempting to explain the laws of life by those of mechanism and chemistry; they considered the body as an hydraulic machine, and the fluids as passing through a series of chemical changes, forgetting that animation was its essential characteristic.' Nevertheless, even while he emphasized the mistakes of mechanistic philosophies, for Darwin matter in motion was nevertheless a crucial factor in defining the organism. Spirit was thus a substance with the power to produce motion, while matter became a sort of transceiver: 'The WHOLE OF NATURE may be supposed to consist of two essences or substances; one of which may be termed spirit, and the other matter. The former of these possesses the power to commence or produce motion, and the latter to receive and communicate it.' Darwin divided motion into two kinds, secondary or mechanic powers, and primary motion consisting of three classes, gravitation, chemistry, and life. A fourth division, not well enough known to be classed, was 'the supposed ethereal fluids of magnetism, electricity, heat, and light.' No longer the animal spirits as subtle little bodies inhabiting the political body, but for Darwin the 'spirit of animation,' a class of mechanic powers that was part of the sensorium, which he defined as 'the medullary part of the brain, spinal marrow, nerves, organs of sense, and ... muscles; but also at the same time that living principle, or spirit of animation, which resides throughout the body, without being cognizable to our senses, except by its effects.' The brain secreted a fluid, perhaps 'more subtile than the electric aura,' for motion and sensation. That it was like the electrical fluid was probable, he explained, because the gymnotus electricus and torpedo (fish that can emit an electric discharge) both had the capacity to voluntarily give out electricity, because an electric shock can stimulate a paralyzed limb, and because the electric fluid needs no perceptible tubes (the nerves having been demonstrated to be solid well before Darwin's time). The head and the spine, the 'central parts of the sensorium constitute a communication between all the organs of sense and muscles.' The governor or musician was no longer present in this materialist explanation, but interestingly the mechanical terms of communication remained in use: 'The *external organs* of sense are the coverings of the immediate organs of sense, and are mechanically adapted for the reception or transmission of peculiar bodies, or of their qualities.'[69] By this point Darwin simply could eradicate the soul from his medical discourse on the body: 'By the words spirit of animation or sensorial power, I mean only that animal life, which mankind possesses in common with brutes, and in some degree even with vegetables, and leave the consideration of the immortal part of us, which is the object of religion, to those who treat of revelation.' What had been called by philosophers an immaterial agent that exists in or with but is distinct from matter could be analogous to 'heat, electricity, and magnetism, [which] can be given to or taken from a piece of iron; and must therefore exist, whether separated from the metal, or combined with it. From a parity of reasoning, the spirit of animation would appear to be capable of existing as well separately from the body as with it.' Moreover, the powers of 'gravity, specific attraction, electricity, magnetism, and even the spirit of animation, may consist of matter of a finer kind.'[70] That subtle spirit, aether, no longer the mechanism of spiritual government over the finite individual body as a soul, was now extended both throughout the enclosed nervous system of animate bodies and beyond into the world at large in a material form like magnetism, heat, light, and electricity. Hence the epigraph translated from Virgil's *Aeneid* Book VI that Darwin chose for *Zoonomia*:

> Earth, on whose lap a thousand nations tread,
> And Ocean, brooding his prolific bed,
> Night's changeful orb, blue pole, and silvery zones,
> Where other worlds encircle other suns,
> One Mind inhabits, one diffusive Soul
> Wields the large limbs, and mingles with the whole.

The governor of the body had become a global soul encompassing the earth and steering motion, sense, and volition through the laws of nature. The science of the body was rather uncomfortably leaving behind the question of whether that governor owed any allegiance whatsoever to an immortal and incorporeal power.

Were Diderot or La Mettrie in France, Lovett or Darwin in England, or indeed McLuhan in Canada or any number of cybertheorists, less 'mad' than Dr William Belcher, who in 1798 wrote the first book on intellectual electricity?[71] 'Intellectual Electricity, Novum Organum of Vision, and Grand Mystic Secret, Namely,' read the title page, 'that the present State of Things is the Consequence of an incipient Change in Human Nature itself: a Revolution that, the more it is thwarted and crossed, the more its Spirit vibrates, kindles and flames; demonstrative of the present Existence of a Sense or Faculty in Man, anticipated or discovered by the Author; whereby Innate ideas are less necessary, and ordinary Optics superseded; the Connection between the material and spiritual World elucidated; the Medium of Thought rendered visible; Instinct seems advancing to Intuition, and Politics assume a form of Magical Intimation.' Citing Newton, David Hartley, and Erasmus Darwin among others as evidence of his theory, Belcher proposed, albeit 'well aware, that the complexity of the human machine is a labyrinth in which the profoundest theorists have been lost,' the following query:

> May not electricity (and consequently lightning) convey ideas in some of the following manners?
> 1. By means of the identity of the electric and nervous fluid.
> 2. By its action on the nervous fluids whether they are identical or not.
> ...
> 5. By its subtil communication with the sensory.
> 6. By its direct access to the soul.
> 7. By its being the vehicle of spirit.[72]

Noting that with the 'excitement of new or forgotten ideas,' he experienced 'flashes' or 'sparkles' at his eyes, Belcher explained that the cause was certainly electricity. 'I cannot but suppose,' he continued, 'that persons of far superior intellect, and quicker imagination, derive ideas from the same subtle source unknown to themselves. My constitution hardly differs from others, except that the eel-like sensibility of my nerves may render me more like an electrical machine, and my thoughts in a manner self-visible, by a kind of *sixth sense* not before noticed.'

Anticipating the academic and cultural furor over the global village by some centuries, Belcher enthused: 'The opinion of Dr Fothergill, formed from the experiments of Galvani and Galli, was that the electric and nervous fluids are specifically the same; which, combined with my actual experience, might suggest an universal communion of ideas, equalling in velocity the solar ray, and the more than figurative meaning of popular *electrification*, importing sudden impression on the public mind as experienced by an army of two thousand men, who being electrified by order of Louis XV all felt the shock at the same instant, as would probably have done a string of men surrounding the globe: for if electric shocks pass with the same velocity as the solar rays, they would in five or six seconds travel a million of miles; more than forty thousand times the circuit of the earth.'[73] Such a realization 'would bid fair for immortality,' he prophesied. 'Certainly human nature would be highly exalted and spiritualized; the body would be transparent, converse intuitive, and disguise odious.' Belcher granted his patent to a

> MYSTIC ADVERTISER, and his Assign or Assigns, for importing in Balloons (Carriage free) this and other his works, into our Mid-region, for the perusal of our loving subjects; and also confer on him an aerial Peerage and Equality, with the name, style, and Title of CHRISTOPHER COLUMBUS SPIRITUALIS, VISCOUNT Sky; and also appoint him one of our *Privy Councillors,* with the privilege of nominating a Deputy to our *Council Chamber* erected on arches of Rainbows, until he shall, by the commixture and leaven of Electric Fire and Vital Air, become sufficiently rarefied to attend in person.
>
> Given at our *Spiritual Court,* holden in a fleecy Vapour, in the night between the Seventh and Eighth of October, Old Style, in the Five Thousand Seven Hundredth and Ninety-eighth Year of our Vice-royalty.
>
> The Devil take JUPITER.[74]

However fanciful Belcher's interpretation of communications between bodies might seem, the notion of an intellectual electricity permeating the body and continuous with the material world would be a not illogical outcome of Willis's and Newton's early speculations on the nature of aether in the nervous system. Moreover, it is these tropes of the unity of human communications with material extension that continue to arise in cyborg speculations.

Tropes of control in the human mind, particularly, consciousness as steersman, governor, or pilot of a mechanical device, have been related to communications metaphors applied to social bodies for centuries. Moreover, the conceptualization of the human body as a mechanistic communications system that governs individual actions and also extends to affect (or effect) the health or disease of the body politic appears in the works of such diverse writers as Hobbes and Willis in the seventeenth century, and Swift, La Mettrie, Sterne, and Belcher in the eighteenth. The supposed convergence of man and machine would seem to be a convergence of metaphors for that subtle entity that makes both man and machine calculate, communicate, move, and act, and the feedback systems that control them. Much as the cyborg construct of networked human-machine represents the health or sickness of a political-social body for cultural theorists today, the man-machine in the seventeenth and eighteenth centuries was figured as a communications system which also represented the governance of the social and commercial hierarchies of the nation. The metaphorical steersman – or lack thereof – remains a powerful metaphor in postmodern discourse when we imagine our souls or consciousness extending beyond the body into networked communications systems. What is both reinforced and destabilized by these systems of communications that have replaced the roadways, communiqués, and pamphlets of Thomas Willis's time, however, is the politicized language of government, materialistic human identity, and ungoverned communications that was composed for the man-machine in the seventeenth and eighteenth centuries.

5 The Woman-Machine: Techno-lust and Techno-reproduction

If the man-machine was characterized as a self-motive thinking machine animated by aethereal or something akin to electrical fluid and regulated by systems of communication between brain and body, was the woman-machine an identical being? Is it possible to identify a gendered precursor to the female cyborg in the mechanistic discourse of the seventeenth and eighteenth centuries? Was the female engine of the eighteenth century a baby machine or a sex machine, with the womb a mechanistic locale of conception in contrast to the brain as mechanism of conception in the man-machine? Such an interpretation would seem to support the feminist theory from the 1980s onward that has suggested such roles in both medical discourse and filmic representations of the female-machine. Gena Corea's *The Mother Machine: Reproductive Technologies from Artificial Insemination to Artificial Wombs* (1985), Jocelynne A. Scutt's collection *The Baby Machine: Reproductive Technology and the Commercialisation of Motherhood* (1990), and Robyn Rowland's *Living Laboratories: Women and Reproductive Technologies* (1992) all explored this theme of the twentieth-century woman as breeding machine in a culture of reproductive technologies. Although scholars have made some cursory connections to the early modern period's tropes of reproduction and of gender, however, the relationship of the contemporary woman-machine to an Enlightenment model is a complex and somewhat contradictory one: most unexpectedly the female, while 'automatic' in a reproductive capability, is rarely depicted as a female machine. Was there a woman-machine? What characteristics, if any, does the mechanistic account of 'woman' share with the modern-day female cyborg? To examine this history, it will be useful to first delineate the salient features of the contemporary female cyborg figure.

In late twentieth-century film and fiction, two dominant versions of the female cyborg emerge. One is the coldly rational and highly sexualized or even fetishized machine who is in control of her own destiny and who may be a disturbingly lethal threat to the male heroes; the other is the horrifying representation of the disembodied and independently reproducing organic-mechanical womb. Extracted from and distinct from the mother, the independent techno-womb is a gruesome exaggeration of monstrous industrial machinery combined with dripping mucal organic tissue. These (predominantly male) characterizations of twentieth-century female techno-bodies contrast tellingly with lived experiences recorded in sociological and anthropological studies of reproductive medicine and technology, where researchers have noted a common image of the pregnant woman as passive vessel carrying the fetus as rational pilot or voyageur. Although many cyborg theorists have suggested a strictly Cartesian or anti-Cartesian identity in terms of body-mind duality, the female cyborg represents a unique formulation of the old trope.

The Female Cyborg in Twentieth-century Fiction and Film, or, Why Do Cyborgs Need Boobs?

Anne McCaffrey's 1961 short story 'The Ship Who Sang' was one of the earliest treatments of the twentieth-century human-machine as a Cartesian rational intelligence piloting a mechanical body. Helva XH-834 is a female brain implanted in the central shaft of a titanium intergalactic scout ship. Her neural, audio, visual, and sensory 'extendibles' are connected and grafted to the ship's shell so that in essence she becomes the ship, her brain piloting and steering the vessel as it would her natural body. Complicating the stereotypical image of the Cartesian rational machine simply as a mind-body dualism, one of Helva's early missions is to deliver 300,000 fertilized ova to the planet Nekkar: the self-directed human-machine here represents female reproductive choice and autonomy. At the story's conclusion, Helva explains to her human scout Kira that she is not interested in reproducing her own child from her parents' stored genetic material, as Kira plans to do herself: '"No," Helva said sharply, then added more gently, "that won't be necessary ... After all," the ship chuckled, "there aren't many women," and Helva used the word proudly, knowing that she had passed as surely from girlhood to woman's estate as any of her mobile sisters, "who give birth to 110,000 babies at one time."' In the closing lines, they speed on 'toward Nekkar and deliverance,'[1] implying not merely delivery of the embryos, but also liberation from carrying the fragile load. This new liberation from the boundaries of 'natural' reproduction ultimately seems to characterize the female cyborg, whether in feminist treatments of the medical discourses of reproductive technology, in feminist cybertheory such as Haraway's 'I'd rather be a cyborg than a goddess,' or in the fetishized depictions of the techno-female in film. Like the male, the female cyborg in contemporary fiction and film invites a Cartesian interpretation of mind-body dualism; unlike the male cyborg, she is a figure consistently defined and problematized by her sexual and reproductive capacities.

Helva, without a gendered human body or avatar, is a rare representation of female human-machines in film and fiction. More typical is the female-gendered cyborg in Mamoru Oshii's anime, *Ghost in the Shell* (1991): narrow-waisted, large-breasted, sexy, dangerous, and rational (therefore unemotional), she reproduces if at all only by 'unnatural' means – in this case, the transfer of organic brain and brain stem, or pure consciousness, to mechanically constructed body (see figures 3 and 25). The fictional female cyborg typically occupies one of the two poles of a spectrum extending from heroism to evil but in either case she effects the coupling of the *femme fatale* with an equally perilous technology. Claudia Springer has identified the female cyborg as a misogynist representation of the 'fetishized phallic woman,'[2] suggesting that the cyborg Eve 8 from the Duncan Gibbins film *Eve of Destruction* (1991), like the Maria-duplicate in Fritz Lang's film, *Metropolis* (1927), is an admonition against amoral quests for technological progress. As Springer explains, Eve and Maria both signify a threatening association of technology with the female body that is explicitly sexual, seductive, and menacing. Eve 8 (suggesting the biblical story of the fall of man through eating the fruit of the tree of knowledge) has had a nuclear bomb installed in her womb by the Defense Department. A cliché of the techno-Fall, the evil female cyborg is regularly situated in the context of a threat to specifically Christian values. In *Metropolis* the fake Maria, representing the Whore of Babylon, provokes lust and desire in the sons of the city and incites the workers of Metropolis to violent rebellion that almost results in the drowning of all their children. The fetish version of the woman-machine apparent in the machine-Maria emphasizes the female android or cyborg as a sign of malevolent techno-sexual desires, in direct contrast to the real Maria who cares for the children of Metropolis, who leads the workers in prayer, and whose name invokes the Virgin Mary, mother of Christ. Springer's two examples are separated by half a century, but they represent perhaps the most common features of the fictional female cyborg in their rendering of simultaneous techno-lust and techno-trepidation. Technically speaking, female cyborgs don't need mammary glands, but in the fantasy world of film they generally have a lovely large pair of them, for obvious reasons.

The razor-girl Molly in William Gibson's *Neuromancer* reiterates the motif of the alluring but dangerous techno-female. Indeed, *Neuromancer* begins with Case running for his life from this woman with retractable razors implanted in the nailbeds of her fingers. Decidedly non-maternal and non-domestic, Molly (evoking the 'moll' or prostitute) develops a sexual relationship with Case but leaves him at the novel's close, her only explanation an uncompromising note: 'HEY ITS OKAY BUT ITS TAKING THE EDGE OFF MY GAME, I PAID THE BILL ALREADY. ITS THE WAY IM WIRED I GUESS ... ' The 'Terminatrix,' in Jonathan Mostow's film, *Terminator 3: Rise of the Machines* (2003), is another version of the lethal and fetishized female cyborg, obviously evoking the dominatrix. Determined to eradicate John Connor on the verge of Judgment Day (saved only by the newly upgraded and morally responsible male T850 unit), the Terminatrix sheathed in skin-tight red latex is again explicitly fetishized and again explicitly non-fertile. Whether she is ultimately on the 'good' or 'bad' side, the fictional female cyborg signifies feminine techno-sexuality on some level as a powerfully menacing force potentially beyond the control of the male hero.

Cyborg 2 (1993), directed by Michael Schroeder, stars Angelina Jolie as the cyborg Casella 'Cash' Reese, a 'peak condition, top of the line,' cyborg who is not only a master of martial arts, but also comes complete with a hyper-sophisticated psycho-emotional program, and 100,000 mg of undetectable bio-explosive chemical implanted in her body. The movie, set in the year 2074, opens with a printed narrative explaining:

CYBORGS HAVE REPLACED HUMANS IN EVERY RESPECT FROM THE SOLDIER IN THE FIELD TO THE PROSTITUTE IN THE BROTHEL.

BILLIONS OF DOLLARS ARE AT STAKE AND AS USUAL THE LOVE OF MONEY PROVES TO BE THE ROOT OF ALL EVIL.

Future Beware.

The first scene shows a number of men in a boardroom viewing a wall-sized video of the live demonstration of a female cyborg 'test unit' and support staff in a tech booth watching the same event through a window.

In what at first appears to be a pornographic peepshow, a well-endowed blonde cyborg sex kitten straddles and enthusiastically fucks an equally gorgeous male cyborg under flashing blue strobe lights. The purpose of the demonstration is to show the capabilities of the 'Osaka operative' (Cash Reese). The demo model has had a bio-explosive injected into her circulatory system and directly linked to a microbeam detonator. As the two cyborgs gasp in pleasure, the techies begin their countdown, and at the moment of what appears to be orgasm the female throws her head back and explodes. Both cyborgs are spewed against the walls and windows in a fiery mess of spattered body parts and fluids – making what one of the technicians callously calls 'cyborg stew.' As the representative of PinWheel Robotics Inc. research and development explains, the cyborg operative is 'aggressively effective ... the cutting edge of counter-espionage technology in a humanoid robotics application.' The rest of the movie entails Cash's escape from this terrible fate. We are expected to sympathize with Jolie as a self-aware cyborg figure, but the depictions in film of the seductive dangers of technological knowledge nevertheless continue to play out the old story of Eve – the conjoined allures of the desirable woman/prostitute and forbidden or prostituted knowledge – as the potential destroyer of humankind. And, as with so many versions of this postmodern myth, humanity is saved ultimately only through the valour of a male hero (here, Colson 'Colt' Ricks with the help of a male rogue fighter named Mercy).

Similarly, in Albert Pyun's original postapocalyptic film, *Cyborg* (1989), it is the 'awesome fighting skills' of the male human Gibson Rickenbacker (Jean Claude Van Damme) who protects and delivers the female cyborg Pearl Prophet, who carries the data necessary to save humanity from a deadly plague. 'It's strange,' Pearl says when she finally arrives, 'but I feel *he's* the real cure for this world.' Neo in *The Matrix* (1999), Case in *Neuromancer,* even the Terminator in *Terminator 3,* all share this responsibility as saviour figures who are ultimately the champions of strength, intelligence, resilience, and bravery, who are the heroes and whose powers ultimately go beyond those of even the most dangerous and powerful sexy female cyborg characters.

Although Mamoru Oshii's anime *Innocence: Ghost in the Shell 2* (2004) to some extent subverts and questions these images, it is perhaps unnecessary to point out that male writers and directors have tended to perpetuate the stereotypes. In contrast, the campy film, *Teknolust* (2002), written, directed, and produced by Lynn Hershman-Leeson, deliberately engages with the apprehensions of the terrifying spectre of the cyborg that is 'one part woman, one part science.' *Teknolust* is the story of biogeneticist Rosetta Stone, 'mother and sister' to three Self-Replicating Automatons (SRAs) that she has created using her own DNA. The SRAs are built to be immortal, but they require infusions of Y chromosomes to survive (their computer spermometer tells them when they are getting dangerously low): the collection of sperm by the beautiful Ruby involves the seduction of unsuspecting men, who subsequently become infected with a dangerous self-replicating virus that renders them sterile, impotent, and branded with an angry-looking rash in the shape of a bar-code in the centre of their foreheads. These cyborgs are only human, however, and instead of annihilation and destruction, this story ends happily, with all the men cured and all six of the main characters in love and in bed. Hershman-Leeson's mocking treatment is a rare critique of the 'teknolust' in mainstream film and fiction that reflects a mistrust of the sexually enticing non-reproductive woman conjoined with the pursuit of artificial organic creation.

In the movies, cyborgs don't get knocked up. The womb as reproductive machine frequently and disquietingly exists, accordingly, in vitro: independent from and extraneous to the 'natural' body, the womb-machine in fiction and film is monstrous both in terms of magnitude and horror. Ridley Scott's film, *Alien* (1979), for example, portrays the human spaceship as 'Mother' carrying fully autonomous offspring within itself as machinic womb, while the alien spaceship is a more monstrous representation of mechanico-organic reproduction, complete with vaginal openings and glistening corridors.[3] Kenneth Branagh's film *Mary Shelley's Frankenstein* (1994) depicts the techno-reproductive process through overwrought imagery of monstrous industrial gears, chains, and pulleys; a seething testicular bulge hanging from the ceiling to release its organic spermlike jism of galvanic electric eels into a series of oversized pipes and channels; a copper 'womb,' reminiscent of a coffin or Robert Fulton's 1798 design for the copper submarine *Nautilus*, filled with an abundance of slippery amniotic fluid collected from women in childbirth, and occupied by a fully grown pilot/offspring awaiting the moment of quickening. In *The Matrix*, while the cyborg Trinity is dangerous, lethal, and sexually desirable, the sinister 'plant' that mechanically reproduces and recycles human beings in individual womb-pods like transparent amniotic sacs incongruously attached to a mechanized metal frame is a grotesque and repulsive system that grows human bodies as 'batteries' for the machine, recycles dead material within a closed energy system, and dupes its complacent citizens/offspring into a false sense of reality. The 'matrix' here is a monstrous womb that circumvents both 'conceptual' rationality and procreativity until Neo the saviour figure is reborn to outwit the machine (see figure 26).

The monstrosities depicted in cyborg films that examine issues of genetic manipulation may be less overtly visible but they, too, react to the evolution from what Laura Briggs and Jodi Kelber-Kaye have termed a 'romanticized natural motherhood.' Briggs and Kelber-Kaye show that Andrew Niccol's film, *Gattaca* (1997), 'located in a science fiction tradition stretching back to the *Alien* movies,' aligns the '"unnaturalness" of genetically modified offspring with homosexuality and communism.'[4] Similarly, as Mary Ann Doane has demonstrated in 'Technophilia: Technology, Representation, and the Feminine,' the conjunction of technology and the female is characterized by both desire and anxiety. The persistence of such contradictions in the genres of science fiction and horror, she argues, manifests 'both a nostalgia for and a terror of the maternal function, both linking it to and divorcing it from the idea of the machine woman.'[5] As I have remarked elsewhere on cyberpunk author Bruce Sterling's *Schismatrix*,[6] the fusion of non-reproductive sexual pleasure and cold rationality in the sexually desirable character Kitsune expresses unease for future gender identities: 'They gave me to the surgeons,' Kitsune explains to her lover. 'They took my womb out, and they put in brain tissue. Grafts from the pleasure center, darling. I'm wired to the ass and the spine and the throat, and it's better than being God.' Her rational consciousness merely 'an amalgam of coldly pragmatic logic and convulsive pleasure,' Kitsune eventually reconfigures herself into a building of voluptuous skin and sphinctered doors, equipped with female pheromones and 'erototechnology' – beds of flesh for male visitors' pleasure – with the stipulation that 'male ejaculations become the property of the recipient ... [which is] an ancient feminine principle.'[7] That Kitsune has almost no emotional sensibility but experiences powerful sexual pleasure, cold rationality, and reproductive control demonstrates once again the anxiety and the desire that cyborg stories register about female intellect, sexual autonomy, and pleasure without 'natural' fertility within the organic matrix of the womb.

Cyborg Reproductive Technologies in the Twentieth Century

The telling counterpart to the fetish sex machine or horrifying reproduction machine in mainstream film and fiction is the representation of the living female within the realm of medical reproductive technologies.

In medical and popular discourses of pregnancy, the fetus is regularly characterized as an autonomous pilot, astronaut, or passenger within the female body. Some writers, such as Donna Haraway, have celebrated the cyborg as representing the possibility of reproductive freedom. Often, however, the technologically manipulated functions of passive female procreativity have been contrasted with the superior creativity of techno-science. Adele Clarke, for example, has delineated cyborg pregnancy as a process mediated by surveillance technologies in 'Modernity, Postmodernity and Reproductive Processes ca. 1890–1990 or, "Mommy, where do cyborgs come from anyway?"' Clarke argues that the modern approaches to reproductive bodies and processes 'were and remain centered on achieving and/or enhancing *control over* those bodies and processes.'[8] Postmodernity, she continues, is witness to 'the moment of the disembodiment of women,' where pregnant women's bodies are 'erased to make way for the one true person – the fetus'[9] (citing Rosalind Pollack Petchesky [1987, 1985], Barbara Duden, and Monica J. Casper [1994]). Janice Raymond has similarly noted in *Women as Wombs* that reproductive technologies 'render women as spectators of rather than participants in the whole reproductive process ... they reduce women to the status of vehicle for the fetus; biologically, they literally sunder the fetus from the pregnant woman. Politically and legally, technological reproduction tends to position the fetus as isolated and independent from the mother but not from the sperm source, the doctor, or the state.'[10]

Female cyborgs have repeatedly been represented as disappearing or invisible subjects and passive containers for techno-fetuses. As Monica J. Casper argues, both fetus and mother become cyborgs through the variety of available technologies in science and medicine, but they are profoundly different types of cyborgs. Casper claims that pregnant women are 'dehumanized by being constructed as cyborgs,' for example, while in ironic contrast, 'technologizing fetuses, turning them into cyborgs, may serve to make them more "naturally" human' since the fetus is given status as an individual patient while the mother becomes construed as 'maternal intensive care units' rather than as patient or actor. Moreover, she suggests, they appear as 'bodies made transparent by imaging technologies ... as living dead technocapsules for uteronauts.'[11] In feminist readings, one of the foremost phenomena associated with visualization technologies such as ultrasound has been the resulting tendency to attribute agency, or personhood, to a fetus. As Rosalind Pollack Petchesky argued in 1987, obstetrical imaging renders the fetus 'not only "already a baby," but more – a "baby-man," an autonomous, atomized mini-space hero.' Here she echoed Barbara Katz Rothman's observation in 1986 that 'the fetus in utero has become a metaphor for "man" in space, floating free, attached only by the umbilical cord to the spaceship. But where is the mother in that metaphor? She has become empty space.'[12] Throughout the 1980s and 1990s, feminist assessments of the culture of reproductive technologies re-examined the image of the female body as a passive machine housing the fetus as autonomous passenger or pilot.

In *Babies in Bottles* (first published in 1994), Susan Merrill Squier corroborated these interpretations, presenting a series of images representing conception and gestation outside of the womb, which once again emphasized the gestating mother as almost completely eradicated from the narrative of childbirth: the image of the mother is notably absent from these images of test-tube babies and reproductive technologies. The cyborg has a history, Squier explains in a later chapter, extending back to the early years of this century: 'the Cartesian fantasy of freeing man (as mind) from the mortal body begins with the notion of freeing the fetus (as proto-man) from the female body. In short, *the cyborg originates in ectogenesis.*' Squier argues that Haraway's model of unmasking the ideologies of organic sexual reproduction in her 'Manifesto for Cyborgs' is untenable and should be re-examined in light of the representation of the fetus itself as the '"perfect man": the cyborg.' It is her contention that 'despite its emancipatory overlay, then, the fantasy/ image of the cyborg is a problematic choice as a resolution of the feminist quandary about birth. We need to consider what it means to think of the pregnant woman as a cyborg, when the cyborg itself has links – via the shared origin in the fantasy of ectogenesis – to a number of troublingly oppressive contemporary representations of pregnancy: from anti-abortion images of the fetus/patient at risk in the hostile, dangerous environment of the pregnant woman, to advertising images that solicit consumer identification with a fantasy-fetus as passenger in a machine-as-gestating woman.'[13] Furthermore, 'foundational to the fantasy of the cyborg, with its denial of the mind/body link,' she writes, 'is an earlier denial of the relationship between fetus and gestating woman.' More recently, Squier has reiterated (without using the term *cyborg*) that the naturalization of reproductive technologies results in the fetus being treated as if it were '*outside,* the rightful subject of medical, social, and legal intervention,' while the mother is 'less a civil, legal, or medical subject than the subject of policing by all three institutions.'[14] The depictions of the machine-woman/cyborg or techno-reproductive woman carrying an autonomous passenger fetus suggests repeatedly the irrelevance of the female body in contrast to the new intelligence carried within.

Donna Haraway, however, has argued in 'The Virtual Speculum in the New World Order' that while visualizing technologies have created the fetus as a public object, the female cyborg does not have to play a largely passive role in the process of surveillance of the fetus onscreen. Haraway comments on the 'technoscientific dramas of origin' where the reproductive imagery of techno-science seems to be 'associated with the organs of cognition and writing,'[15] noting, however, that the image of the extrauterine fetus floating in cyberspace is a sign of an entity established through variously related communities of practice: the woman viewing the technologized fetal image may have a number of options at her fingertips. More recently Rayna Rapp has documented the tendency of expectant mothers to describe the foetus as the stereotypical floating space traveller after viewing a sonogram image of the embryo. 'Fetuses visualized as persons loom large in the contested politics of feminism and antifeminism that have become a central aspect of our social landscape,' she writes.[16] In any case, the prevalence of the representation of fetus in the womb as pilot or astronaut within the female cyborg body is striking, given the historical association of the pilot as soul or rational intelligence in the man-machine, and the womb of the irrational female as analogous to the brain of the rational male.

Female Cyborg Origin Stories

Like other cyborg origin stories, female cyborg origins are regularly situated in the changing medical and philosophical climate of the Scientific Revolution and the Enlightenment and, also like other cyborg origin stories, these histories are incomplete. Petchesky's article, for example, indicates a brief history in the early modern period with the claim that the free-floating autonomous fetus 'extends to gestation the Hobbesian view of born human beings as disconnected, solitary individuals.'[17] Haraway more recently has postulated a history of the postmodern reproductive-surveillance technologies in Renaissance techniques of perspective and heightened realism in art. The 'salvation stories' or 'key techniques for the realization of man' rendered through new visual techniques in the sixteenth century, she reasons, share 'viewing conventions and epistemological assumptions' with late twentieth-century computer-generated realism. Thus, Haraway considers Albrecht Dürer's

Draughtsman Drawing a Nude (1538) as a metonym for 'the entire array of Renaissance visual techniques,' the story of an apparatus for turning 'disorderly bodies into disciplined art and science,' and pointedly notes the marked screen between woman and artist as a not-so-subtle divider between art and pornography.[18] Haraway's association of gazing upon a sexually appealing image with all the negative implications of the word pornography is, *as* always, a problematic one – as is her anachronistic association of Dürer's perspective machine with gynecological examinations, which she acknowledges did not become common practice until the nineteenth century. Nevertheless, Haraway's argument that technoscientific representations of pregnant and fetal bodies have been influenced by 'iconic exemplars of early modern European art/humanism/technology' demands further exploration. The image of the female reproductive body in early modern medicine and anatomy, however, is surprisingly missing from this short history of fetuses and mothers in the 'new world order.' Robbie Davis-Floyd and Joseph Dumit have also addressed the history of female cyborgs, to a limited extent. Situating the female cyborg's origins in seventeenth-century western European philosophy in their 1998 book, *Cyborg Babies: From Techno-Sex to Techno-Tots*, they argued that the metaphor of the female body as a defective machine 'formed the philosophical foundation of modern obstetrics.' Floyd and Davis suggested that the cyborgification of both women and fetuses began with the Cartesian model of the body: 'The practical utility of this metaphor of the body-as-machine lay in its conceptual divorce of body from soul, and in the subsequent removal of the body from the purview of religion so it could be opened up for scientific investigation.' They suggest that 'Catholicism' in western Europe in the seventeenth century held woman inferior to man, with little or no spirituality, resulting in a conception of 'the male body as the prototype' of the human machine. This in turn, they claim, resulted in a conception of the female machine as 'abnormal, inherently defective, and dangerously under the influence of nature, which ... was itself regarded as inherently defective and in need of constant manipulation by man.' Along with the 'wide cultural acceptance' of the man-machine, they conclude, was a 'demise of the midwife and the rise of the male-attended, mechanically manipulated birth,' resulting in an 'accompanying acceptance of the metaphor of the female body as a defective machine – a metaphor that eventually formed the philosophical foundation of modern obstetrics.'[19] These summaries, none of them longer than a page or two, are evocative while somewhat cavalier: they compel a more thorough examination of both medical treatises and cultural assumptions in the complex history of the early modern human-machine.

The woman-machine in the early modern period is actually quite an elusive figure, and certainly was not inevitably imagined as inferior (the man-machine, after all, was also treated in medical tracts as subject to severe malfunctioning and breakdown); but she was, generally speaking, a body in which the mechanical apparatuses of sexual attraction and reproduction served to re-inscribe old gendered hierarchies. The woman-machine was the default man-machine with custom components to enable breeding, which, in turn, rendered her in some representations as more shallow, nervous, emotional, sensitive, and less rational than the male. As Thomas Laqueur has argued, while 'Scientific advances ... did not destroy the hierarchical model that construed the female body as a lesser, turned-inward version of the male ... the political, economic, and cultural transformations of the eighteenth century created the context in which the articulation of radical differences between the sexes became culturally imperative.'[20] That difference was portrayed in large part in the mechanical terms of reproductive and nervous components. If the problematic philosophical conundrum for medical conjectures of the man-machine was conception in terms of rational thought or the soul, the analogous problem for fully understanding the woman-machine was conception in terms of the mechanism of reproduction, which materially influenced both sexual and emotional functions. The nervous sensibility and 'hysteria' of the female through the eighteenth century was a mechanical interface of mind and womb where sensibility or sensuality was constantly at war with rationality.

To recapitulate, in the postmodern myths engendered by fiction and film, the female cyborg is dangerous in her intellectual and physical facility, and in her overtly seductive but non-reproductive sexuality, while the mechanical womb or matrix is fearsome, monstrous, and grotesque. In the medical practice of reproductive technologies examined by feminist scholars, the reproductive female is portrayed as passive, mechanical, and irrelevant in the new technologies that focus our gazes upon the autonomous fetal passenger as patient-pilot. The histories that are hinted at in these critiques are centred once again in revolutionary philosophy and medical techniques of the seventeenth and eighteenth centuries. As with the origin stories posited for the cyborg described in previous chapters, however, the succinct origin stories for the female cyborg are also unsatisfactory. Dürer's drawing machine (known to Alberti, Leonardo da Vinci, and Bramantino before him) was used for male *and* female bodies, and lutes as well for that matter, and while Dürer's famous woodcut from his treatise on measurement, *Underweysung der Messung* (1527), certainly calls attention to a rather obvious heterosexual male fantasy, it seems inadequate to define or even historicize contemporary reproductive imaging techniques using this most superficial of comparisons. Moreover, the human machines of both Descartes and Hobbes, which are used to tell us about the history of female cyborgs, were mostly male, while the reproductive female body in philosophy of the period has largely been ignored by most female cyborg origin stories. Where was the woman-machine in the early modern period? Is it possible to locate a history of these deeply ingrained patterns of depicting female cyborgs somewhere in the seventeenth- and eighteenth-century philosophies and anatomies of the human machine?

The answer is yes – but it is a surprisingly qualified yes because the female machine is almost entirely absent from the anatomical, philosophical, or fictional literature of that period. Paradoxically, the female identity was a mechanical product òf her weaker body and nerves, and typically discussed in terms of distinctly 'female' attributes of domesticity, desire, and reproduction. The proximate cause of acute hysteria in women, wrote the Edinburgh surgeon John Aitken (d. 1790), 'seems to be a condition of the nerves giving morbid sensibility' most frequently arising from 'strong passion or violent emotions of the mind, such as results from *disappointed love.*'[21] In contrast, as James Grantham Turner has pointed out, the mechanick philosophy that explained human physiology in terms of matter in motion presented a paradox for the male libertine or 'Man of Sense,' who both celebrated and rejected a mechanical emobodiment independent of traditional morality. 'By generating such ambivalence around the issues of method, mechanism and imitation, libertinism effectively brings into question the very assumption it is supposed to uphold – the absolute autonomy of the individual,' Turner comments, 'and yet, according to the materialist philosophy of the libertine, the individual is a product of blind drives and random circumstances.' Mechanical behaviour was no ideal for the man-machine. Furthermore, 'libertine sexuality cannot be understood simply as a surrender to spontaneous physicality; it is inseparable from the cerebral triumph over the opposite sex, from mastery exercised through tactical reason.'[22] Turner sees Samuel Richardson's character Lovelace in *Clarissa* as a space on which 'contradictory discourses leave their trace.' I would add to this that active reason and free will in no way contradict the image of the rational man-machine; it is the female who is automatically and mechanically determined by the conformation of her body. As Raymond Stephanson argued in his important essay on 'Richardson's Nerves,' 'Clarissa dies because of her nervous sensibility.'[23] Men, too, were sensible creatures, but they were generally conceived to be stronger in body and nerve as well as potentially more rational because of their physiology – and, as in Lovelace's case, they could be most wilful in their pursuit of pleasure.

The paradox or contradictory discourse would seem to be that wrong choices and ill-considered wilfulness can render the man-machine purely mechanical: Seeing that his own proud spirit has been bested from 'below,' Lovelace exclaims: ''tis poor too, to think myself a machine in the hands of such wretches ["these *women*"]. – I am *no* machine. – Lovelace, thou art base to thyself, but to *suppose* thyself a machine.'[24] However, when he admits to Belford that he has plotted himself into a wretched position, Lovelace is forced to acknowledge that though he hates to be 'compelled' even by his own choice, he has become 'a machine at last, and no free agent.'[25] As Richardson's friend George Cheyne remarked, 'the *Males* having stronger, but coarser, both Bodies and Faculties, by their own Licentiousness and Luxury, often bring on their own Sufferings, and bodily Misfortunes; whereas the *Females* in general, having weaker, but more delicat and pliable Bodies and Spirits, and tied down by Custom, and the *Tyranny* of Men, to many Restraints (which Men insolently despise) are more *temperat, abstemious, and modest.*'[26] The distinction here, and commonly throughout the period, is that the male 'Animal Machin' (as Cheyne repeatedly designates the human body) makes active and self-willed choices, whether pursuing or denying mechanical sensual pleasures, while the female (even if – paradoxically – spiritually superior) is largely predetermined by the mechanisms of her body. The female mechanism is also a site where contradictory discourses leave their trace.

Oddly enough, the 'woman-machine' did not exist as common terminology, even though the female body as mechanism was implied by the mechanistic discourse in a variety of categories: the 'artificial' woman was a most dangerous form of femme fatale; the womb, apart from the woman, was mechanical and, with a growing interest in the medical government of the birthing process, a mathematically defined system of independent forces that, unless controlled and manipulated by male rationality, risked catastrophic failure; the womb was also a passive automaton in contrast to the active and autonomous embryo machine that it carried. In other words, the relationships between the female cyborg and the female machine are quite clear, although the female machine herself is virtually invisible.

Where's the Woman-Machine?

London physician and natural philosopher Walter Charleton's *Enquiries into Human Nature* (1680) begins with a preface defining his subject as 'the most abstruse Oeconomy of Nature in the body of Man; a System of innumerable smaller Machines or Engines ... fram'd and compacted into one most beautiful, greater Automaton.' Charleton here endorses the image of the man-machine, but also emphatically redeems human nature with the immediately following clarification that all these disparate mechanisms are 'adjusted to one common End, namely, to compose a Living Ergasterium or Workhouse, in which a Reasonable and Immortal Soul' may do its divine work.[27] Like Thomas Willis, Charleton represents the soul in the man-machine as feminine, but also as in Willis, the representative human for his discourse on human nature is categorically male. Almost a century later, Laurence Sterne's Yorick exclaims in *Sentimental Journey* (1768), 'What a strange machine is man, framed with such nice mechanism by Nature's hand, that every element impedes his perfect motion!' thus introducing a riotous dialogue between his sensual body and disapproving soul. Charleton's earnest and Sterne's playful invocations of the mechanistic philosophy in these texts call attention to two basic assumptions about the human-machine in the seventeenth and eighteenth centuries. First, that it was by default male; second, as we have already seen, that it inevitably problematized the connection of the rational, immortal soul to the purely mechanical motions of the human body. In the latter part of the eighteenth century, the man-machine (that is, the living body defined by mechanicks or the laws of motion) also came to represent a calculus of labour and appeared alongside various other moving contraptions in works such as the second edition of William Emerson's engineering manual, *The Principles of Mechanics* (published in 1758). Emerson cites Borelli who 'computes the force of the muscles to bend the arm and the elbow': 'a strong young fellow can sustain at arms end, a weight of 28 lb. taking in, the weight of the arm. and he finds the length of CB to CK to be in a greater proportion, than that of 20 to 1. Whence he infers the strength of these muscles to be so great, as to bear a stretch at least of 560 lb.' (see figure 15). Furthermore, he concludes 'It is evident that all animal bodies are machines.' The example following shows the mechanics of the 'motion of a *man*, walking, running &c.' A 'good footman,' he obverves, 'will run 400 yards in a minute.'[28] Similarly, in his *Survey of Experimental Philosophy* (1776), Oliver Goldsmith writes that the force the shoulder muscles exert to keep the hand and arm 'extended, while the hand holds a weight of about twenty pounds, is astonishing ... this natural machine, thus fashioned by the Great Workman, is infinitely more powerful than any artificial machine that man could form.'[29] As we have seen, the male human-machine, if not completely ubiquitous then certainly acceptable as a model for the functions of the body, continued to appear in a variety of discourses both specialized and popular through the end of the eighteenth century as 'animal machine,' simply 'machine' or 'engine,' and in descriptions of numerous body systems as 'mechanisms.'

The obvious question that arises from the persistent terminology of the man-machine is: Was there a woman-machine? Predictably, there seems to be no equivalent language for the laws of motion governing a working woman-machine, nor is there notable debate about the materiality of the soul in the rational woman-machine. While the female may fit within the category of human machine in a non-gendered way – as in the words of Richardson's Clarissa to Anna: 'O my dear, what a poor, passive machine is the body, when the mind is disordered!'[30] – the *woman-machine* would seem to be a term that is not used in English literature of the period. How can we account for this paradoxical absence of the woman-machine, given that the 'fair sex' is consistently defined in the material terms of nervous and reproductive embodiment?

We might turn to the infamous author of *L'homme machine* to determine whether *la femme machine* was an entity in itself; however, La Mettrie has very little to say about women in his notorious text: 'In the fair sex,' he wrote with no great originality, the soul and temperament arise from the delicacy of the female nervous system. Hence, he argues, 'the tenderness, affection and lively feelings based on passion rather than on reason, the prejudices and superstition whose deep imprint can scarcely be erased' in women resulted from the mechanism of their inferior material. 'In men, on the contrary,' he writes, 'whose brains and nerves have the firmness of all solids, the mind, like the features, is more lively. Education, which women lack, further reinforces [men's] souls ... how could they not be more grateful, generous, constant in friendship, firm in adversity, etc.?' Women should not envy 'the twofold strength' of the nervous system and education given to man, he concludes, 'for the one exists only in order to be all the more affected by the attractions of beauty, and the other only to minister all the better to their pleasures.' For the rational male, significantly, art is not the deceptive or transitory material vanities of women (described in the following section of this chapter). For the male, art as a product of education makes possible the highest order of creative or intellectual genius elevating humanity above all other creatures. The man-machine implies a complex hierarchy of conception: the rational, incorporeal ideas of the male represent a 'natural' ascendancy over the corporeal conceptions of women with their weaker nerves and irrational passions. Pregnancy and the young girl's disease known as chlorosis, he argued, both can accordingly wake 'depraved tastes,' and make 'the soul execute the most atrocious plots' because 'the brain, the mind's womb, is perverted ... together with the body.'[31] Certainly, the man-machine for La Mettrie is a superior mechanism, which, aided by superior matter and by education benefited from a livelier mind. The crucial point is that for La Mettrie, the active, engaged, thinking and creative man-machine is an *ideal* – however material.

La Mettrie of course was not unique in suggesting women were more susceptible to the natural infirmity of their delicate anatomy: woman's representation in the literature of sensibility as a being governed by bodily mechanisms is well known.[32] Indeed, the female consciousness was in many formulations more often subject to the weakness of the body's natural material nervous mechanisms than the male's and as many scholars of history, literary criticism and gender studies have demonstrated, women throughout the period were consistently represented as being influenced by their particularly inferior embodiment, their weaker nerves, and by their wombs. Nevertheless, the explicit terminology of female machine or woman-machine is almost completely absent even in the most overtly mechanistic fictional, medical, or philosophical texts: rather, 'woman' is more usually associated with nature in a mechanic world-view where the autonomous rationality of the man-machine is opposite to the automatic reflexivity of the mechanical female.

Typically, the machine-woman is at most implied, and in this instance is characterized by natural passive mechanisms that make her categorically distinct from the active and rational man-machine. Moreover, the mechanical woman would seem to be irrelevant to the man-machine debate over mind and matter; where the female machine is active and self-moving, she is characterised by her artifice or vanity. Although the term *mechanical* is only rarely stated explicitly, the mechanical woman is suggested by a number of well-known representations not surprisingly

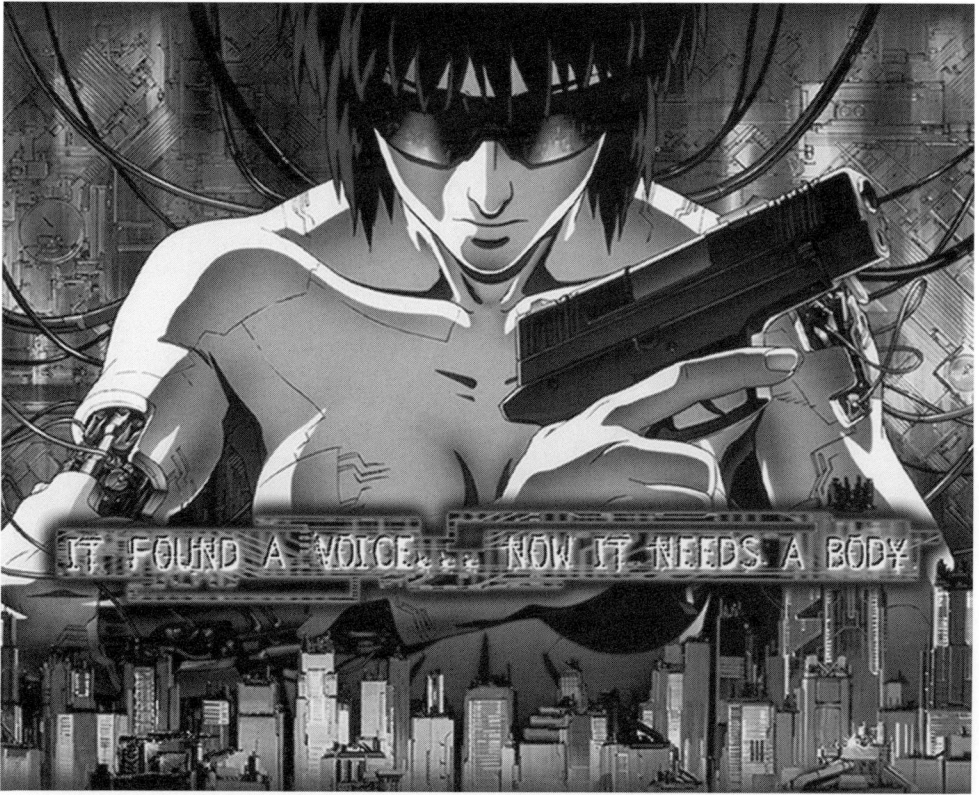

25. Promotional image from Mamoru Oshii's *Ghost in the Shell* (1996) featuring the cyborg Motoko Kusanagi.

26. The womb as monstrous mechanical-organic 'plant' (*The Matrix*, 1999).

BREAST-WORK or FEMALE FORTIFICATION.

And now the dear Creatures appear, | *With BREAST-WORK projecting in Front,*
As if for HOT ACTION inclin'd Sir. | *And extended BUMBATTERIES behind Sir.*

Published March 15, 1786, by S.W.Fores, at the Caricature Warehouse, N°3, Piccadilly. *vide Paddy O'Brien.* 1.6

27. The first fembot? 'Breast-Work or Female Fortification' signed R.T. (March 15, 1786). Etching with stipple, with hand coloring, 23.6 x 19.6 cm (design).

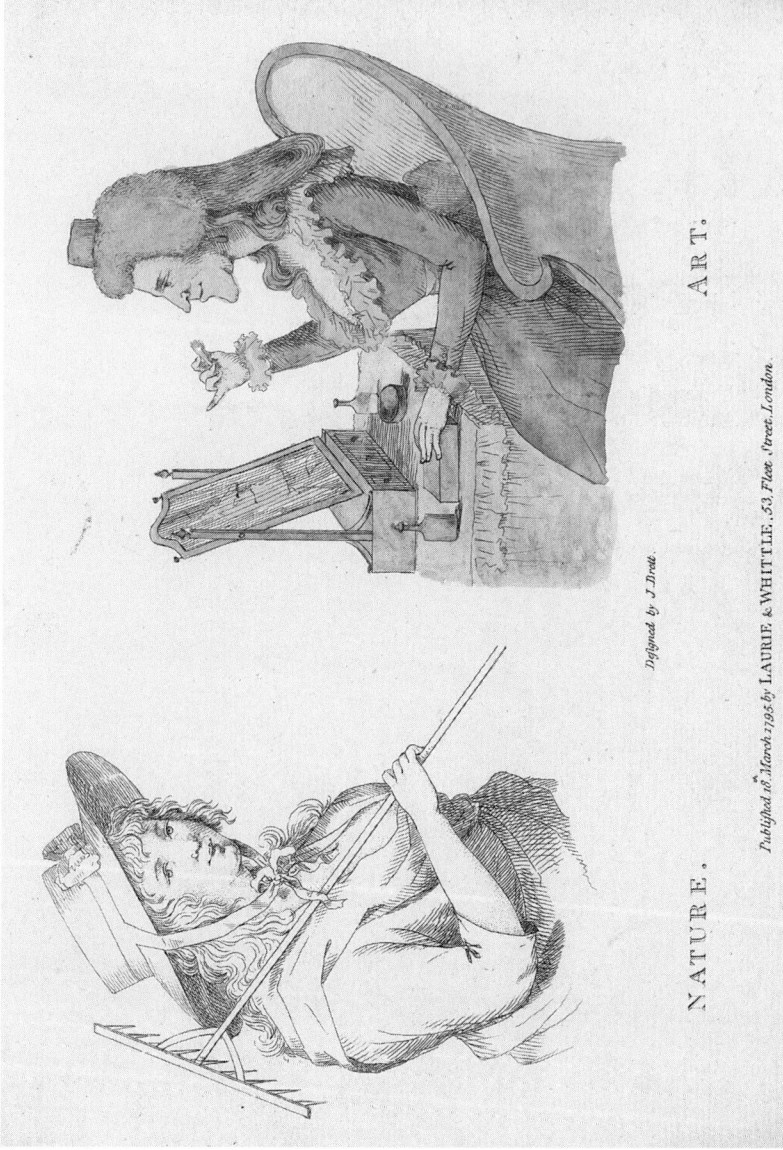

NATURE.

ART.

Designed by J. Brett.

Published 18 March 1795 by LAURIE & WHITTLE, 53 Fleet Street, London.

28. 'Nature. Art.' by J. Brett (March 18, 1795). Etching with hand coloring, 18.5 x 24.8 cm (sheet). Published by Laurie & Whittle.

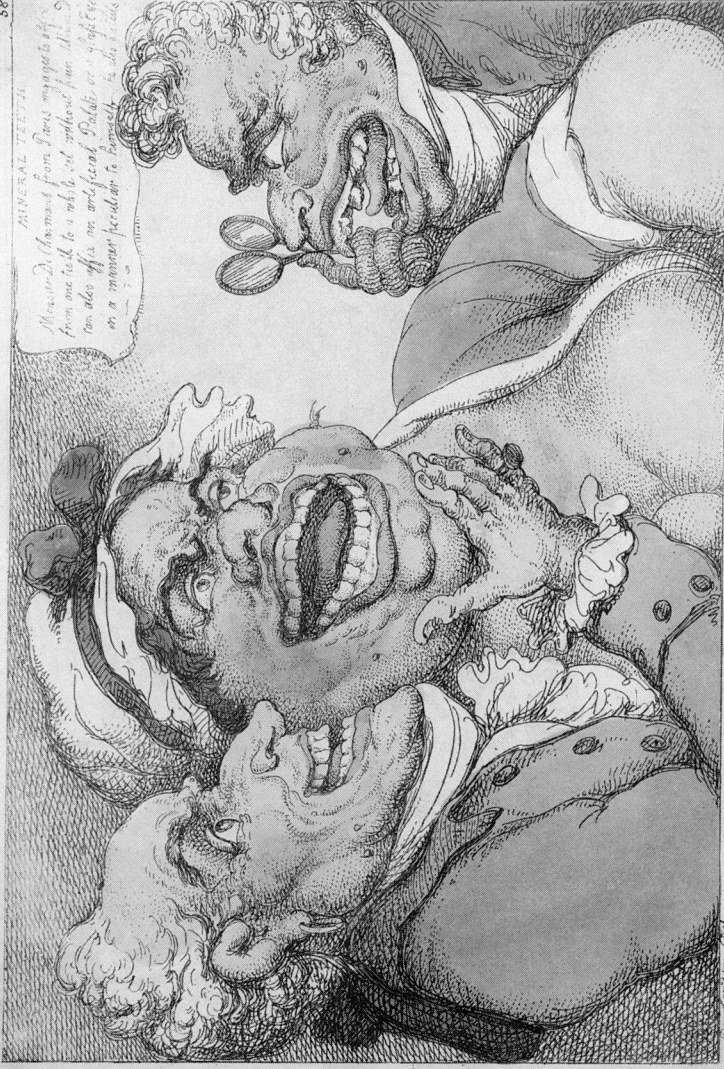

29. 'A French Dentist Shewing a Specimen of His Artificial Teeth' by Thomas Rowlandson (February 27, 1811). Etching with hand coloring, 21.7 x 32.8 cm (design). Published by Thomas Tegg.

30. 'Moll Handy. With a Letter of Recommendation to a Service,' probably by George Bickham, in *Bowles and Carver's Caricatures* (1740).

31. The womb as flower. Giulio Casserio's *De formato foetu* ... illustrated by Odoardo Fialetti (Frankfurt, 1631). A crude version of this drawing, retaining the flower covering the genitals and the floral-womb image, appears in Jane Sharp's *The Midwives Book* (1671).

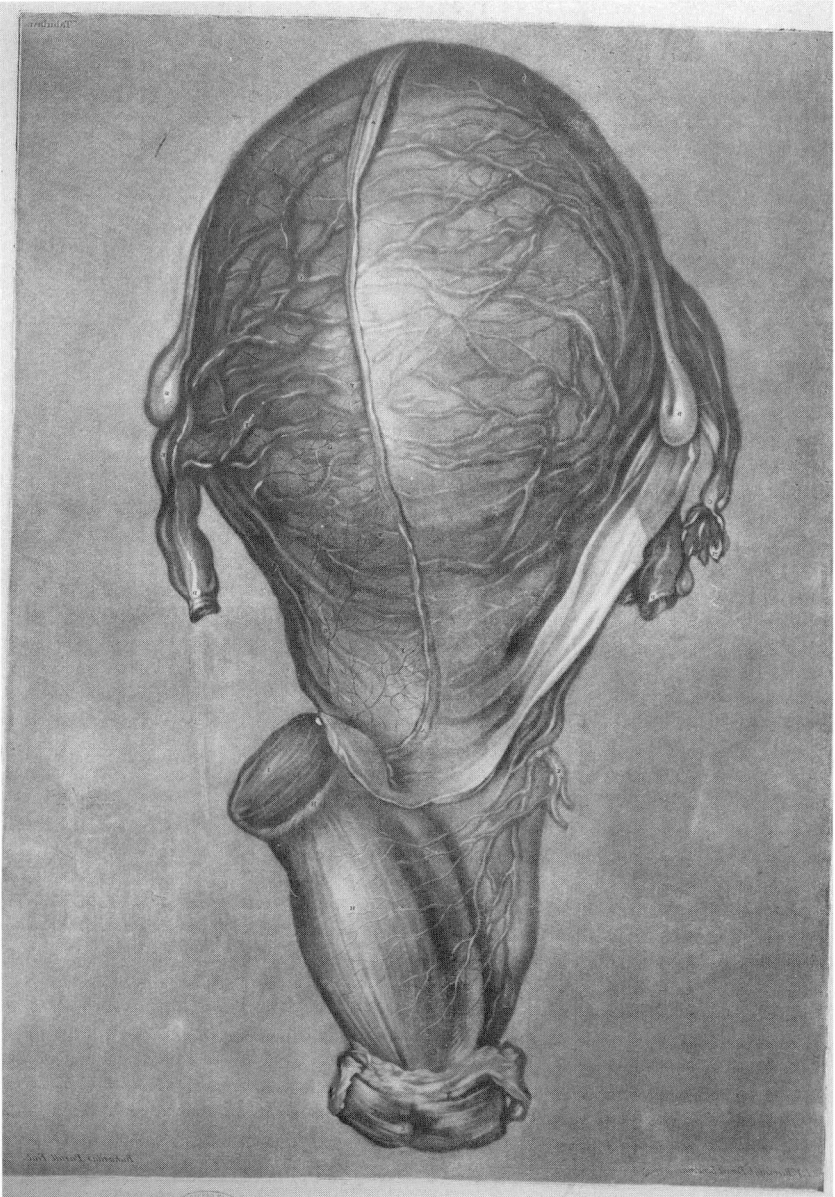

32. The pregnant uterus. Charles Nicolas Jenty, *Explicatio demonstrationis uteri praegnantis muleris* (London, 1757).

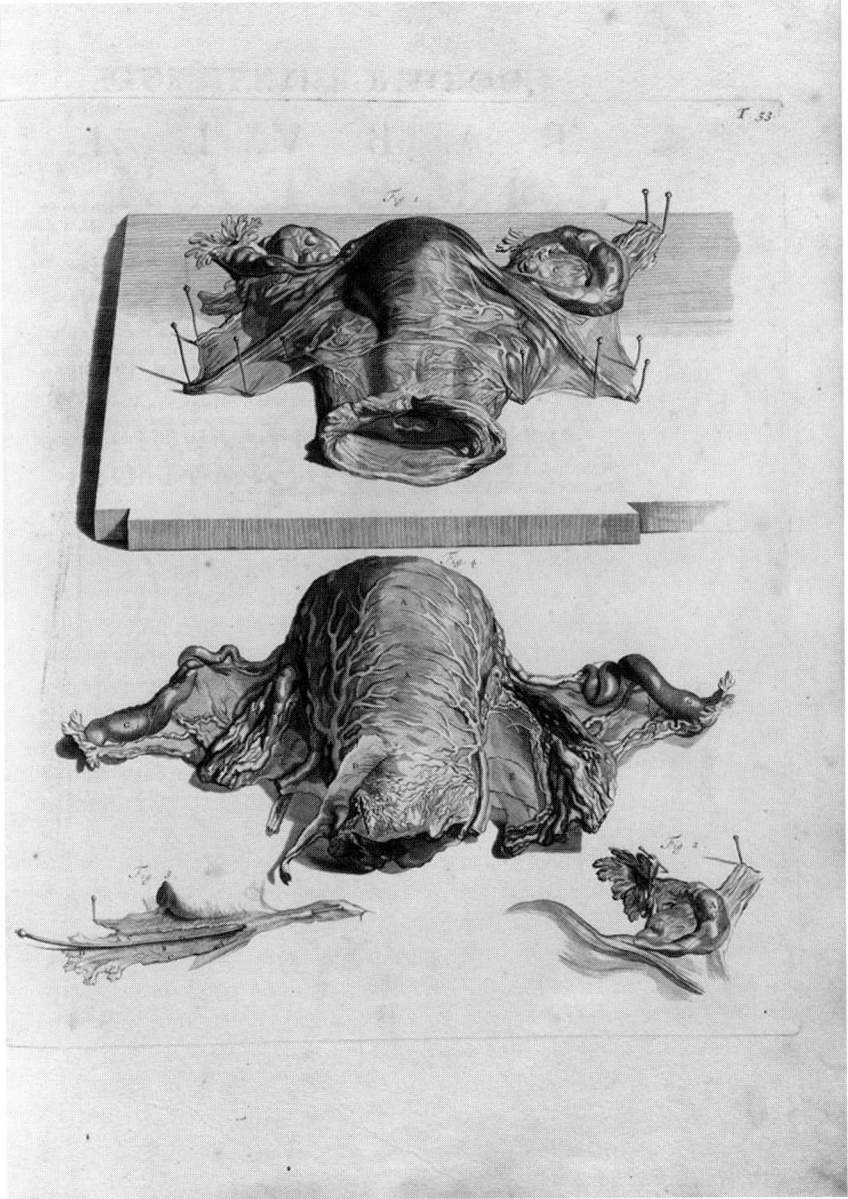

33. Uterus, ovary and Fallopian tube, engraving/etching by Gerard de Lairesse, in Govard Bidloo's *Anatomia humani corporis* (Amsterdam, 1685).

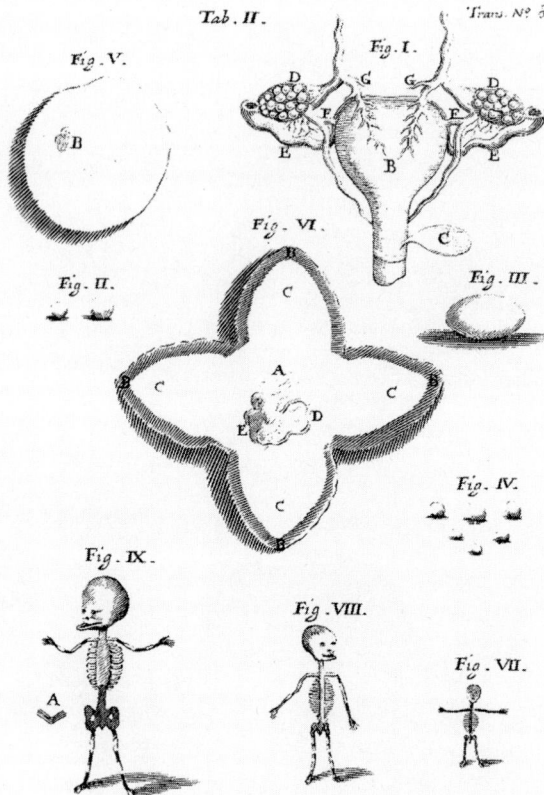

34. The excised matrix and the foetus as little engine. 'An Account of What Hath Been of Late Observed by Dr. Kerkringius Concerning Eggs to be Found in All Sorts of Females, by Dr. Kerkringius' (*Philosophical Transactions*, 1672). Fig. I is 'the *Matrix*, with its chief dependencies'; Fig. II shows the eggs from 'the testicles of a Woman'; Fig. III shows a bigger egg as was found in women 40 and 18 years of age; Fig. IV shows smaller eggs found in the testicles of a cow; Fig. V is an egg 'after it was fallen into the *matrix* of a Woman, and in which he saw that little *embryon* marked *B*, whereof he found the head begun to be distinguished from the Body'; Fig. VI is a bigger egg showing 'a Child of 14 days after Conception'; Fig. VII represents 'the *Sceleton* of an Infant found by the same in one of these Eggs three weeks after Conception'; Fig. VIII 'exhibits the *Sceleton* of another Child, found also by him in an Egg, a moneth after Conception'; Fig. IX 'represents the Sceleton of an Embryon found by him six weeks after Conception.'

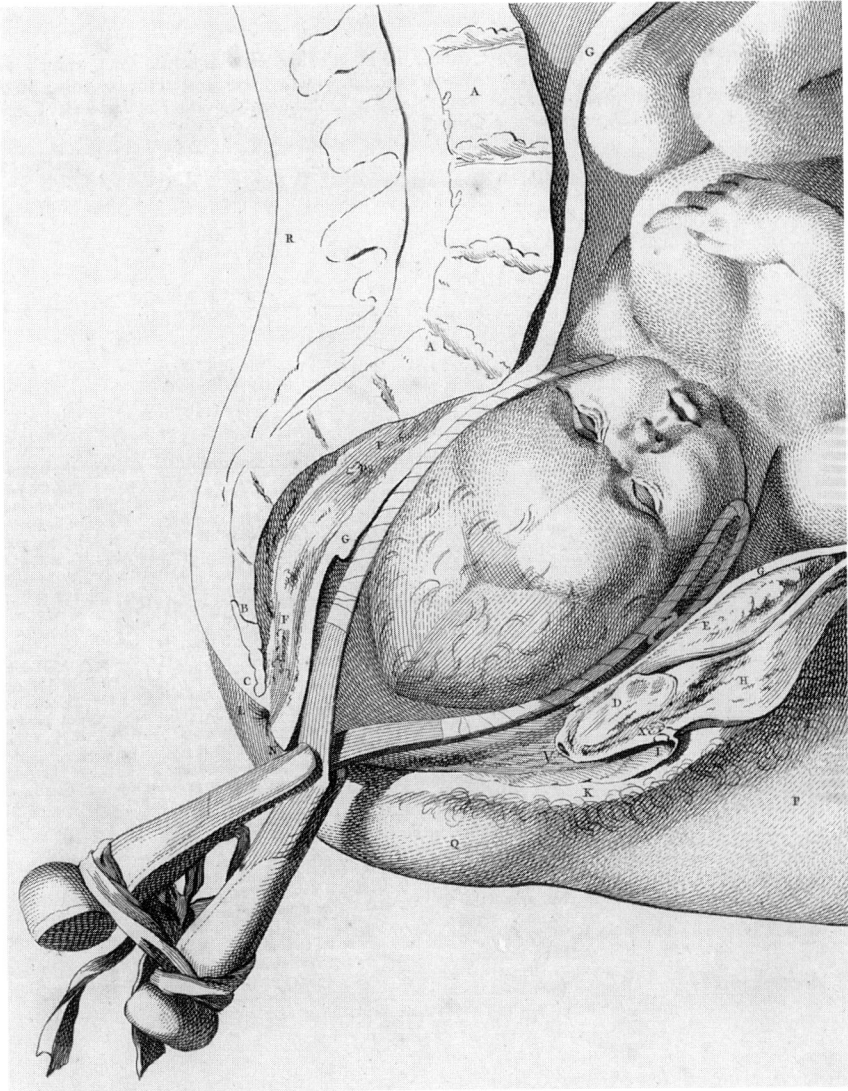

35. Delivery of the fetus using forceps, shown in situ. Engraving/etching by Jan van Rymsdyk, in William Smellie's *Sett of Anatomical Tables* (London, 1754). The child and forceps here are rendered whole and lifelike, while the mother's body is cut in two and abstracted as representational outlines.

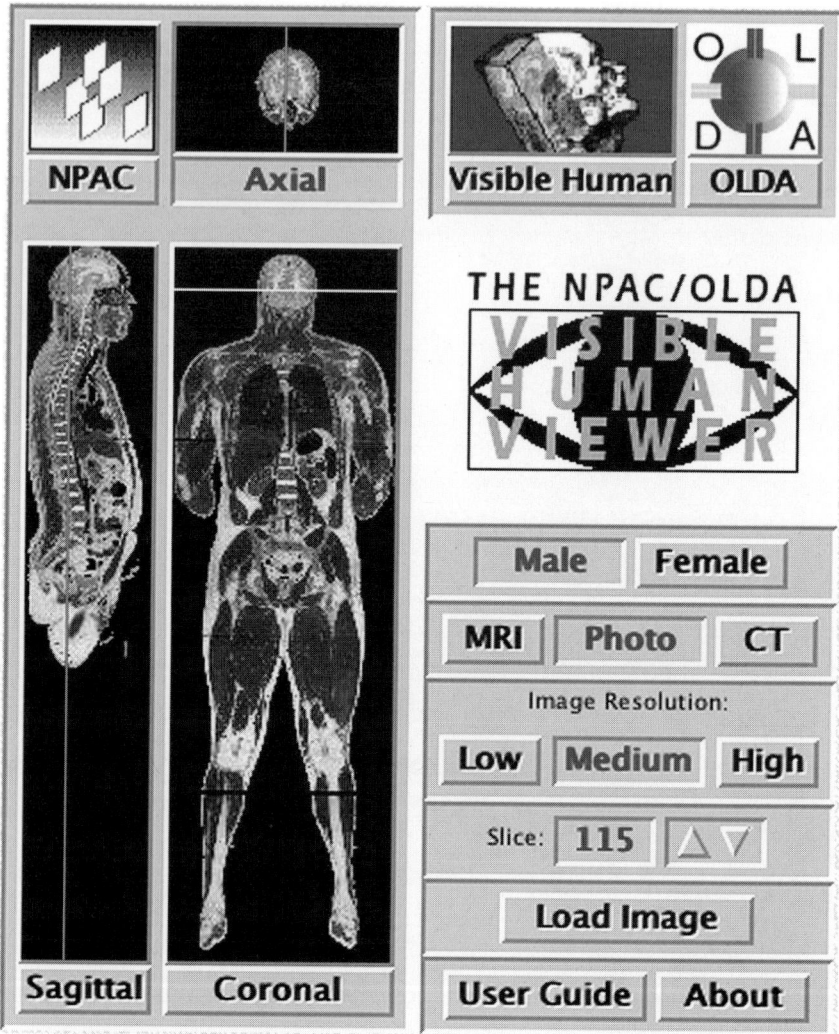

36. NPAC/OLDA Visible Human Viewer. (http://www.dhpc.adelaide.edu.au/
projects/vishuman)

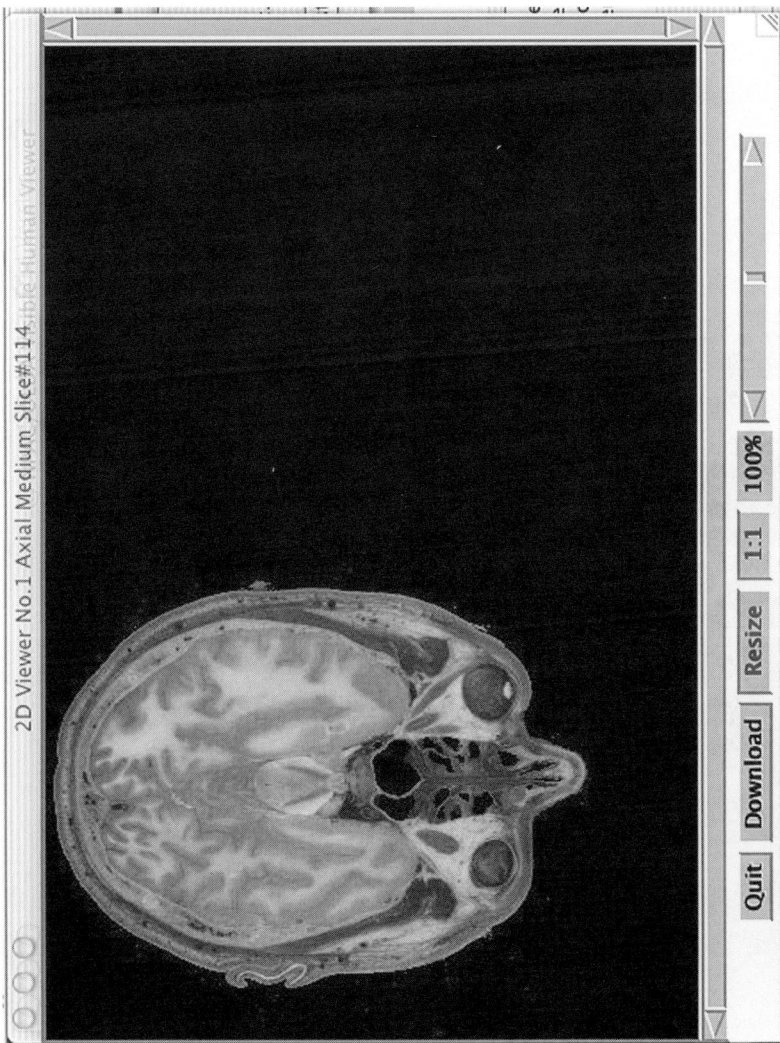

2D Viewer No.1 Axial Medium Slice#114 Visible Human Viewer

Quit Download Resize 1:1 100%

37. Brain page from the Visible Human Viewer. (http://www.dhpc.adelaide.edu.au/projects/vishuman)

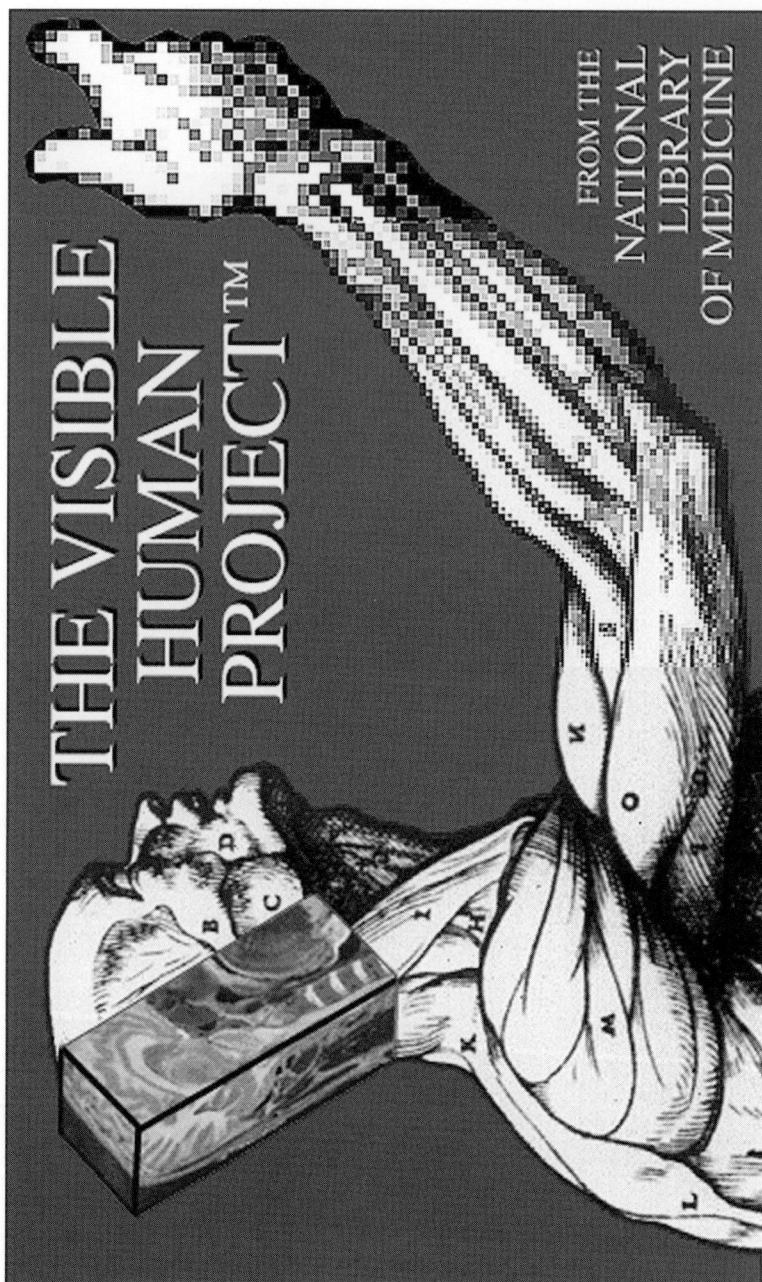

38. Visible Human Project logo.

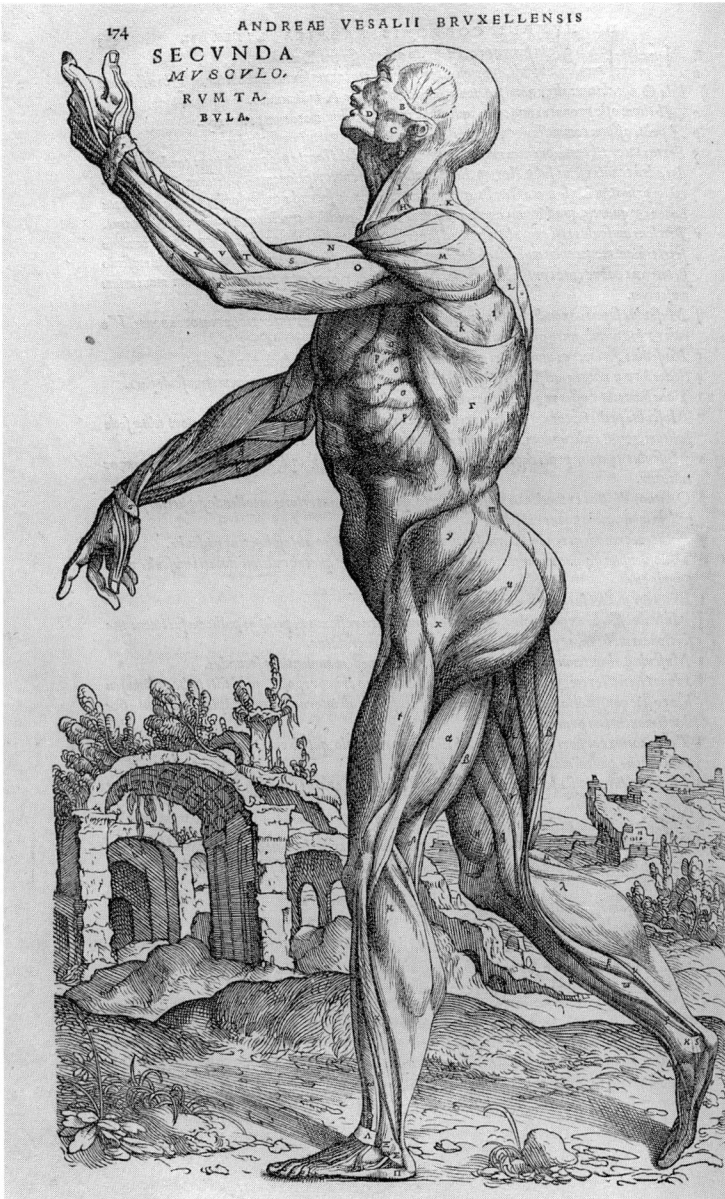

39. Second plate of the muscles, from Vesalius' *De humani corporis fabrica* (Basileæ [Basel], 1543).

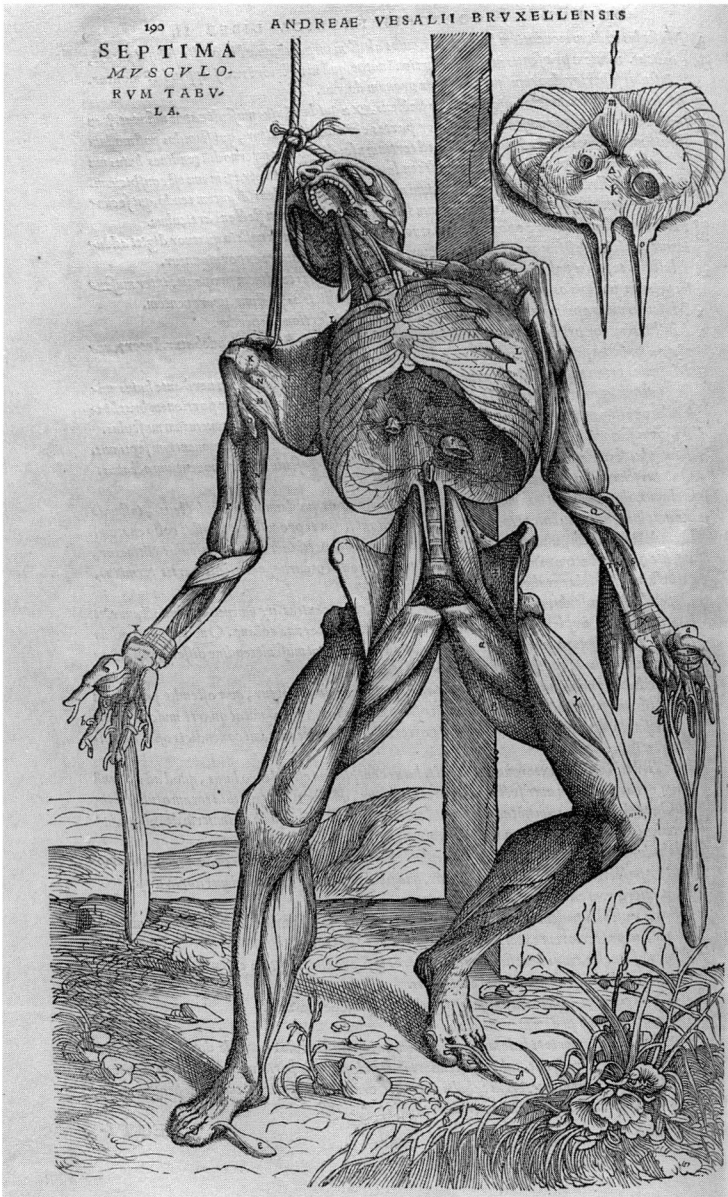

40. Seventh plate of the muscles, showing also the excised diaphragm, from Vesalius' *De humani corporis fabrica* (Basileæ [Basel], 1543).

limited to either domestic or a sexual economies: she is the grotesque or sexually desirable woman who is the disturbing product of artifice and prosthetics; she is the mechanical body that comes like (or with) machines; she is the female whose mind is governed by the material structure of her nerves (in the literature of sensibility and pornography); occasionally, she appears as a mechanism in a domestic household; finally, in medical-philosophical literature she is not present at all as an active, thinking subject, but subsumed by the auto-womb that is characterised by the natural and non-rational forces of reproductive conception and creation.

Female Vanity and Mechanick Art

The vanity of feminine artifice, in contrast to a preferable 'natural' state, is one version of the female as a counterfeit creation wherein she is humorously, occasionally grotesquely, modified by prosthetic intervention. The artificial woman of the eighteenth century signifies female fashion, extravagance, and vanity, and she evokes responses simultaneously erotic and disquieting. The comical rendering of womanly artifice in a print entitled 'Breast-Work or Female Fortification' (1786), for example, registers the sexual titillation inherent in the excessive bodily 'extensions' of feminine attire that are both inviting and defensive (figure 27). The caption reads:

> And now the dear Cratures appear,
> As if for HOT ACTION inclin'd Sir,
> With BREAST-WORK projecting in Front,
> And extended BUM-BATTERIES behind Sir.

A more cruel jest upon women's artificial bums is the eccentric London bookseller and hack writer John Dunton's *Bumography: Or, a Touch at the Lady's Tails* (1707), which jokingly derides the monstrosity of prosthetic extensions to the female anatomy in a statement ultimately registering distaste for fashion, commodity, and the marriage market-place. 'LADYS, ... I did this year (being at Turnbridge-Wells),' he begins, 'Indulge my self the Liberty of viewing your MONSTROUS (or New Fashion'd) Tails.' The lampoon is inspired by 'Lady's Tails ... being grown so High and Mountainous, that Three Lady's walking A-Breast wou'd Baricado the Walks in the broadest Place.'[33] The 'Bumographical Jest' on women's vanities is particularly aimed at the artificially sexy bodies created to 'trap' men into marriage:

> As soon as you arriv'd at the long wish'd for TEENS, what Arts have you not used to Engage our Rambling Affections? How have you made it your Business to spruce up, and FINIFIE your selves, that you might appear to the best Advantage? Witness your High Top-knots, Ridiculous Furbelows, Perpetual Washings, Curlings, Powderings, Various Garbs ... Affected Postures, Tempting Smiles, and Amorous Glances And when all this will not Wound our Hearts, you then try what your Tails will do; and by these Battering Guns you even sho your very Souls at us.

1

Leave these *Deluding* Tricks and Shows,
   Be Honest and Down-right;
What *Nature* did to View Expose,
   Don't you keep out of Sight.
The Novice Youth may chance Admire
   Your Dressings, Paints, and Spells;
But we that are *Expert* desire
   Your *Sex* for somewhat else.

2

In your Adored *Face* and *Hair,*
   What Virtue could you find,
If *Women* were like *Angels* fair,
   And ev'ry Man were blind?
You need no *Pains* or *Time* to wast,
   to set your *Beauties forth,*
With *Oyls,* and *Paint,* and *Drugs,* that cost
   More than the *Face* is worth.

3

Nature her self her own Work does,
   And hates all needless Arts,
And all your *Artificial* Shows
   Disgrace your *Nat'ral* Parts.
You're *Flesh* and *Blood,* and so are we,
   Let *Flesh* and *Blood* alone,
To Love all *Compounds* hateful be;
   Give me the *Pure,* or none.[34]

The aversion here expressed to the 'compound' creature whose allure arises from extensions and modifications to the natural body – and whose danger is emphasized by combative aggression ('battering guns') – prefigures the simultaneous desire and anxiety provoked by the gun-toting female cyborg (figure 25). Cyborgs, the antithesis of purity, often register anxiety or elation for the breakdown of gender boundaries; but 'pure' flesh and blood – 'pure' nature – clearly have historically signified a more comfortable version of woman's 'natural' domestic roles.

A jest this may be, but the horror of the 'monstrous' falsity of fashionable women conveys misogynistic anxiety about the inversion of the neat ideals of female-nature / male-culture binaries and perhaps also material/spiritual ones: distressingly the woman's sexual appeal might seem 'natural' to the aroused male but to his chagrin he discovers she is deceptively an artificial construction. She is adulterated, then, not only by the debris constituting her simulated body but also by the wicked dishonesty of using cultural and materialistic accoutrements for fooling men into sexual desire:

> Then you who wou'd not love do this
> Learn of me what *Woman* is:
> Something made of Thread, and Thrumb,
> A meer Botch of all, and some.
> Pieces, Patches, Ropes of Hair,
> Inlaid Garbage ev'ry where.
> Outside Silk, and Inside Lawn,
> Scenes to cheat us neatly drawn.
> False in Legs, and false in Thighs,
> False in Breast, Teeth, Hair and Eyes.
> False in TAIL, and false enough,
> Only true in Shreads and Stuff.[35]

Do not fall in love, Dunton warns his male readers at length, offering a recipe that starts with boiling Cupid alive. He acknowledges there is such a thing as an honest woman, but 'The Misfortune is in the CONJUGAL LOTTERY, for One Prize there's a thousand Blanks.'[36]

Dunton concludes with a 'Bumographical Description of a Good Wife' who is characterized by industry, frugality, and chastity: she remembers that 'tho' her *Tail* does Conquer his weak Side, yet that God and Law have appointed him to be her HEAD.' Moreover, she 'considers she is call'd A HOUSE-WIFE, and endeavours to make good the Title'; she refrains from 'Gadding Abroad' and is not 'Sluttish' at home; she 'provides Liberally for her *Family*, but has an Eye that nothing be wasted, and remembers that an ill manag'd Kitchen has destroy'd many a Noble Hall'; she 'spends more Time in Prayer and Exercises of Devotion than between the Glass and the Dressing-Box'; she 'suits her Dress and Apparel (but more especially her *Rump* and top-knots) to her Husband's Quality, rather than the Fashion; yet loves *Neatness*, and cannot endure any Paint on her Cheeks, but the Natural Vermilion of Modest Blushes'; she is trustworthy with her husband's money. Finally she 'is not perpetually draining his Purse for Modish Vanities.'[37] The good wife characterizes all that the artificial creature is not. In this Dunton's satire registers discomfort with a self-determined female whose supposed ambitions of finding wealthy husbands at such fashionable spas as Turnbridge Wells – driven by their vanity and greed – make it all the more difficult to find a domestic and obedient family provider who would let her husband do her thinking, and moreover is not rendered some bastard compound by the new commercial materialism.

Jonathan Swift's devastating caricature of female vanity in his scathing 'Progress of Beauty' (1728) also makes an explicitly moral statement on the binary of spirit and 'mechanick matter,' in this case in an aging whore. The transformation of Celia's 'cloudy wrinkled face' by degrees to her 'artificial face' serves as a vile reminder of both the mortality of the flesh and ugly sin of women's artifice for commercial gain:

> But, Art no longer can prevail,
>   When the Materials are all gone;
> The best Mechanick Hand must fail,
>   Where nothing's left to work upon.

> *Matter*, as wise Logicians say,
>   Cannot without a *Form* subsist;
> And Form, say I as well as they,
>   Must fail, if *Matter* brings no Grist.[38]

'Rotting' Celia who strolls the street when all sober folk are in bed is defying her mortality in vain, Swift chides, and the materials of her outward beauty are merely transitory. The poem rather callously concludes with a request for *new* nymphs each month presumably for the narrator's own enjoyment, but the artificial female is nonetheless a moral admonishment against female vanity and material concerns.

The subject of satiric prints and poetry alike, the deep distrust of physical artifice forms a principal focus of repugnance. 'Nature. Art.' designed by J. Brett and published in 1795 calls attention to the ludicrous vanity of female artifice in a cruel representation of an aging and homely woman applying colour to her face as laughably inferior to the buxom and charming figure opposed to her, prettily flushed with the bloom of presumably undadulterated youth (see figure 28). Rowlandson's devastating caricature of 'A French Dentist Shewing a Specimen of his Artificial Teeth and False Palates' (see figure 29) renders a woman's artifice as grotesque as the unhealthy 'natural' state of the man beside her. Although this depiction seems more a xenophobic comment on the French than particularly targeting feminine artifice, the image captures once again the distrust for the artificial augmentation of the natural body: the dentist here who can also according to the caption, 'affix an artificial Palate or a glass Eye in a manner peculiar to himself,' seems to have constructed nothing so much as a disturbing monstrosity.

However material or mechanical, the presence of prosthetic beauty aids in these examples does not result in a definitive woman-machine for the age – even if these representations foreshadow gendered assumptions addressed by and reincorporated into certain female cyborg figures as a disturbingly alluring artificial construct. The distinction between nature and artifice distends the category of human-machine beyond the boundaries within which the man-machine has been defined by the physics of mechanical philosophy – that is, as matter in motion. What moves the woman-machine? Notably, of the few references I have been able to find to the specific term *female machine* or *'female engine'* in eighteenth-century literature, vanity is recurrently the feminine motive force.

The trope is used for humorous effect as in the mischievous poem by an anonymous author, 'The Sigh Revers'd,' where the body too tightly corseted results in an embarrassing mishap of compressed gas:

Friendly Air, mis-call'd a Vapour,
    Lovely *Silvia's* Ease and Shame;
For when e'er thou dost escape her,
    All her Face is in a Flame.

Wholesome Crack of Female Engine,
    Skrew'd for Shape a Peg too high,
Which th'imprison'd Air revenging,
    Forces *Silvia* to let fly.

Hateful Messenger to Lovers,
   Pressing thro' a Crowd of Foes;
For tho' none thy Sound discovers,
   Sav'ry Fumes inform the Nose.

Shapeless F——t, who e'er has shewn thee,
For an outward Form thou'st none
Tho' in Publick few will own thee,
All caress thee when alone.[39]

In Frances Brooke's novel *History of Lady Julia Mandeville* (1763) the female is explicitly contrasted to the rational man-machine with regard to 'petticoat politics' in a passing comment: 'Vanity is the moving spring in the female machine, as interest is in the male.'[40] Similarly, in William Combe's satirical *The Devil Upon Two Sticks in England* (1790), the demon Asmodeus characterizes vanity as the moving force in the female machine. Explaining to Don Cleofas that 'human virtue is sometimes too strong for us all,' Asmodeus describes his unsuccessful attempt to exert his influence in getting a beautiful young widow to open her door to a young man who has fallen in love with her. The young man had determined to fall from his horse in front of her house as if by accident, Asmodeus continues, and 'though I contrived to draw her to the window at the moment this even took place, when I imagined that common politeness and humanity would operate in favour of our stratagem, her pride suppressed it all; for the moment she saw what was going forward in the street, she sent her *femme de chambre* with the most peremptory orders, that no one should be admitted into the house, on any pretence whatever.' The reason she rejects the young man, he explains, is that she is of a very good family and 'therefore has a certain dignity to support. ... And as for vanity, which is such a leading movement in the female machine – that, at present, operates wholly against us. ... but let us once get her vanity on our side and I will soon contrive that her conscience shall give us very little trouble.'[41] Another example occurs in the anonymous *Louisa Mathews. By an Eminent Lady* (1793). In a gossipy conversation between Mr Abercrombie and his neighbour Lady Susan Forrestor, Abercrombie wonders aloud what might have made Lady Elenor Palmington despise the beautiful Miss Mathews: '– stole her lovers? been praised for a finer complexion, or shape? or admired for a more graceful minuet?' Upon Lady Susan's light chastisement he replies, 'For female injuries, these are the greatest possible ones!' Lady Susan confides that

Lady Elenor has determined to marry a particular gentleman since the two estates united would be superior to her hated neighbour's: 'The Cliffords and Palmingtons were always at daggers-drawing,' she explains, 'they are now rival beauties of equal consequence in the same county, but an additional weight of five thousand a year would turn the scale decidedly in Lady Elenor's favour.' Once again vanity, here accompanied by domestic competitiveness, is the force that moves the female mechanism: Abercrombie replies, 'What intricate wheels put the female machine in motion!'[42] In these few cases it would seem that the construct of female machine is invoked to reinforce a none-too-subtle condemnation of women's less than virtuous conceits: the machine or mechanick in these cases implies a force outside of any rational or spiritual ideal. Consistent as they are in their representation of the specific terminology of 'female machine,' these three texts cannot represent a compelling argument for a presence of the moving female machine or female engine: the inventory here creates an impression of a prevailing image when in fact such representations are few and far between, and none of them appear in mainstream texts.

Domestic Machines?

One might imagine that, if the later version of the man-machine is defined by his labour, the woman as mechanism might be defined within the larger machine of a household economy, but these images are also quite rare, perhaps partly because of a reluctance to depart from what Ludmilla Jordanova has characterized as 'the naturalness of the family,' which in turn legitimized the sexual division of labour and the separation between public and domestic life.[43] When mechanical images of housewives do occur, they seem predictably calculated to emphasize a familiar moral framework of appropriate activities in the realm of domesticity. In 1711 Joseph Addison, without using the term *machine*, contrasted the ideal wife in a regular household economy to the woman of artifice attracted to 'glittering Gew-Gaws ... so many Lures to Women of weak Minds or low Educations.' Addison describes the good Aurelia's family as an ideal microcosm of the larger commonwealth, 'under so regular an Oeconomy, in its hours of Devotion and Repast, Employment and Diversion, that it looks like a little Common-wealth within itself.' They are the 'Envy, or rather the Delight, of all that know them.' Fulvia, in contrast, is one of those women of 'light, fantastical Disposition [who] considers her Husband as her Steward, and looks upon Discretion and good House-Wifery, as little domestick Virtues, unbecoming a Woman of Quality.' Although Aurelia's household seems to function like clock-work, it is Fulvia who is mechanical, living 'in a perpetual Motion of Body and Restlessness of Thought.'[44] Samuel Richardson's *Pamela: or, Virtue Rewarded* (1740), presents a more thoroughly articulated version of the female as mechanism in a domestic machine. The household economy here is analogous to a mechanical body with two governors. The woman-mechanism is a governor (directing what might be called animal ser-vants rather than animal spirits) but ultimately subservient to the husband or sovereign:

> But consider, Child, the Station you are raised to, does not require you to be quite a domestic Animal ... You'll have nothing to do but to give Orders. You will consider yourself as the Task-mistress, and the common Herd of Female Servants as so many Negroes directing themselves by your Nod; or yourself as the Master-wheel, in some beautiful Piece of Mechanism, whose dignify'd grave Motion is to set a going all the Under-wheels, with a Velocity suitable to their respective Parts. – Let your Servants, under your Direction; do all that relates to Houshold Oeconomy: They cannot write to entertain and instruct as you can. So what will you have to do? ... In the first Place, Endeavour to please your Sovereign Lord and Master; and, let me tell you, any other Woman in England, be her Quality ever so high, would have found enough to do to succeed in that ... May your Time be filled up with Reputation to yourself, and Delight to others, till a fourth Employment puts itself upon you; and that is, (shall I tell you in one Word, without mincing the Matter?) a Succession of brave Boys, to perpetuate a Family that has for many hundred Years been esteemed worthy and eminent, and which, being now reduced, in the direct Line, to him and me, expects it from you.[45]

That is to say, the woman's most important role in the properly function-ing domestic machine is explicitly reproductive. Finally, the comic image attributed by the British Museum catalogue to George Bickham suggests a female machine constructed by the tools of domestic labour represent-ing a domestic mechanism from the servant's perspective (see figure 30). The punchline here primarily concerns the working girl as repro-ductive and sexual: the text below is a letter of recommendation for Moll Handy asking if Lady Crosspatch might offer employment for a servant whose only fault is that 'she had the Misfortune by a fall to be Crack'd & is become Pot Belly'd, but as this Small fault is so Common in Our Sex I hope ... that it will be overlook'd by you. P.S. She will come for small wages.' These images both present a version of the woman as mechanism in the domestic economy: while they associate the mechanical arts of good housewifery with the female, they designate her most significant role not in terms of the actively working mechanism but rather in terms of her 'natural' sexual and maternal imperative.

Such an interpretation is supported by Jane Brereton's poem 'Pastora to Captain Fido on his last Epistle.' Brereton, who had left a tempestuous marriage in 1721, started in 1734 to publish pseudonymously some of her verses in the *Gentleman's Magazine.* These prompted some comical exchanges with 'Fido' (Thomas Beach):

> 'Gainst the whole Sex, you open War declare,
> And subtly urge, that *we* have no pretence
> To raise our Faculties, and aim at Sense;
> Gravely affirm, that all *we* ought to do,
> Is to inspect a Family – – and few.
>
> Content in Ignorance to drag our Chain,
> And blindly serve our haughty Tyrant Man,
> Who, vainly swell'd with his imperious Rule,
> Thinks Nature destin'd Woman a – tame Fool;
> A meer Machine, devoid of Reason's Guide,
> And like the Brutes design'd to sooth his Pride.
>
> Your just Preheminence *we* all allow,
> But boast aspiring Souls, as well as you;
> Indu'd with Reason, active Pow'rs, and Will,
> And can like you distinguish Good from Ill.
> [...]
>
> But if by barb'rons Laws we are confin'd,
> Nor dare reform or cultivate our Mind,
> Our upstart Passions will assert their Force,
> For nought but Reason's Check can stop their Course.
> For if by Nature these should be suppress,
> We're mere domestick Drudges at the best.[46]

Brererton's imagery of domesticated woman-machine as beast machine presumed to be distinct from the rational man epitomizes the political and indeterminate quality of human-machine terminology: few literary works characterize woman as a 'meer machine,' however condescending the writer may be about the female's position and duties.

Sex Machines: The Mechanical Operation of the Slit

Then there are the 'mechanick Jades [who] frig themselves with candles of about four in the pound,'[47] as the knowledgeable Frank explains to a wondering Katy in the anonymously published *The School of Venus* (1680). Somewhat at odds with the representation of woman in the domestic sphere is that of the independent sexual female who defies naturalized categories of home and family. Women who manually operate their own cunts represent a small number of 'mechanical' females. As the legendary Fanny Hill describes it, 'I ... squeezed, and compres'd the lips of that virgin-slit, and following mechanically the example of *Phoebe*'s manual operation on it ... brought on at last the critical extasy, the melting flow.'[48] Can we at last find a definitive woman-machine in the realm of mechanical sex? One fascinating case is that of a book by a supposed Physician of Bath calling himself Mr Lobcock (i.e., large relaxed cock or dull fellow). In *A Safe-Conduct Through the Territories of the Republic of Venus. Containing, a Practicable Proposal for the Prevention and Final Eradication of a Certain Disease* 'Lobcock' recommends prudent regulations and a wise government of prostitution, since it is, after all, 'interwoven into the very nature and web of human society.' We can, he claims, 'no more prevent girls from being manufactured into whores before they are twenty, than we can prevent whoring itself.' Manufacture notwithstanding, the writer's argument rests upon a shifting foundation of that old standby for prescribing appropriate codes of human behaviour – nature: 'They who abound in the riches of this world will expend themselves in the gratification of their desires, whether that be in the purchase of a maidenhead at thirteen, or of a pottle of green peas at Christmas,' he explains: 'Without vainly striving to chain the winds, or to counteract the decrees of Providence, or grieving till our guts ache, because we are unable to work miracles; let us attend to the performance of our *real* duty, that of recommending a just and philosophic moderation to others, and of observing it ourselves. Nor let it be supposed that even the pains of purgatory, or limited damnation, ought to be inflicted for satisfying a natural appetite when it craves, out of a common stock which Nature herself seems to have reserved, and set apart for that purpose; – for a venial trespass, which, if *properly committed*, is just as harmless as eating flesh in Lent.'[49]

Indeed, 'Lobcock' argues, plain reason and common sense show us that female virtue, family peace, and general safety depend upon sexual release for passionate men: 'Carnal lewdness is deaf and furious as the waters: – You may regulate, but you cannot impede its natural course. There *must* be Whores, that there *may* be chastity,' he concludes with an eloquent flourish. 'Solitary drains,' however – in what might seem to the distrustful reader a somewhat self-serving argument – renders the healthy natural human mechanical and ill; in the case of the male, an automaton:

> The excessive loss of the seminal fluid, by slight of hand (even if a Hand-maid be employed) doubles all the usual ill effects of too much venery in the regular way, inducing fatal symptoms, and incurable maladies peculiar to itself.
>
> Nature, friendly to those who rationally obey her impulses, will not suffer a deviation from her established rules, or a violation of her laws. Venereal excesses enervate and debase both soul and body; they yet admit of remedy, and from sedulous endeavours, we may hope for a renovation of the im-paired faculties, at some period or other. But any material damage derived to the constitution from solitary drains, is to a certain degree irreparable; for the functions, both mental and bodily, will be ever after wavering and uncertain; like those of an imperfect automaton, which at some times perform its destined action, at others be entirely useless.[50]

The passage is a variation on the libertine figure described above in terms of Richardson's *Clarissa*: while the mechanical dictates of bodily desires are a natural function of the man-machine, man must never allow himself the self-indulgence to become merely automatic. Further-more, 'The pallid and sallow Onanist shews unmanly backwardness, and base timidity.' The automaton is not machine-manly: if he comes in contact with a real woman expecting sexual relations, he is rendered 'forever incapable of enjoying the genuine and lawful pleasures of na-ture,' his penis reduced to a 'mere drooping lilly, as relaxed and useless to all the purposes of delight and harmony, as a wetted catgut.' The female engaged in such 'illegitimate gratification' is similarly rendered incapable of 'natural' pleasure, although she does not become machine-like: 'It disqualifies them for the purposes of generation, by a gradual exsiccation and debasement of the creative juices; and by rendering nature's storehouse too loose and flaccid to retain her genial treasures. And, analogous to what happens in the like case to the male sex, the woman becomes incapable of enjoyment in the natural way, and can experience no pleasure but in her own abuse.'[51] The assumptions of the bodily functions of creative-female-womb-storehouse as counterpart to

creative-male-brain-storehouse are manifest in these passages. Most obviously, in addition to the impairment of 'natural' sexual pleasure, the
disease supposedly wrought upon the male by mechanick operations is a
drain of mental capability; the disease wrought upon the female is a
drain of procreative capability. What this out-of-the-way example encapsulates, then, are a number of the conventional characteristics delineating the functions and government of male and female bodies. The
man-machine, as opposed to the effeminate automaton, is active, virile,
and rational; the eighteenth-century female is again associated with
sexual titillation or gratification, and legitimate breeding. Her naturalness, not surprisingly, is left intact.

At the opposite extreme of the gendered definitions for the sexual-
mechanical woman is Catherine Macaulay's arch mechanistic argument
defending her marriage at the age of forty-seven to twenty-one-year-old
William Graham. Less than two years after moving into Alfred House in
Bath, the home of the wealthy seventy-three-year-old Reverend Thomas Wilson, she announced her marriage to Graham (younger brother
of the infamous quack James Graham with whom it was rumoured she
had had an affair). The public censure at such an apparently scandalous act prompted a satirical defiance in her poem 'A Remarkable
Moving Letter!':

... from a cloud prolific Fancy drew,
Astride my Quarto Volumes Cupid flew;
Gr-h-m was perch'd behind, and in his hand
A portly syringe stood, or seem'd to stand
[...]

Such were the portents of my love and fate,
And climacteric widows cannot wait.
I rose, forgetful of the storm that frown'd
Of Alfred-house, of 17,000 L.
Of brother S—br-dge, continence, and pelf,
Thy much-lov'd peruke, and thy dearer self!

Material atoms, variously combin'd,
Form all the vice or virtue of the mind.
Souls and Spa-water differ but in phrase,
So preach the Doctors of degen'rate days;
Give then this female, frail machine the blame,
That fan's a widow's furor into flame.[52]

The explicit reference to the woman-machine serves to appropriate and invert the expectations for the function of female as mechanism.

The most renowned sex machines of that time might be in Cleland's *Memoirs of a Woman of Pleasure* (1749), a novel in which we encounter one of the few direct representations of the woman-machine in this period.[53] Here we find Louisa rendered a 'meer' machine by the fabulous fucking of the simpleton Good-natur'd Dick. Fanny herself reaches orgasm mechanically more than once, but as we will see, there is a significant difference between the reactive or non-rational 'mechanical' responses of the women, and the active, puissant, and rational man-machine. Even while casting her as a mentally active and intelligent being, Cleland describes a clitoral-mechanical imperative over rational will in Fanny's enjoyment of her insalubrious profession: 'with women, of our turn especially, however well our hearts may be dispos'd, there is a controuling part, or queen-seat in us, that governs itself by its own maxims of state, amongst which not one is stronger in practise with it, than, in matter of its dues, never to accept the will for the deed.' Fanny is governed by a ruler seated at her sensual centre – not in the brain where Descartes and Willis placed their governing monarch or engineer in the man-machine – but in the clitoris, which directs her will and passions. Fanny elsewhere describes animal spirits rushing 'mechanically' to her own 'center of attraction.' These are the two instances where Fanny is characterized as a mechanism. Her body is, interestingly, nowhere described explicitly as a machine. Fanny's sensibility, of course, functions finally to confirm traditional constructs of matrimony, love, and morality: 'I could not help pitying,' she comments upon her marriage to her true love Charles, 'those who, immers'd in a gross sensuality, are insensible to the so delicate charms of VIRTUE, than which even PLEASURE has not a greater friend, nor than VICE a greater enemy.'[54]

In the two-volume edition of *Memoirs* (1749) printed by Thomas Parker, 'machine' appears twenty-two times (all but once to refer to the penis); 'engine' appears six times (once or perhaps twice to describe Fanny's orgasm, and all others to describe the penis); 'mechanically' appears four times (all referring to female masturbation and sexual responses); 'instrument' as an organ (always male) appears nine times. The penis is 'a stately piece of machinery,' an 'active force,' a 'weapon,' a 'battering piece,' and it is 'animated,' whereas the female sexual organs are controlled, governed by animal spirits (and ungovernable by Fanny's rationality), often 'attacked,' enclosing, or receiving, and never – as mechanical as they are – an active or energetic force in the capacity of the male engine. Compare, for example, the following excerpts where Fanny's response is mechanical reaction to the male's 'attack' or 'driving' forces. She describes her first experience: 'my petty-coats and shift were soon taken up, and their stronger center of attraction laid open to their tender invasion: my fears however made me mechanically close my thighs; but the very touch of [Charles's] hand insinuated between them, disclosed them, and open'd a way for the main-attack.' Later, she says: 'I had it now, I felt it now: and beginning to drive, he soon gave nature such a powerful summons down to her favourite quarters, that she could no longer refuse repairing thither: all my animal spirits then rush'd mechanically to that center of attraction, and presently, inly warm'd, and stirr'd as I was beyond bearing, I lost all restraint, and yielding to the force of the emotion, gave down, as mere woman, those effusions of pleasure, which in the strictness of still faithful love, I could have wish'd to have held up.'[55] As to her own arousal in another encounter, Fanny informs us: 'You will ask me perhaps, whether all this time I enjoy'd any perception of pleasure? I assure you, little or none; till just towards the latter end, a faintish sense of it came on mechanically, from so long a struggle, and frequent fret in that ever sensible part.'

The unwilled bodily responses – if not absolve then seem at least formulated to excuse – Fanny from the guilt of willing corporeal sin or the betrayal of her beloved Charles, but they also contrast markedly and importantly with the animated male's 'mighty,' 'in-driving,'[56] or 'terrible spit-fire' machine. Where the women are active or dynamic, it is in pushing the appropriate buttons to turn on the powerful man-machine. Even a bashful young country boy becomes manly and vigorous after Fanny, playing with his buttons, 'which were bursting ripe from the active force within' reveals 'not the play-thing of a boy, not the weapon of a man, but a may-pole of so enormous a standard ... such a breadth of animated ivory ... it stood an object of terror and delight.' Fanny, the instructor, is now 'determined' by her bodily reaction to her novice: 'he feels out and seizes gently the center-spot of his ardours: oh then! the fiery touch of his fingers determines me, and my fears melting away before the growing intolerable heat, my thighs disclose of themselves, and yield all liberty to his hand.'[57]

The female mechanism is decidedly automatic in contrast to the energetical organ of the male. The rollicking episode of Good natur'd Dick and his awesome 'battering piece' in volume two confirms this construction again. My readers will remember how Louisa and Fanny set to work on Dick to 'put the principles of pleasure effectually into motion, and to wind up the springs of its organ to their supreme pitch.'

Being so provoked, the simple-minded 'natural' transforms into the vital machine: 'For now the man-machine, strongly work'd upon by the sensual passion, felt so manfully his advantages, and superiority, felt withal the sting of pleasure so intolerable, that maddening with it, his joys began to assume a character of furiousness which made me tremble for the too tender *Louisa:* he seem'd at this juncture greater than himself; his countenance, before so void of meaning, or expression, now grew big with the importance of the act he was upon. In short, it was not now that he was to be play'd the fool with.' The idiot boy now becomes manful, superior, and furious; his eyes 'shooting sparks of fire, his face glowing,' his cock becomes an industrial engine forceful and smoking: 'now, by dint of an enrag'd enforcement, the brute-machine, driven like a whirlwind, made all smoak again, and wedging its way up, to the utmost extremity, left her in point of penetration nothing to fear, or to desire.'[58]

Louisa, in contrast to her earlier aggressive seduction of the young man, becomes passively reflexive as her body is in turn worked upon by the active and energetical force of Good-natur'd Dick. To every action there is an equal and opposite reaction, said Newton: overwrought with bodily pleasure, the loss of her self-control is what makes Louisa a machine, reactive and quite mindless: 'catching at length the rage from her furious driver, and sharing the riot of his wild rapture, she went wholly out of her mind into that favourite part of her body, the whole intenseness of which was so fervously fill'd, and employ'd: there alone she existed, all lost in those delirious transports, those extasies of the senses ... In short, she was now as meer a machine, as much wrought on, and had her motions as little at her own command, as the natural himself.' Louisa rendered 'wholly out of her mind' specifically contrasts the machine-Dick actually losing his mindless vacuity. These two cases, along with the recurring language of Fanny or her cunt as a vessel at sea encourage a reading of the female as inverse and weaker version of the mechanical male: the 'driver' for the machine-woman is the male organ of pleasure inside her cunt, as opposed to the rational pilot in the mind of machine-man (Fanny is regularly 'driven' by her ardour for a cock). As Peter Sabor has shown, the metaphor of the vessel is sustained throughout the two volumes: among numerous other such references Fanny makes her 'launch into the wide world' as an orphan without guidance; on her maiden voyage to London she is as a ship tossed in a whirlwind, she struggles to stay above water with the town's current at the end of volume one; she witnesses Louisa get 'out of compass,' and finally gets 'snug into port'[59] when her true love Charles returns.[60] That the 'port' is 'the sovereign authority which love had given him' over Fanny to insist upon marriage and thus finally to raise her above the vice of mere sensual pleasure in addition to bestowing 'a legal parentage on those fine children you have seen by this happiest of matches'[61] shows even the notorious Cleland upholding a traditional morality: the ideal steering of the female mechanism would seem to be by a male navigator, and toward the safe haven of love, virtue, marriage, and domesticity.

Reproductive Machines: Knowledge, 'Geometrical Certainty,' and the Automatic Womb

To consider present-day invocations of fetus as passenger in a machine-as-gestating woman alongside the rise of mechanistic thought through the early modern period might seem to suggest a reasonable hypothesis that the eighteenth-century woman-as-machine was the origin of 'cyborg' reproductive machines. In fact, what we find is that the eighteenth-century woman was rarely characterised directly as a machine, although her automatic womb was featured as a distinct mechanism within which the fetus was a passenger. Medical and popular treatises alike habitually associate the womb with mechanical growth processes of nature, and even with the relatively sophisticated medical advances in the period, the uterus might be characterised simultaneously as the root or the soil out of which the fetus grows like a plant and as a bottle or vessel that contains the fetus. In either case, the fertilization by the male seed tended to be characterised as a creative process using the tools of tilling or of distilling (the tools implying thereby an act of deliberate creation). The representation of the fetus as a man in miniature, often autonomous, and often an entire 'engine' or 'machine' distinct from the mother, contrasts the representation of the mother herself, whose wholeness or rational personhood is eclipsed by descriptions of discrete mechanical systems such as the ovaria, the tubes and tissues that promote fluxions of fluids and nutrients, or the geometry of the womb. Toward the end of the eighteenth century, and in the specific context of midwifery manuals, the reproductive woman might be an 'animal machine,' but in such contexts not surprisingly the debates over rationality, knowledge, and understanding that had dominated earlier dialogues on the man-machine are absent, except insofar as any argument about knowledge and understanding is relegated to doctors and man-midwives defending and promoting their own expertise. In general terms from the late seventeenth to the end of the eighteenth century, the womb undergoes a revision from a mechanical substrate for the plant-like fetus to a mechanical apparatus that is mathematically determined by and manipulated according to properties of force, angles, curves, and torque. While the roles of both male and female seed in the reproductive process were contested and indeterminate, the dominant version of the male partner as active, life-giving, and creative force gradually gives way to a new hierarchy where active intervention and knowledge, even conception itself, is the terrain of the medical professional. It would be wrong to suggest the reproductive woman is increasingly characterised as mechanical; nor is she overtly and consistently imagined as a woman-machine. The definitions of mechanism, however, become increasingly mathematical, and therefore establish the territory of the new medical industry, in which the woman is again mechanical or automatic rather than an active and self-motive machine.

At the end of the seventeenth century the woman's reproductive body is mechanically fruitful like nature, her automatic womb acting as the unconscious motions of matter constrained by the design and purpose of an omnipotent engineer. Thus Robert Boyle compared the world itself to a pregnant automaton or a ship steered or maintained by the natural laws of motion and God's will:

> I consider the frame of the world already made, as a great, and, if I may so speak, pregnant automaton, that, like a woman with twins in her womb, or a ship furnished with pumps, ordnance, &c. is such an engine as comprises, or consists of, several lesser engines; and this compounded machine, in conjunction with the laws of motion, freely established and still maintained by God, among its parts; I look upon as a complex principle, whence results the setled order or course, of things corporeal. And that which happens, according to this course, may, generally speaking, be said to come to pass according to nature, or to be done by nature, and that which thwarts this Order may be said to be preternatural, or contrary to nature. And indeed, though men talk of nature as they please, yet whatever is done among things inanimate, which make incomparably the greatest part of the universe, is really done but by particular bodies, acting on one another by local motion, modified by the other mechanical affections of the agent, of the patient, and of those other bodies, that necessarily concur to the effect, or the phaenomenon produced.[62]

Here nature, a ship, or a pregnant woman alike are automatons which an outside active and vital force directs: 'For, whatever the Structures of these Engines be,' Boyle writes, 'they would be as little, without the Co-operations of external Agents; such as the *Sun, Æther, Air*, &c. be able to exercise their own Functions, as the great Mills, commonly us'd with us, would be to Grind Corn, without the assistance of Wind or running Water.' As we have seen, the brain of the human-machine also exercises its function by active agents compared to aether or light, but because that engine is rational, the forces are the agent of the governing monarch or rational soul. For passive nature or automatic womb, there is no such governing pilot – only the mechanical forces put in place by God.

As with all such generalized claims, there are exceptions. The poem 'The Water-Engine,' which appears in the anonymous miscellany *Pleasure for a Minute: or, the Amorous Adventure* (1730), makes a female engine the brunt of a racy joke:

> A Female Engine, 'tis, you'll say,
> That I thus lively here display;
> All other Engines it exceeds,
> It does those mighty wond'rous Deeds;
> It's Water, mix'd, is of such worth,
> It procreates, and Man brings forth:
> But when you well this Engine try,
> It draws the Lover's Water dry,
> It has a Well that ne'er has been
> Home-fathom'd yet all venture in;
> Tho' sometimes this will burn like Flame,
> When in the Hands of vicious Dame:
> It is the Mover of Desire,
> First kindles, then puts out Love's Fire.[63]

Although the image draws on the stereotypically hazardous femme fatale, this portrayal of the female organs of generation as active and 'moving' engine is unusual and arguably even more uncommon outside of humorous or satirical literature. Certainly, the only specific reference to the term female machine or woman-machine I have been able to find in English medical literature outside of midwifery manuals during this period is in George Wallis's *The Art of Preventing Diseases, and Restoring Health*: on menstruation he comments that much time has been wasted in endeavouring to account for how nature performs 'this operation in the female machine.'[64] While the woman is included in the general category of the human machine in medical literature, the default tends to be the man-machine unless the topic under consideration is the uterine system. In the specifically gendered discussions of corporeal conception the female body is not so much machine, as it is passively mechanical as nature. Thus, the physician Peter Chamberlen (1601–1683), relying heavily on Galen in his *Midwife's Practice* (1665), observed that both male and female have testicles or 'stones' to produce the seed that is equally necessary for generation, but he described the male member as the active (and purposefully implemented tool) – the 'instrument, Agent or Tiller of mans generation' – while the woman's womb is the passive 'field it self of generation.' The seed is partly poured, and partly drawn in by the active womb, but it is the male seed that has the 'forming faculty' while the womb or matrix has only the '*vis plastica*,'[65] the term also applied by medieval and Renaissance scholars to the shaping force within the earth that formed stones into the shapes of fossilised bones.

Chamberlen also, it might be added, had reason to proclaim female passivity or inferiority. He was of the infamous Chamberlen family who had invented and kept secret over several generations the short forceps they used to assist delivery (for which they charged handsomely), and who purportedly had instigated the 1616 and 1633 attempts to incorporate midwives – aimed at granting monopolies to the Chamberlens rather than the midwives. Chamberlen had earlier attempted to gain control over the London midwives; they in turn petitioned the College of Physicians in 1634, complaining that he expected them to come to his house for lectures, and 'for ought as they can discern by his carriage would monopolize the whole practice among Child-bearing Women'; the college upheld the midwives' claim, not because they were averse to licensing but rather because he 'is not otherwise able to instruct them than any other the meanest Fellow of our College, unless he understood it by the use of Iron instruments.'[66] Dr Chamberlen represents the contested sites of conception and knowledge that would continue through the eighteenth century: his book not only reinscribed the old version of the female body as passive recipient of the male's active and vital seed; it also proclaimed the physician's authority over female midwives. A 'great many Women (not wanting ignorance, nor impudence), presume to take upon them this Mysterious Office' thus resulting in mortal danger to both mother and infant, he wrote in his two-page dedication to 'the English Ladies & Gentlewomen, Especially to the more Studious in the ensuing Subject.' In order that 'both Root and Branch, tree & fruit may flourish,' he continued, he would undertake to 'unlock the most abstruce

Cabinet of nature, to untie her gordion knots in her most principal Fabrick ... For a *Midwife* that is ignorant hereof, is no more fitting for that Faculty, then a blind man to judge of Colours.'[67] Chamberlen's asserted control over and knowledge of the secrets of the female-gendered nature functions to reinforce his claims of superiority over women's bodies as inert and fruitful plants.

The writings of both quacks and legitimate doctors over the course of the following century demonstrate the auto-womb as a different category entirely from the active and self-directed (by default male) fetus-machine. The specific advantage in describing a woman's reproductive functions as passive mechanical processes and the fetus as self-sustaining machine would be to gain licence to mechanically manipulate the body carrying the little engine. Paradoxically, although generation and parturition would be increasingly described as mechanical rather than spiritual or 'occult' processes through the eighteenth century, the woman herself is not inevitably designated a woman-machine, which would imply a complete and self-directed engine.

As theories of generation changed, the descriptions of women's passivity and affinity with mechanical nature were revised, reinscribed, and reincorporated to assert male authority. Angus McLaren has noted that both animalculist and ovist theories in the eighteenth century tended to present women as passive recipients of active male seed. If the egg contained the fetus, commonly the sperm was what shook it into life. McLaren quotes William Smellie: 'The *Ovum* being impregnated, is squeezed from its *Nidus* or husk into the tube by the contraction of the *Fimbria,* and thus disengaged from its attachments to the *Ovarium,* is endowed with a circulating force by the *Animalculum,* which has a *vis vitae* in itself.'[68] Smellie's version of conception avoids old-fashioned metaphors of fields and plants, but the image of the natural female mechanism as passive ground nourishing the fetus as if it were a plant persisted. As John Arbuthnot in 1731, and Ephraim Chambers in his *Cyclopedia* of 1741–3 described it, 'A Foetus in the womb is indeed nourish'd like a Plant.' John Cook in 1762, after Blondel, described the female uterus as 'soft mud' or even thick stones and walls in which the roots of the fetus, like those of herbs, shrubs, and trees, strongly adhere. John Quincy's *Medical Dictionary* retains the definition in the improved and corrected eleventh edition of 1794 under the definition for embryo. The female body as fertile ground for plant-like growth provided a figure of natural beauty and femininity in the reproductive process (see figure 31). At its most mechanical, the eighteenth-century womb is a geometri-

cal calculus of force and angle, but such characterisations often resisted
the image of the woman herself as a machine. This appears most strikingly in images: earlier depictions of the pregnant woman as a desirable
subject give way to representations of the womb as a disembodied (and
disturbingly meatlike) measurable structure (see figures 32 and 33). The
womb at the end of the century is a separate structure upon which
mechanical calculation and instrumentation may be imposed. At the
same time, new mezzotint techniques and anatomical practice render it
simultaneously 'organic' and a mechanism in isolation from the woman
herself. Indeed, the woman as subject simply disappears as womb, fetus,
and mechanical delivery assume greater importance (compare figure 31
with 34 and 35).

Some earlier images of the womb might suggest a mechanical process
of distillation, similar to that Willis described occurring in the brain: the
late seventeenth- and early eighteenth-century womb as a field sown with
seed is simultaneously a vessel (Chamberlen notes that the uterus is so
named in Latin since it is 'hollow like a Bottle, and is filled or distended
like a leather bag or bottle with the infant contained in it'[69]). The womb
could be also characterized as a distillation apparatus for material chemical processes in generation, menstruation, or parturition. In the process
of generation humours or spirits are purified and infused into the
material that will become the embryo, but this distillation, however
mechanical, is nevertheless as natural and undirected as water filtered
through sand and clay (in contrast Willis's analogy for distillation in the
brain is the alembic, a process of actively directed separation by the
analytical philosopher). The iatromechanist James Keill suggested a
mechanical distillation process with an organic-agricultural analogy: 'when
the Animalcule gets into the Womb,' he writes in the second edition of
Anatomy of the Humane Body Abridg'd (1703), 'the Humours which distill
through the Vessels of the Womb, penetrating the Coats of the Egg,
swell and dilate it, as the Sap of Earth does Seed thrown into the
Ground.'[70] James Handley in his *Mechanical Essays on the Animal Oeconomy*
(1721) alternatively suggested that both semen and ova participate in
the formation of an entire embryo which, once in the uterus is 'tumified'
and 'dilated' by the humour 'distilling thro' the Vessels of the Womb.'[71]
The conflation of the womb as distillation apparatus and substrate appeared much later in the racy sexology of Nicholas Venette, translated
into English and published in 1750 as *Conjugal Love; Or, the Pleasures of the
Marriage Bed*. Venette described the womb as an apparatus whereby the
male's 'virile seed' is diluted and rarified so that the seminal spirit might

be separated by the gross material of the seed and through a process of 'superfoetation' result in a new conception. The egg itself, however, is like an acorn contained in a 'membranous cup' and, once fecundated, 'being turgid and rarefied by the seminal spirit, can no longer contain itself in its cup, but falls from it *of its own accord* into the fallopian tube and from thence, by a successive constriction of all the fibres, it is forced into the bottom of the womb.'[72]

Even within the logic of a mechanistic tradition, the womb as mechanical purifying apparatus is imagined as a natural vegetative ground for seed. The woman/womb was both natural and mechanical but not a machine. In these cases the active and motive powers that characterize the fetus as human machine are often in sharp distinction from the woman carrying it. Accordingly, what I examine here are two subsets of mechanical reproduction that have a bearing upon later cyborg images of the overly rational and non-reproductive female and of the mechanical womb existing separately from the thinking and feeling woman: first, a much older anatomical paradigm of the 'natural' function of the female, which comprised not only the obvious function of the female organs of generation but also metaphors of male versus female creativity; and second, the new expositions of the mechanical womb by man-midwives eager to assert their authority as qualified and educated experts.

Metaphors of incorporeal versus corporeal conception carried the influence of Aristotelian semen-centred theories on generation. Aristotle postulated that the semen contains the soul, while the woman's menstrual blood provided the embryo with a body and material mass. The imagery is of plant growth until the point where the fetus becomes motive: 'When the embryo is formed, it acts like the seeds of plants. For seeds also contain the first principle of growth in themselves, and when this (which previously exists in them only potentially) has been differentiated, the shoot and the root are sent off from it, and it is by the root that the plant gets nourishment; for it needs growth ... Since the embryo is already potentially an animal but an imperfect one, it must obtain its nourishment from elsewhere; accordingly it makes use of the uterus and the mother, as a plant does of the earth, to get nourishment, until it is perfected to the point of being now an animal potentially locomotive.' The parts of the fetus exist only in potential until 'the active and the passive come in contact with each other ... in the right manner, in the right place, and at the right time'; what is necessary for the female material to form a living being is the male 'principle of motion,' where the imagery is of the artist, like God himself creating Adam from clay: 'The female, then, provides matter, the male the principle of motion. And as the products of art are made by means of the tools of the artist, or to put it more truly by means of their movement, and this is the activity of the art, and the art is the form of what is made in something else, so it is with the power of the nutritive soul. As later on in the case of mature animals and plants this soul causes growth from the nutriment, using heat and cold as its tools.' Conceding that, if the woman has a nutritive or generative soul in addition to the material of the embryo's formation it might appear that she does not need the male at all, Aristotle explains that the male provides the sensitive soul, without which 'it is impossible for face, hand, flesh, or any other part to exist; it will be no better than a corpse or part of a corpse.'[73]

Seventeenth-century anatomists still accepted that semen was a distillate from the male brain, and contained the male soul, which was then transferred to the material substrate in the woman's body. The ovist William Harvey's *Exercitationes de generatione animalium,* or *Disputations Concerning the Generation of Animals* (1651), for example, disputed the Aristotelian theory that the menstrual blood supplies the matter for the fetus, but he emphasized the male contribution to conception as active, and as analogous to creative thought. The male is the artist, the engineer; the female provides the ground. The hierarchy of rational creativity over mere mechanical corporeality is explicit: Harvey's premise is that the male seed has an incorporeal power or virtue. The woman is made 'fecund by no perceptible corporeal agent' but 'sperm confers fecundity by some kind of contagion' of this power in the same way that 'iron touched by a magnet is immediately endowed with the virtue of the lodestone and draws other iron bodies to itself.' The incorporeal virtue, the spirit of a new being, is transferred to the corporeal egg from which all life begins in a process corresponding to the transcription of sensation and creation of thought or memories in the brain. The woman is impregnated by an idea: 'Does a female conceive in her womb in the way in which we see with our eyes or think with our brain?' he asked. The 'virtue proceeding from the male in coitus' is a 'power communicated to the uterus' by the male, which changes the inner surface of the uterus for future conception, making it correspond 'in smoothness and softness to the inner parts of the ventricles of the brain.' Since the substance of the uterus is so very much like the brain, Harvey argues, 'why may we not justly surmise that the function of each of them is also alike, and that what imagination and appetite are to the brain, that same thing, or at least something analogous to it, is awakened in the uterus by coitus and from this proceeds the generation or procreation of the egg?' The immaterial spirit that is the rational soul in the brain is analogous to the spirit of life in the uterus: 'For the functions of both are called conceptions and both are immaterial though they be the principles of all the actions of the body.' The male is like a 'skilful artificer' who conceives a child by the same process as he would conceive art. The woman's role is automatic mechanism. The artist, Harvey explains, 'accurately portrays things ... which he did see in the past,' but the bird in a cage can sing its summer song without practice, or can build a nest without ever having seen a model – from instinctive and passive imagination but not memory (that is, the rational knowledge of the artist is of a higher order than the mechanical and corporeal knowledge of brutes). So also a spider 'without either a model or a brain, weaves its webs by the help of imagination only.' In terms of this analogy, argues Harvey, surely his readers would not 'deem it a most absurd and monstrous thing that a woman should become the artificer of generation, being impregnated by the conception of a general and immaterial idea' from the male.[74]

In 1672 Thomas Willis also emphasized the metaphor of the brain as womb in terms of a distinctively male creative wisdom: 'The Poets feigned *Pallas* to be formed within the Brain of *Jupiter*, and from thence to be born. In truth, within the Womb of the Brain all the Conceptions, Ideas, Forces, and Powers whatsoever both of the Rational and Sensitive Soul are framed; and having there gotten a species and form, are produced into act.'[75] Willis would later write about how 'when the Desire is vehement, almost the whole Soul is drawn into Parties, and by a certain going out from the Body, wanders towards the desired thing, or at least emits a Portion of it self ... in that mad affection of Lust, in which the genital Humor, containing Fragments picked from the whole Soul is poured forth.' Reiterating Harvey's notion of the male begetting another individual as an idea formed within the testicles and then transferred to 'a convenient womb,' Willis's synechdoche relegates the womb to functional storehouse for the male idea/conception, much as the brain was characterized as the storehouse for memory in the human engine. For the male there is an inverse relationship between ideas in the head and those in the genitals: if there is 'too great Expence' of the animal spirits in venery, the result may be depletion of that which would ordinarily be reserved for the brain and nerves; or the spermatic vessels might even 'snatch' the spirits from the brain. The corporeal soul in both man and woman is feminine, and because she is so easily 'incited to Lust' and 'inclines herself towards the Genital Members,' she must be shown the

'Commands of Reason and Religion' by the 'Superior Mind' (the male rational soul) so that her powers are not expended on 'Disorderly Excursions,' but rather are 'reduced into Order, the Acts of Sobriety, Prudence, and of other Science, and Discipline.'[76] This was the new science: bodies might be explicable machines, but conception, hovering between materiality and immateriality, touching on human souls as well as gender roles, was regularly reinterpreted to assert cultural hierarchies.

From Willis's scholarly medical treatise to the books intended for a popular audience, the assumptions about male creative conception and lower order mechanical production in the womb prevailed well into the eighteenth century. For example, the popular sex and pregnancy manual from the 1680s throughout the early decades of the eighteenth century, *Aristotle's Masterpiece,* also imagined the fetus as an artistic creative endeavour of the male. Here the male seed has a 'forming Faculty & Vertue in the Seed ... it being abundantly endued with Vital and Etherial Spirit, which gives shape and form to the *Embryo.*' The embryo is then transformed in the womb into a 'decent & comely Figure of a Man' from this initial 'rough draft':

> The second time of forming is constituted when Nature and the force of the Womb, by the use of her own inbred forces and Virtue, makes a manifest mutation in the Conception, so that all the substance seems Congealed Flesh and Blood, which happens about the 12 and 14 day after Copulation; and though this Concreation or Fleshy Mass abound with hot fiery Blood, yet it remains undistinguishable without form of figure, and may be called the rough draught, or *Embryo,* and well likened to Seed which is sown in the Ground, which through kindly Heat and Moisture grows up by degrees, into a perfect form, either in Plant or Grain; or as when a Potter Fashions a Vessel out of a rude Lump of Clay.[77]

Although the tract specifically provides instructions for getting a male or female child, the default embryo here as elsewhere is male; as elsewhere, the woman's role is at best the artist's mechanick tool for shaping the material in order for the conceptual idea to be given form.

The Dutch physician Regnier de Graaf's 'discovery' of ova in viviparous animals in the late seventeenth century marks a juncture where the older representations of male creativity acting upon lifeless material in the womb begin to lose medical validity. His writings also act to reinforce a new hierarchy wherein male scientific intelligence is asserted over the bodily mechanisms of female reproduction. Four years before he published his work on the female organs of generation, De Graaf published his *Tractatus de Virorum Organis Generationi Inservientibus,* or *A Treatise Concerning the Generative Organs of Men* (1668). In the preface to this work, he describes the. creation of the universe as a collaboration between a female nature bringing forth men and Almighty God as architect making them immortal by providing them with reproductive organs. This would, at first, seem to invoke a version of the old Aristotelian theory wherein the male provides the 'form' while the female provides the 'matter,' thus explicitly reinforcing a hierarchy of creation that contrasts conscious, creative intelligence with mere mechanical labour: 'Since, because of the material at hand, Nature could not bring forth a man who was immortal, the Maker of the Universe, so as to succour the frailty of the human race, contrived what he could in the way of immortality. He thought not to sustain a man for a few thousands of years but rather to prolong his existence for eternity in the form of ever new offspring. With this as His end He constructed divers organs by which the work necessary might be performed and He did such a workmanlike job that with full justice we can proclaim that here, if anywhere, almighty God left a great testimony not only of His providence but also of His marvellous wisdom.'

De Graaf, however, defends the female, who is not, he asserts, merely an inferior male: 'Quite ridiculous therefore is the opinion ... that Nature Fully equipped the male sex and added the female only as a sort of appendage like a peacock's tail. No less insultingly does Aristotle call the female an incomplete male or, as the barbarian philosophers say, an animal which just happened.' The analogy slews almost imperceptibly at this point. Suddenly, it is the female nature who is actively fashioning immortality and moreover, she is thinking intelligently about her work: 'We ourselves think that Nature had her mind on the job,' he says, 'when generating the female as well as the male, for without females there would be no generation of any animal.'[78] De Graaf's description of the creative union of male God and female nature underscores familiar assumptions about male and female creativity, but the indeterminacy between who makes and who brings forth, between what is designed and what is not – who's in charge, so to speak – suggests a more complex and contradictory imagining of the human generating machine.

Four years later in his *De Mulierum Organis Generationi Inservientibus Tractatus Novus,* or *New Treatise Concerning the Generative Organs of Women* (1672), de Graaf would explicitly state the woman's active, participatory, and creative role in the process of generation. William Harvey had already inscribed the frontispiece to *De generatione animalium* with the famous words '*Ex ovo omnia*' and the Danish Niels Stensen and Dutch Jan van Horne had already identified eggs in the female, but de Graaf claimed finally to have 'made bold to strip off Nature's robe and to reveal the first threads of our nativity, the whole workshop of human manufacture and its tools.' The concealed mystery that his new discovery had revealed, in short, was that 'all animals without exception, man included, take their origin from an egg, not from one formed in the uterus by semen (as Aristotle says ...) or the virtue of semen (as Harvey ... seems to hint) but from one which exists in women's "testicles" before coitus.'[79] De Graaf's language simultaneously reinforces the new mechanistic approach to human physiology, acknowledges female action and creativity in human reproduction, and – by 'stripping off Nature's robe' – reclaims the seemingly perpetual hierarchies of male intelligence over female-as-sexualized-body and female-as-passive-nature.

Jane Sharp's *The Midwives Book Or the Whole Art of Midwifry Discovered* (1671) represents a more equitable version of conception, insofar as Sharp stresses that women and men both produce seed; but as elsewhere man 'is the agent and tiller and sower of the Ground' while the woman is 'the Patient or ground to be tilled,' and her womb a 'Field.' Sharp explains that male seed is more 'active' than the female seed, but she does not idealize the male's creative role: the soul comes not from the semen, but 'is infused from above.' Moreover, the womb actively moves itself to embrace, stir up, and bring the seed to action. 'Conception,' she writes, 'is the proper action of the womb after fruitful seed cast in by both sexes ... it is not copulation, but the mixture of both seeds is called conception, when the heat of the womb fastens them ... The seeds of both must be first perfectly mixed, and when that is done, the Matrix contracts it self and so closely embraceth it, being greedy to perfect this work, that by succession time she stirs up the formative faculty which lieth hid in the seed and brings it into act, which was before but in possibility, this is the natural property of the womb to make prolifick Seed fruitful, it is not all the art of man that setting the womb aside can form a living child.'[80] The gradual acceptance that the female seed was not merely material substrate but played a role in active creation of the embryo did not, however, translate into a new gender equity in reproductive matters.

Microscopic studies from the 1670s on would also challenge the notion of material human conception beginning as an immaterial idea, but the metaphors of the hierarchy of male genius and the lesser female corporeal imagination continued in various guises. Venette, for example, would cavalierly reduce the woman's role to receiver of impressions or communications from the male mind: 'man ... acts with more firmness, feeds more heartily, defends himself with more courage and presence of mind, reasons with more strength, and contributes towards the getting of children with more alacrity. He acts particularly in generation, where he communicates himself, and by other actions of the body and mind, gives proofs of this strength and heat; whereas the woman only suffers the impressions a man makes upon her, and often is not ready so soon as he to furnish withal to form a man. In short, she is only to conceive, to give suck, and to breed up children.'[81]

La Mettrie similarly equated the womb to a 'lower' brain: 'what fruit would the most excellent school produce without a womb perfectly open to the admission of the conception of ideas? It is as impossible to give a single idea to a man deprived of all the senses,' he wrote, 'as to give a child to a woman in whom nature was absent-minded enough to forget to make a vulva, as I have seen in one who had neither opening nor vagina nor womb.' La Mettrie's point obviously was not to explain the reproductive mechanism, but nevertheless reinforced the stereotype of conception in the womb as a process akin to conception in the brain. These metaphors gave rise to a series of metaphors of child as a sort of mechanically produced text, via 'communications,' 'imprints,' and 'impressions.' La Mettrie – although he was no authority on reproduction – agreed with Nicolas Malebranche that the fetus could be imprinted like the *tabula rasa* of the mind, suggesting that there are evident communications of vessels between mother and fetus, and that there may be also communications by the nerves: 'Since there is obviously communication between the mother and child and it is hard to deny the facts produced by Tulpius and by other equally trustworthy writers ... we believe that it is by the same means that the foetus feels the effect of its mother's impetuous imagination, as soft wax receives all sort of impressions, and that the same traces or desires of its mother can be imprinted on the foetus.' He goes on to confirm the old relationship of male and female creativity, citing the 'famous A. Pope': 'the efforts and nerves of his genius are etched on his physiognomy,' he argues. 'It is totally convulsed, his eyes are starting out of their sockets and his eyebrows are lifted by the muscles of his forehead. Why?' he asks. 'Because the source of his nerves is in labour and all his body must feel the effects of such a difficult birth.'[82] As Raymond Stephanson has argued, the brain-womb and imagination-birthing tropes functioned to establish warring versions of intellectual potency for creative male authors.[83] In an inverse relationship that is ripe for metaphorical critique, as Aristotelian portrayals of the masculine semen as creative intelligence in the physical processes of conception, the new print culture engendered a new space for male authors to assert their creative supremacy.

The hierarchy of male and female intelligences in the reproductive process would be renegotiated by the new professional medic, who asserted in print his superiority over the supposedly shoddy and ill-educated female midwives. If the analogy of male intelligence as architect or artist in the act of conception would lose favour as eighteenth-century medical tracts developed sophisticated physiological explanations for conception and fetal development, the assertion of licensed intelligence and art in the symbiotic relationship of print culture and professional medicine was gaining ground. The frequent trope used by man-midwives comparing their authoritative treatises to offspring born of parturition and difficult labours would seem to suggest a shift from vital seed to medical institution in the understanding of males as authors of pregnancy. Medical texts assumed the role of the children of creative genius as man-midwives jostled for authority in their newly established profession. Robert Erickson has called attention to the metaphors of the book-as-infant in the English surgeon James Wolveridge's *The Speculum Matricis* (1671).[84] Daniel Turner also used the trope of book-as-baby in his pamphlet war with James Blondel concerning the powers of female imagination on the unborn child, writing of the *'New Imagination'* promoted too precipitously by Blondel's publisher: 'it was near twelve Months that the *Mountain* was thus *pregnant*, whether a *Mouse* or a *Monster* is now brought forth, let the Reader judge.'[85]

In 1724 the London man-midwife John Maubray depicted his weighty tome *The Female Physician* (more than four hundred pages long) as both an independent and self-directed vessel and as the offspring of his dedicatees, 'all Learned and Judicious Professors of Physick, as well as Ingenious and Experienced Practicers of Midwifery.' Maubray boasted that his was 'a *substantial* Work [that] will stand securely upon its own Bottom, and make its way into the World without any *secondary Helps*, whereas a slight and Defective Piece will fall and be quash'd.' Self-importantly he offers his issue to his dedicatees' '*August societies* in general, and in this manner to render some small account of the TALENT, which some of your selves have entrusted me with for *Improvement*.' Maubray concludes his dedicaton with: 'The *undertaking* then being yours by *Birth*, it has a Natural Claim to your *Care*; And it being the *Duty* of Men to provide for their *Off-spring*, it has a peculiar Right to your *Patronage*.' Maubray's book as healthy child – conceived from the commixture of these various male 'talents' and delivered without any 'secondary Helps' – is qualified to distinguish the superior qualities of its fathers. In the chapter on 'the Qualifications of the Ordinary MIDWIFE' Maubray admits that men and women both may practice the art of midwifery, and he details the necessary qualities of body and mind for the female practitioner. However, in the subsequent chapter on 'the Qualifications of the Extraordinary MIDWIFE' (the 'Andro-Boethogynist' physician or surgeon who is categorically male) he cautions:

> [He] ought not only to be endued with all the *Qualities* and *Qualifications* mentioned in the *Two* preceding *Chapters,* but also to excel the WOMAN-MIDWIFE in many special *Particulars,* and ingenious *Points*; which noways belong to her *Female Province.*
>
> FOR it is not enough that *He* knows how to relieve and *lay* the *labouring Woman,* however difficult or preternatural her *Case* may be; nor is it sufficient that he understands how to *help* and succour both the MOTHER and the INFANT after such a *Deliver:* No, so much of his Business might be easily learn'd and enhanced by *old Women,* were they but *Docile,* and not such obstinate *Creatures.*
>
> BUT *He* ought farther also to know (*first*) how to prevent preternatural *Disasters* incident to both the *one* and the *other,* in their respective *States* of CHILD-BED and INFANCY: And, *Secondly,* how to administer Relief and perform the CURE, in *Case* of any dismal *Accident* whatsoever to *one* or *either* of them in their dangerous Condition.

Here the conversion of pregnancy to 'dangerous condition,' notably, requires the direction of a male steersman: these are the cases, he explains with rhetorical flourish, 'which try his *Skill* and *Knowledge* most, as the *Tempest* or *Storm* best discovers the *Judgment* and *Capacity* of the MASTER-MARINER.' The mother herself is hardly a subject here other than to be provided with direction, instruction, and treatment, while the author's book, like the fetus who receives his rational soul in the third month and 'becomes MAN,'[86] is a self-steering or fully autonomous child: Maubray is clearly and powerfully positioning male authority and direction over female bodies clearly lacking that all-important steersmanship.

The infamous case of Mary Toft of Guildford also calls attention to these newly contested territories of conception. The confirmation in 1726, after much ado and many examinations by John Howard, midwife and surgeon of Guildford, and Nathanael St André, Surgeon and Anatomist to the Royal Household, that Toft was giving birth to rabbits placed professional authority in a particularly awkward situation. Toft claimed to have been startled by rabbits while working in a field early in her pregnancy, and over the course of some weeks delivered seventeen of the creatures in pieces that later turned out to be parts of a cat, an eel, and young or skinned rabbits. According to Dennis Todd the Toft family, victims of a decline in wool manufacturing due to foreign competition and increasing tariffs, had apparently decided on the hoax as a means of attracting paying viewers. That the illiterate Mary Toft could so gull the professional man-midwife not to mention the Royal Surgeon proved to be a volatile issue once the investigation into the fraud became public. As Todd comments, 'It was no longer simply a question of the truth of Mary Toft's monstrous births. At stake now were reputations.' The affair was investigated by Ciriacus Ahlers, Surgeon to His Majesty's German Household; by Sir Richard Manningham, Fellow of the Royal Society and Licentate of the College of physicians; by the highly respected anatomist and midwife Dr James Douglas, by Thomas, second Lord Onslow, Lord Lieutenant of Surrey and Justice of the Peace; and by Sir Thomas Clarges, Justice of the Peace. Toft was eventually charged with being a 'Notorious and Vile Cheat' and imprisoned at Bridewell. The case against her was eventually dismissed; St André would fall into disfavour at court, and Howard himself was charged with being involved in the 'Cheat and Conspiracy of Mary Toft,'[87] although this charge too was ultimately dropped.

But why should a woman be legally charged and imprisoned for pretending to give birth to rabbits? One could interpret this entire course of events as male reactions to the wrong kind of creative conception for a female whose responsibilities were generally supposed to be domestic and reproductive – and deferential to an educated male hierarchy. The imaginative conceptions of Mary Toft challenged the higher order male genius in all too obvious ways. Such gullibility came under intense scrutiny, of course, but the dialogue was unmistakably a contest of male authority over the processes of reproduction and childbirth. Even the fictitious Lemuel Gulliver felt the necessity to publish an authoritative pamphlet, *The Anatomist Dissected: or the Man-midwife Finely Brought to Bed*, wherein he knowledgeably critiqued the philosophical and anatomical misunderstandings and the conduct of St Andre: 'Suppose,' he submitted, 'that one were to see a letter from *Battersea*, importing that a woman there had been deliver'd of five Cucumbers, or indeed a hundred Letters, would that lead a Man of Sense to believe any Thing, but, either that the People who wrote those Letters had been grossly impos'd upon themselves, or intended to impose upon him.'[88] As we have seen in these contests of learned authority it was no strange oddity to imagine letters as the offspring of male intelligence and the sign of superior scientific knowledge, but one could only stretch the analogy so far: to imagine a woman being delivered of a hundred letters was indeed as preposterous as any other monstrous birth. (Admittedly, I stretch my own comparison somewhat beyond plausible proof here; nevertheless, these masculine contests of authority over letters and births rendered the participation of women virtually irrelevant and occasionally inappropriate.)

These examples would seem to provide ample evidence for Davis-Floyd and Dumit's claim that the 'demise of the midwife and the rise of the male-attended, mechanically manipulated birth,' accompanied a wide acceptance of the metaphor of the female body as defective, but we must here re-examine the assumption that the mother was imagined a 'machine – a metaphor that eventually formed the philosophical foundation of modern obstetrics' as Davis-Floyd and Dumit have postulated. For we can see convincing evidence for a prevalent trope of the female body in need of professional male guidance and authority, and even, as we shall see, of the fetus as a machine, but the mother as machine is somewhat harder to discover.

Whether he arose from animalcule or ovum, the fetus as a little male passenger in the female body assumes a new importance with the mechanical understanding of human reproduction. If the father's role as architect in 'creative' conception was diminished by the new science, the male fetus assumed right of way over the female body: 'If *Leewenhoeck's* ... notion be true ... by what Right,' exclaimed James Blondel, 'has the Mother's Fancy any Influence upon the Body of the *Foetus*, which comes from the *Semen virile*, and which is, in respect to her, but a *Passenger*, who has taken there his Lodging for a short time? If the Father could not cause, by the Strength of Imagination, any Change in the Animalcule which was originally in his Body; I desire to know, why the Mother should plead that Priviledge in Exclusion to the Father?'[89] That Blondel should subtitle his tract against 'vulgar error' with his own credentials as 'Member of the College of Physicians, London' is an indication of the stakes in these arguments: authority inscribed by institutional license was the territory of mechanical physiology and the most interesting machine in the debate was not the woman-machine but the baby-machine.

Indeed, throughout the late seventeenth and eighteenth centuries, it was consistently the fetus and not the mother who was the machine. Walter Charleton's *Natural History of Nutrition, Life, and Voluntary Motion* (1659) renders the pregnant woman visible primarily as womb. 'Lest the *Chorion* should at any time be corrugated or shrieveled up together, and so streighten or compresse either the Liquor, or the Infant,' Charleton explains, 'Nature hath affixed the same to the *Placentae Uterina*, to the end, that adhering to the bottome or upper part of the womb, it might hang fast, as an Apple hangs by its stem, or as our Globes of Glass are hung up by strings to the Seeling of a room. So that the *Chorion* thus adhaering to the *Placenta Uterina*, which is fastned to the bottome of the womb, and the *Amnios* in like manner adhering to the *Chorion*, in the same upper part; and the lower part of each membrane being depressed by the weight of the Infant, and of the Humors contained in them: it thence comes to pass, that this Natural Machine both of the Child, and Membranes ... is afterward made of an oval figure.' The instability of Charleton's analogy of apple and stem mingled with the comparison of glass globes emphasizes the dual nature of the reproductive female as repository, both soil and vessel. The 'Natural Machine' is not the woman as vessel, however, but the child together with the membranes of the womb. The male fetus is active, independent, and self-governing: '(for he sits with his leggs crossed, his heels drawn up to his buttocks, his elbowes resting on his knees, one hand held up close to his ear, the other to his cheek, for the more firme and easie sustentation of his head) yet, in that situation he hath need of a Mansion of an Oval Figure, that swimming in liquor, he might keep his head above water, and at his pleasure take in his nourishment by his mouth, and also inspire the temperate aer surrounding his head, in the void space of the Secundines.'[90]

Similarly, the *Philosophical Transactions* in 1672 published an explanation by the ovist Dr Kerkringius, who claimed to have witnessed in a month-old fetus a 'whole humane shape' (see figure 34). 'Behold the Figure well,' he notes, 'which represents this little Engin in its natural size. It already in a manner sustains itself.'[91] The role of the woman here is not overtly mechanical: her body carries the ova, and there is some question regarding whether the vesicles or eggs from which the embryo is formed are fastened in one place and then loosened to form the embryo elsewhere. As is typical, the fetus and not the mother is described as an active engine. Decades later, the animalculist physician Daniel De Superville ('Privy Counsellor and Chief Physician to His Most serene Highness the Margrave of Brandenburg-Bareith, President of the

College of Physicians, Director of the Mines and of All Medicinal Affairs in the Margravite, Member of the Imperial Academy Naturæ Curiosorum, and of the Royal Society of Berlin,' as he was designated in *Philosophical Transactions*) argued: 'According to the most general Opinion, there must be at every Conception a new Creation of a Soul: Or, according to others, there is always a Legion of created Souls fluttering about in the Air, and watching the Minute for entering into a fruitful Egg as soon as it is impregnated. What an Extravagance is this! Would it be as absurd to believe, that every *Animalculum* has already its Soul, which waits only for the little Machine's unfolding itself in order to exert its Function?'[92]

Similarly, George Cheyne comments in 1742 that 'it is now pretty evident' that 'the *Principle* of Generation, or the *infinitesimal Corpuscle* of the *Animal,* is in the *Male*; and that the *Female* is only, as it were, the primary *Nurse* to it, furnish'd with a proper *Nidus* and the *specific* Nutriture for it, till it has acquired Strength enough to bear its own proper *Element;* which State it no sooner attains to, than it breaks its Prison, and escapes.'[93] The female is not described as a machine in any of these cases, whether the author ascribes to ova or semen as the source of life: in fact, she is barely present, represented principally as the womb holding the active and self-motive machine.

John Stedman's *Physiological Essays and Observations* printed in Edinburgh in 1769 unequivocally represents the ideal and active machine as the fetus not the mother. Women, he argues, suffer a variety of wasting diseases but frequently deliver children whose 'full and healthy appearance bears no proportion to the emaciated state of the mothers.' Stedman explains that this phenomenon is due to the fetus's superior circulation even when the mother's is weak: 'This more natural circulation in the body of the foetus, than in that of the mother, may be accounted for, in some measure, from the moving power in the child's body being greater than that of the mother's, in proportion to the size of the two bodies, according to the universal law, which obtains in mechanics: for, of similar machines, the smaller do more, in proportion to their magnitudes, than the larger.' The woman's body as mechanism here is singularly passive: her blood may be tainted but this is not passed on to the fetus since 'perhaps the blood, which is conveyed to it, may undergo such a percolation, in passing from the uterine vessels to those of the *placenta,* as to render it purer than that circulating in the mother's body.'[94]

Well over a century after Charleton's treatise, in the promotional pamphlets of the quack Dr James Graham in the 1780s, the woman is represented again as passive fertile ground and the embryo as 'animal machine.' In Dr Graham's version, the embryo, a child in miniature, is like a plant in the ovarium until excited into motion: 'the complete future child, (like the plant with its leaves, flowers, fruit, and seeds, wrapped up to us invisible in the seed), actually subsists in a dormant or inanimate state in the ovarium, or in the egg-bed of the mother, each ovum containing a child sketched out in miniature, in all its parts, long before the commerce between the sexes; – and that what is styled the act of generation is only the means intended by nature to animate it or set the animal machine a going – or in motion.' Once the germ has been fecundated, it 'escapes from the ovarium' and becomes an autonomous new life:

> the motions of its inherent fluids and faculties commence; for though the embryo or fœtus is supplied with fluids from the mother, they are circulated by its own powers; and on this circulation of the fluids, LIFE depends; for the germ when once endowed with life, is fully possessed of the means of continuing it; and that chiefly by the influence of its own admirable mechanism.
>
> This is what is called conception; and with this view the vagina, and the womb, are most plentifully furnished with nerves; and during the connection of sexes, those parts seem to be endowed with a tenfold portion of sensibility. The state of nerves which occasions this increased sensibility is no doubt communicated to the fallopian tubes, by which their ragged ends are erected, and applied to the germ in such a manner as to facilitate its escape. The germ finds a ready access into the open extremity of the tube, and through the tube itself, assisted by the increased secretion of mucus, a convenient passage to the womb; to which it soon adheres, and is nourished during nine months by the mother's blood.[95]

Drawing on traditional Aristotelian conventions to sell his modern treatments, Graham's pamphlets characterize the motion characteristic of the tiny new human machine in terms of male energy added to passive material. Here again, the woman in her reproductive capacity is not a machine and in fact is present only as vagina and womb in Graham's description of the process of generation. Graham's use of the passive voice to describe the woman's role further diminishes her responsibility: the state of her nerves 'is communicated'; the fallopian tubes' ragged ends 'are erected and applied.' That the woman's body acts without her conscious knowledge or effort in reproductive processes is not noteworthy in itself, but Graham's language to describe the germ shifts markedly to the active voice. Once endued with vitality from the semen, the germ changes from plantlike seed to a tiny animal machine in motion, circulating its own fluids by its own powers and, still no more than a germ, it 'finds ready access'; 'it adheres.' The exceptional sensibility that the vagina and womb have during sexual relations would seem to be the woman's only function, serving to create the opening for the germ's escape. Furthermore, what gives the germ its amazing power of motion and vitality is the 'exquisitely penetrating seminal liquor' of the male, which 'excites or rouses' the germ 'and communicates to it a new or more sensible activity; or that which we call LIFE!' However, in order to promote his treatment (thus augmenting the male semen with the doctor's application of electrical energy), Graham characterizes the 'prolific liquor of the male' as 'stimulus' and 'nutritious fluid, exquisitely adapted to the extreme delicacy of the parts of the germ.' While both celebrating and undermining the important role of the male semen, Graham reiterates the fetus's self-sufficiency: 'motion being once impressed on the little mobile, or moving body, is there preserved, I believe solely by the energy of its own mechanism.'[96] The new terminology of the fetus's own 'mechanism' and 'energy' replacing the older vital spirit infused by the father's creative idea insists on a hierarchy of female function distinct from fetal autonomy and doctor's authority.

Dr Graham is, of course, selling the virtues of electrical energy as an equivalent to older versions of vital spirit (he provides solutions not only for infertility, but also 'Receipts for the Preservation and Exaltation of Personal Beauty and Loveliness,' as his subtitle claims, emphasizing only too clearly his opinion of the female's place in society). He boasts of curing a beautiful young lady who had undergone a regimen of being 'electrified for an hour or two daily, under ethereal and magnetic influences,' bathed in cold water every other day, and along with her husband given a glass of green sage and licorice root juice every morning. Through the ministrations of the good doctor, 'assisted, no doubt, by an electrical touch or two from her husband, the lady became effectually pregnant.' All this was, essentially, a pseudo-learned advertisement for the doctor's 'celestial or magneto-electrical bed' to cure impotence and sterility which, he informs his audience, he has placed in a spacious room to the right of his orchestra. The bed, supported by six glass pillars, covered in Saxon blue and purple satin, and perfumed with Arabian spices, supposedly produced the necessary 'celestial fire and vivifying influence' for immediate conception – all for the price of a mere £50.[97]

While Dr Graham is beyond the pale of what might be considered as legitimacy, having studied at Edinburgh University without taking a degree, and derided by some of his peers as a mountebank and charlatan, his treatments and lectures attracted a following. Graham's treatments are an extreme example of a growing confidence in the authority of doctors and their prosthetic mechanisms to enhance the mechanical workings of natural bodies.

Particularly in descriptions by the promoters of the new profession of man-midwife, delineating the uterus as a combination of geometric forces and spatial arrangements granted authority to the doctor, with his superior knowledge of mechanics and geometry. Towards the second half of the century we see the occasional occurrence of the woman as animal machine but this is no iteration of the rational human machine: the labouring woman is mostly automatic. *The Midwife's Companion* (1737) by Lancaster physician, surgeon, and man-midwife Henry Bracken (1697–1764), for example, is a an outspokenly mechanistic treatise that stops short of using the term woman-machine but advertises forcefully the superiority of mechanically trained doctors over those 'ignorant Pretenders to this Art, meer Old Women.' Bracken makes one reference to the woman as a generic machine in describing the mechanism of jaundice: 'Thus may all the various Operations of the whole Animal Machine,' he writes 'be made plain ... without having recourse to that subtle, tho' ignorant, Subterfuge of *occult Qualities.* I esteem a Person who practises Physick (without a thorough Knowledge of the Mathematicks) a mere Empirick, or a Practiser upon *hear-say* only.' The use of the term 'animal machine' here functions not so much to categorize woman-as-machine as it does to assert Bracken's advanced knowledge as a mathematically trained physician well versed in the mechanics of bodies. In contrast to his singular use of the term *animal machine,* Bracken repeats the superiority of his training in mechanics throughout the treatise: 'I am well convinced that a great Part of the Practice of Physick (and the Force and Operation of Medicines) may be mathematically and mechanically accounted for, yet I own there are many hidden Mysteries undiscovered.' The marginalia here specifying the subject, 'Mathematical and mechanical Knowledge necessary in a Physician,'[98] underscore the validating education of the male practitioner.

London man-midwife William Smellie (1697–1763) notes as well how his success unfolded when he 'began to consider the whole [process of assisted birth] in a mechanical view, and reduce the extraction of the child to the rules of moving bodies in different directions'[99] (see figure

35). For teaching he used mechanical models of a woman in childbirth, which students would use for practice. Practising on a mechanical model, however, was not a widely accepted technique. A year after Smellie's first edition of *A Treatise on the Theory and Practice of Midwifery* in 1752, York physician and man-midwife John Burton (1710–1771) publicly chastized him in *A Letter to William Smellie, M.D. Containing Critical and Practical Remarks Upon his Treatise on the Theory and Practice of Midwifery*. Particularly annoying to Burton was the general misconception that Smellie had invented the mechanical theory of midwifery: 'as the Art of delivering a Woman of her Child and Secundines is entirely a mechanical Operation, whether it be done by turning the Child in the Womb, to extract it by the Feet, or by Assistance of Instruments,' he opined, 'so the mechanical Laws or Rules are to be our general Guide. In which View you tell us *(c)*, you began to "consider the *Art of Midwifery*, and reduce the Extraction of the Child to the Rules of moving Bodies in different Directions;" which your *Eccho*, the *Review* Writer, Number III. has repeated, and mistaken you to be the first Writer on this Subject that has gone upon such Principles.' However mechanical the man-midwife's theoretical and practical knowledge, the woman herself was not to be considered a machine: Burton rebuked Smellie for his description of moving the child's head with forceps: 'Upon a Machine you may do such a Thing, but for many Reasons it cannot be effected on a human Body in Labour; the Difference betwixt these two being very great.' Burton phrases his objections in terms of the superiority of his own mechanical expertise, at one point disputing Smellie's description of a breech delivery with the dismissive comment that 'whoever understands the Mechanism and Bulk of a Child, the Size and Shape of the Pelvis, and the Laws of Mechanics, will easily see that this Case *can never happen*.'[100] All disputes of method aside, the authority of the man-midwife in these cases resided in his knowledge of mechanics.

Mechanical knowledge secured for the trained physician a right of entry to the birthing body. As Jean Astruc (1684–1766) put it in his *Elements of Midwifry*, 'the art of midwifry wants but little of having attained its utmost perfection, and its operations arrived to an almost geometrical certainty: and this is not surprising; for, after all, the art of midwifery is reduced to the following mechanical problem: *"An extensible cavity, of a certain capacity being given, to pass a flexible body, of a given length and thickness, through an opening dilatable to a certain degree ..."* which might be resolved geometrically.'[101] In 1766, Thomas Cooper in like manner argued that 'THE Art of Midwifery depends on a few simple and obvious

Principles, chiefly deducible from the Laws of Mechanics. It is a Branch
of the Medical Science.' As a branch of medical science, it would increasingly exclude the uneducated practitioner from childbirth. 'The Study
of Midwifery therefore, as in itself an Art, or as a Branch of Surgery only,'
Cooper argued, 'should always be preceded by the Study of Mechanics
and Medicine, for we cannot too diligently search the Means of Perfection, as far as it is attainable by Human Nature, in an Art of such
Importance; since the Errors that may be committed in the Exercise of
it, are of the utmost Consequence, as not only the Health and Welfare,
but the Life itself of Mankind, is interested therein.' Cooper's description of parturition based on mechanical principles is worth including at
length for its detached analysis of childbirth as mere geometry:

> Mechanics is a mixed, mathematical Science, which considers Motion and
> moving Powers, their Nature and Laws, with the Effects thereof. That which
> considers the Motion of Bodies, arising from Gravity, is called Statics, in
> Contradistinction to that Part, which considers the mechanical Powers, and
> their Application, properly called Mechanics.
>
> Mechanical Powers denote the six simple Machines, to which all others,
> how complex soever are reducible; these are, the Balance, Wedge, Lever,
> Wheel, Pulley and Screw, but may, however, be all reduced to one, viz. the
> Lever, the Principles whereon they depend, being nearly the same in all.
>
> Statics is the Branch of this Science, which has the nearest Analogy with
> the Study of Midwifery, and the Lever, the only adventitious Member
> required, to assist the Operations of the animal Machine, in the Act of
> Parturition.
>
> There is a Power inherent in all Bodies, whereby they tend to some
> Common Point call'd the Centre of Gravity; and that, with a greater or less
> Velocity, as they are more or less dense, or as the Medium they pass through,
> is more or less rare ... The Action of Nature in the Birth of a child, may be
> compared to a Weight, descending, or a Body being moved to a given Point
> or Centre of Gravity, assisted by the adventitious Powers of a propelling
> Lever, through an irregular cylindric Passage, the Diameter of which is
> barely sufficient to admit of such Protrusion; and in this View, the Aperture
> of the Pelvis, is the Centre of Gravity, and the Muscles set in Action by the
> Labour-pains, are the propelling Lever.

Therefore, Cooper concludes, 'an Idea of the mechanical Powers' will
enable midwives to manipulate the 'propelling Force from within' or
apply the 'proper Assistants or Levers, from without.'[102] In this most

extreme version of women's labour, we see the woman as 'animal machine.' Others, however, would present the process as mechanical while reminding readers to recognize the real bodies involved (even if only to establish their own authority).

The *Course of Lectures on the Theory and Practice of Midwifery ... Fully Explain'd, and Clearly Demonstrated; Particularly, those Operations Which are Conformable to the* Principles *of Mechanical Motion* (1767) by John Leake (1729–1792) was another defence of the benefits of medical mechanical science over untrained amateurs. Here, in his proposal to teach midwifery to both male and female students, he promised to teach the '*Art* of *Injecting*' and making anatomical preparations to 'illustrate the Science of *Midwifery*' but also emphasized the necessity of '*real Practice*' before one was deemed proficient; but to practice upon real bodies could do harm to both the patient and the pupil's reputation, he explained. Therefore, they needed to learn their art with '*artificial Bodies, so mechanically* and *naturally constructed,* as nearly to correspond with the several Parts concern'd in *Parturition.*' However, Leake cautioned, 'the bare Name of an *Apparatus*, which is much easier obtain'd than the Apparatus itself, is not sufficient to give it *Value.*' Indeed, he continued,

> *Machines badly constructed*, are so far from being useful, that they do much *real Harm*, by misinforming the Judgment of the *Student*, and giving him a false Idea of Nature ...
>
> This induced me to bestow no small *Labour* and *Expence* on that *Apparatus*, on which I propose to *demonstrate* the *Practice of Midwifery*, having executed all the principal Parts of it with my own Hands; and having also, for several Years past, been constantly making *Alterations* and *Improvements* in it ... in order to render it as complete and useful as possible; being thoroughly convinced, that what is commonly call'd *Machinery*, (made and sold by *Mechanics* as it generally is) would prove totally useless; for, they are not only unacquainted with the *anatomical Structure* of those Parts which it ought to resemble, but also with the Uses to which it will afterwards be apply'd.

His own machine, Leake concluded, could be 'look'd upon as the most direct Road to *practical Knowledge*, that can possibly be pointed out for his [i.e., the pupil's] Improvement and future Success.' The terminology of commerce – value, labour, expense, reputation, utility, improvement, practical knowledge, success – compounded with the naturally *and* mechanically constructed 'several parts' involved in parturition implicates Leake in a complex deconstruction of the pregnant mother into the manipulated organs of a lucrative industry.[103] Founder of London's first charitable Lying-in Hospital for those women who are 'reduced to real want,'[104] Leake approached his own profession with an astute business sense. Appointed physician at the hospital, he also taught courses to a series of pupils who practised their new skills on patients in the hospital and in the surrounding district. Leake charged his students four guineas for the first lecture course, two for the second, and ten if the pupil wanted to attend the practice of the hospital and district.

He would later defend the need for medical training both practical and theoretical in his *Introduction to the Theory and Practice of Midwifery ... to which are Added, a Description of the Author's New Forceps* (1787). His impractical three-bladed forceps ultimately would be scorned by his peers, but Leake based his authority on the proposition that midwifery is a science, and no arena for those lacking in the knowledge of the mechanisms of childbirth:

> Whosoever professes to teach a Science, ought to present those intending to study it, with a clear and comprehensive View of all its branches, and the means most conducive to a perfect knowledge of the whole. *Midwifery,* respecting its operative part, may be term'd an Art; but as it comprehends the nature and treatment of Diseases, it may more properly be considered as a *Science,* divided into Theory and Practice: Theory consists in a competent knowledge of *Anatomy* and *Physiology,* particularly what relates to Generation, and the menstrual Flux; œconomy of the gravid Uterus, the nature of Parturition, and Doctrine of *Diseases incident to Women and Children.* The method of assisting with dexterity in *laborious* and *preternatural Labours,* and acting with judgment in Cases of danger or difficulty, constitutes the practical part of this Art. Without a previous and distinct knowledge of all these, no one deserves the name of *Accoucheur,* for if he ventures to give advice, or assistance not founded on rational Theory and the established Rules of his profession; he will act like a bungling Artificer, who vainly attempts to repair a complex Machine, without being acquainted with the several wheels and springs which compose it, or the principles upon which its motion depends.[105]

To relegate the art of midwifery without knowledge to the mere bungling operations of an 'artificer' is a claim founded upon new distinctions between the forms of valid scientific knowledge as opposed to merely manual 'mechanics,' and, while Leake does not make the mistake of categorizing women as machines, he denigrates their roles as rational and self-governing participants in the process of labour. Dismissing the supposition that imagination can affect the fetus as ignorant superstition in his earlier *Lecture Introductory to the Theory and Practice of Midwifery*, Leake had argued: 'A woman's mind, from the delicacy of her bodily frame, and the prevalence of her passions, is liable to so many excesses and inordinate motions that had such causes been productive of marks or monsters, they would certainly have been much more frequent. Besides, it ought to be remarked, that conception does not depend upon the will of the mother, but results from the nature and disposition of the several animal functions, and therefore happens whether her imagination be for or against it: Surely then it would be extremely irrational to suppose its influence over the fœtus greater than that which prevailed over her own body.'[106] As medical science gained authority over reproduction, its authors reinscribed women's supposed incapacity for the self-governing rationality of which doctors were now claiming to be masters.

In this context, all the more fitting is the epigraph to *Principles of Midwifery, or Puerperal Medicine* (1784) by Edinburgh surgeon John Aitken (d. 1790), MD, Fellow of the Royal College of Surgeons, Member of the Royal Medical Society, Surgeon of the Royal Infirmary, and Lecturer on Anatomy, Surgery and Midwifery in Edinburgh:

> To me be Nature's volume broad display'd
> And to peruse her all-instructive page,
> My sole delight.

'Gentlemen,' he writes, in the prefatory 'The Author to His Pupils,' 'I have the honour to present you with these *Principles of Midwifery*.'[107] The womb or woman as book of nature to be read and interpreted by male authority seems a fitting summary to this cycle of communications and control in the female mechanism.

To conclude, while there are some provocative similarities between the eighteenth-century auto-female mechanism and the female cyborg today, to suggest any kind of origin story would be problematic. The medicalization of conception and childbirth today has roots in the eighteenth century, as do cyborg representations of the mother, the womb, and the autonomous fetus-as-passenger, but there is no pervasive image of woman-machine in the eighteenth century except in some few midwifery manuals where the womb is treated as a geometrical entity of forces and angles (and even in these, the machine is quite organic in the images accompanying the text). Woman remains at the end of the eighteenth century both nature and passive mechanism upon which the man-machine's rational conceptions are inscribed. She is deceptively artificial when her sexual appeal is conscious and deliberate; without domesticity and marriage, the artificial female mechanism is at her most threatening; when her non-reproductive sexuality is given free reign, as in the case of Fanny Hill, her sexual desires are still governed by mechanical response and still in need of male direction, government, and ideally marriage; when pregnant, she is regularly the automatic vessel in which the active and autonomous fetus is carried and while she is occasionally an animal-machine in this capacity, she is a different creature than the active engine that is the fetus. The fetus-passenger is the human-machine, by default male and from conception active and vital, sometimes even rational and autonomous. What this history tells us is that there are clear relationships between representations of the woman-mechanism and the female cyborg who is simultaneously threatening, dangerous, and sexy while non-reproductive, and utterly threatening as disembodied and rational womb over which male heroes must reassert their control (the mechanical plant of *The Matrix* – conflating networked mechanical intelligence and mechanical reproduction without the participation of a human or spiritual steersman – might be the clearest version of this anxiety). These competing and unstable histories of the non-existent woman-machine figure from which the contemporary cyborg not so directly descends, however, do not prove a causal relationship. Sometimes, however, in some cases, they evidence an evocative and compelling similarity to images for women's roles and rationality in the twentieth and twenty-first centuries.

6 Cyborg Conceptions: Bodies, Texts, and the Future of Human Spirit

Whence, I often asked myself, did the principle of life proceed? It was a bold question, and one which has ever been considered as a mystery; yet with how many things are we upon the brink of becoming acquainted, if cowardice or carelessness did not restrain our inquiries.

Mary Shelley, *Frankenstein, or The Modern Prometheus*

In 1817 Mary Shelley wrote the story of what might be the first cyborg in literature: a nameless 'monster,' created out of fragments from charnel houses, the dissecting room, and slaughterhouses, and animated into life by galvanic current. Drawing on themes of the Biblical Eden, the Fall, and the myth of Prometheus punished for giving the knowledge of the gods to humans on earth, Shelley's novel is a moral tale of dangerous knowledge: the ambiguous morality in Victor Frankenstein's attempt to 'unnaturally' create a human being is in part framed by the disparity between his obsessive knowledge of science and the literary knowledge that is enjoyed by other characters. 'I delighted in investigating the facts relative to the actual world,' Victor explains of his youthful studies, while his cousin Elizabeth 'busied herself in following the aërial creations of the poets.' The contrasts between the material pursuits of natural philosophy or science and the spiritual ones of poetry and artistic creativity are thus made explicit (although later problematized) from the outset. Victor's playmate Clerval's favourite study consists of books of chivalry and romance: he composes plays about Orlando, Robin Hood, Amadis, and St George. It is not poetic genius that Victor embodies: 'Natural philosophy is the genius that has regulated my fate,' he declares. Later, only the aesthetic uplift of seeing nature's beauty and hearing Clerval (his imagination 'too vivid for the minutiae of science') invent 'tales of wonderful fancy and passion' can draw the haunted Victor out of his science-induced suffering and depression. Frankenstein's creature itself formulates an admiration of 'virtue and good feelings,'[1] as well as an individual identity that is all too aware of its own inhuman difference, by reading the works of Goethe, Plutarch, and Milton.

Indeed, Maureen N. McLane has explored the novel as proposing, 'a remedy for the horrifying body which science has produced: the humanities.' That is, she proposes, the humanities serve to 'acquire human being' in the novel. Quoting from R.S. Crane's *The Idea of the Humanities*, (1967) in which he had stated that 'the peculiar dignity of these arts is said to lie in the fact that their cultivation and pursuit differentiates the activities distinctive of man from those of animals,' McLane argues convincingly that 'the monster's *bildung* makes explicit the problematic status of "human being" (a species category) and the "humanities" (the cultivation of the "good arts" ... eloquence and reason, belles lettres, or the "liberal arts").'[2] What makes a human? If materialism and mechanism were increasingly pervading the body, taking the place of the rational soul and its spiritual hierarchy, then the literature of sensibility and later of the Romantics helped to establish a new hierarchy and morality where sensitivity, creativity, and literature were considered the stuff of humanity. If Willis's influence had established that the rational soul distinguished man from animal and purely material mechanism, then Frankenstein's creature registers a cultural anxiety attendant upon the new formulations of electricity as a vital energy. Mary Shelley's work considers the possibility that even a creature without a God-given soul, constructed by man entirely from base material, could possess sensitivity, reason, and even a morality and sense of justice no worse than the problematic morality and justice of human beings. The nineteenth century witnessed new formulations of the human machine identity arising directly from eighteenth-century natural philosophy that described humans as machines. Frankenstein's creature is perhaps the most evocative, remaining today a powerful symbol of the moral implications of medical-technical scientific hubris, but other continuities of man-machine and machine are evident as well.

Bruce Clarke's *Energy Forms: Allegory and Science in the Era of Classical Thermodynamics* discusses the 'cultural resilience' of classical thermodynamics throughout the nineteenth century that still relied upon a Galilean-Newtonian synthesis of mathematic and dynamic principles for matter and motion. With the 'hardy resilience of cultural anachronisms,' aether continued to inform the world-picture in new versions of energy, body, and society, and this too would 'leave its trace on field theory and the Ethernet' in postmodernity. For example, in H.G. Wells's *The Time Machine*, Clarke explains, 'a full-blown cyborg fantasy leaps directly out of raw mid-nineteenth century materials – early thermodynamics, social evolutionism, and the concurrent stirrings of cultural response to electromagnetic apparatuses and theories.'[3] Laura Otis examines how nineteenth-century technologies of communications such as telegraph networks and railway systems informed our understanding of the communications within the nervous system, and how in turn physiological studies have influenced communications engineers. 'From the 1850s onward,' she writes, 'as more and more people began communicating through telegrams, the public ... began to understand themselves as "connected" and to envision themselves as cross-points in a net.' The supposedly clear boundaries beween human bodies and their external mechanisms of communication continued to be as porous as they had been in earlier centuries. Otis suggests that 'drawing a distinction between organic and technological systems has become increasingly problematic. No mind can be abstracted from the information system that feeds it, and rather than passively receiving data, the mind controls the circuits that monitor its own environment. If a "medium" is a substance that allows communication, then nerves are undeniably a "medial apparatus." They perform the same functions as technological communications devices, and they can be studied with the same scientific methods.'[4]

The cyborg nexus of techno-bodies, techno-minds, and technologies of communications in *Frankenstein*, in theories of energy and thermodynamics, and in new machines for communicating continued to build on older metaphors and analogies for the human-machine. The following section examines in closer detail how these connections have continued to inform the understanding of human consciousness and identity in cyber-theory and fiction through the late twentieth century.

Virtually Human: The Electronic Page, the Archived Body, and Human Identity

Disputants, many of them writers, say to me, 'Words are still words – on a page, on a screen – what's the difference?' ... The changes are profound and the differences are consequential. Nearly weightless though it is, the word printed on a page is a thing. The configuration of impulses on a screen is not – it is a manifestation, an indeterminate entity both particle and wave, an ectoplasmic arrival and departure. The former occupies a position in space – on a page, in a book – and is verifiably there. The latter, once dematerialized, digitalized back into storage, into memory, cannot be said to exist in quite the same way. It has potential, not actual, locus. ... And although one could argue that the word, the passage, is present in the software memory as surely as it sits on page *x*, the fact is that we register a profound difference. One is outside and visible, the other 'inside' and invisible. A thing and, in a sense, the idea of a thing.

Sven Birkerts, *The Gutenberg Elegies*

What changes when we no longer think of the page as 'real'? Since the arrival of television and the personal computer, the presentation of our ideas via configurations of electron beams rather than fixed upon a more palpable page has inspired repeated commentary upon not only how our texts will change, but also how *we* will change. Whether a lamentation for the loss of the texture and substance of the book, or an exultation for a supposed new traversing of boundaries and freedom from hierarchical structures, a predominant conclusion has been that we are altered by our media. For many writers the 'terminal' page signals profound changes to our historical perspective, to our understanding and experience of community, to our cultural and democratic values, to our ability to teach our young to be moral and engaged citizens, to human agency and subjectivity, and even to human identity itself. Certainly, the material form of communication can shape cultural paradigms, what Harold Innis characterized as 'grooves which determine the channels of thought of readers and later writers.'[5] But does our media actually 'restructure consciousness,'[6] as early media theorists such as Adorno, Ong, or McLuhan assumed and contemporary theorists have continued to do? (Adorno, for example, pessimistically claimed that the technologies of film and radio, which produce 'technicized forms of modern consciousness,' result in imagination's being 'replaced by a mechanically relentless control mechanism.'[7]) *Is* the material of our communication indeed such a powerful determiner of human consciousness or identity as various writers following these early media theorists have claimed? Is the electronic page a technology that can *effect*, as some have argued, freedom and democracy or cultural decline? Is the page's influence on our culture more important than our culture's influence on the page? And to what extent are such queries determined by our own culture as professors and students of the humanities who are, not coincidentally, responsible for most of the speculation upon the materiality of our texts? Whether the materiality or immateriality of the page is a causative force in shaping human identity and society would seem impossible to determine conclusively. Indeed, it would seem impossible to ever arrive at a conclusion so profoundly informed by metaphors for the page as a body presenting a visible, material form of knowledge, which in turn both reflects and moulds the invisible, immaterial entity that is our rational self or consciousness. Speculation upon the immateriality of our electronic texts, therefore, is speculation about knowledge, morality, and education, and also about human embodiment, an ever-present debate about the old philosophical definitions of the physical body and its relationship to human 'spirit,' identity, or rational intelligence.

The page or computer network (material housing of the text) and the body (material housing of the self) are understood to be permeable, almost equivalent, in certain theories of electronic textuality. The so-called fusion of machine and human in networked communications systems has been imagined as a sign not only of a transition from book culture, but also of the impending redundancy or evolution of the human body. Sven Birkerts for example, has conflated book, body, and humanity in his discussion of 'the fate of the book': 'Maybe we are ready to embrace the pain of leaving the book behind;' he writes, 'maybe we are shedding a skin; maybe the meaning and purpose of being human is itself undergoing metamorphosis.'[8] For Arthur Kroker and Michael Weinstein, human intelligence is reduced to 'a circulating medium of cybernetic exchange'[9] in the networked text. The desire to become virtual, they suggested, promotes 'a radically diminished vision of human experience and of a disintegrated conception of the human good: for virtualizers, the good is ultimately that which disappears human subjectivity.' Sadie Plant has described 'the cyborg you become' when jacked into the computer as 'more or less directly connected to your central nervous system; more or less hooked up to its own abstraction and the phase space in which you are both drawn out.'[10] Whatever this new entity is, she concluded, it will be post-human. More optimistically and just as speculatively, Christopher Keep, claiming that there is 'more than a simple metaphorical relationship between hypertext and the body,' suggests that by reading electronic texts we are 'engaged in a border experience, a moving back and forth across the lines which divide the human and the machine, culture and nature.' The reader thus 'becomes an extended space of production, a series of flows, energies, intensities, discontinuities, and desires which refuse the (en)closure of the normative body.'[11] Surely there is a little irony in some of these musings but nevertheless my rude question is this: why is it assumed that human identity is so dependent upon the material page that when texts becomes electronic, consciousness or subjectivity flows out as if by osmosis?

Our texts, like our human identities, it seems to me, are unlikely to undergo radical revolution to reflect a new 'posthumanity': our texts, electronic or otherwise, are still engaged in very human themes of life, love, sex, and death; we still, as ever, make war and make love; there are still those who traverse boundaries, and those who erect them. There are and will continue to be good and bad citizens, good and bad students, avid and apathetic readers. Whatever difficult choices the computer age will necessitate, electronic textuality will not change these very human traits. What *has* changed is our analogies for the page as body and the text as thoughtful reflection: the simulacra produced by computer program and displayed upon the ethereal and immaterial page subverts conventional tropes for body, mind, and human spirit. The electronic environment has enabled us to understand our bodies as computational configurations of atoms and electrons, the mind as electrochemical charge and the body as DNA: the body no longer is the cathedral for the soul. The body is a biomechanical system, a complex machine, which can defy death when updated with artificial components. The electronic environment has also destabilized a centuries-old system of inscribing and disseminating principles of critique, judgment, and morality through the stable and material texts sanctioned by our educational and religious institutions. The page is a tool for teaching – but in the electronic environment, anyone can be an author. And so, if we imagine the electronic environment as virtual not real, as fleeting and malleable not permanent and canonical, as technological and commercial not literary and artistic, or as permeable and rhizomatic not fixed and hierarchical, we imagine we might predict the future: mind-body and mind-text meet figuratively at the locus of the page, and if the materiality of the page changes – so the theory goes – so must we as humans. However, a significant and generally unacknowledged part of this conversation about the incipient changes to humanity is concerned with the incipient changes to the humanities and liberal arts, and with the increasing influence of the sciences of material technologies.

Some Conceptual Frameworks: The Electronic Page and the Book of Life

What do we mean by 'electronic page'?[12] *Page,* of course, comes from the Latin *pagina,* a column of writing in a scroll. The display of text and image on a computer screen in a horizontal or vertical format similar to the dimensions of our traditional paper pages, which we read by 'scrolling' up or down, corresponds quite closely to the traditional page. Indeed, the style sheets and html markup used to create Web pages specifically replicate certain visual aspects of the printed page such as colour, text formatting, and image placement. Virtual reality displays could be considered a form of page as well, since the viewer scrolls through the apparently three-dimensional text of imaginary space, viewing images and reading texts that are written into the program. The electronic page is both database and bits. Potentially, it is far more malleable and responsive to user input than the paper page, but it is nevertheless a form of page that displays the artefacts of human creativity. The term *electronic* describes the flow of electronic charge that results in generating, sending, receiving, storing, and displaying the data that comprises a computer text: the text is in electronic form when bombarded onto certain types of screens to form images, or when it is transferred through circuits, or stored temporarily in random access memory (the use of *electronic* to describe computer-mediated texts, however, is slightly problematic since such a text may be transferred as sound waves, or stored as physical bumps etched into the surface of a CD). The electronic page is both a visible display legible to humans, and invisible machine-readable bits that are the information for the transmission and storage, content, layout, markup, and programmed functional capabilities of the visible page. This invisible text written on the main store of a computer is divided into manageable sections called pages, and a similar amount of data or part of a program is also called a page. The electronic page is markedly different from the paper page, but whether in fact a page is a 'thing' if the data is not printed on a material substrate is a philosophical question with which I will not trouble myself here. When I write of the computer-mediated text as page I refer to a text that may be displayed visibly onscreen in a familiar format, but which is also necessarily invisible binary data written onto the surface of a computer disk or tape. The electron beams patterning the screen through a shifting flux of energy may be rewritten, lost, or recovered in moments, it may be stored as database and generated dynamically for display depending on the reader's input, but the electronic page is nevertheless a text inscribed through human ingenuity onto the physical elements of the world in which we live.

This very invisibility, changeability, and seeming instability creates the illusion of a new form of human consciousness that is permeated by, and permeates, the computer network. There is a history to this image of the mind, however: the material page has been traditionally understood as re-presenting the human form as repository for knowledge, memory, creativity, and imagination; likewise, the human mind has traditionally been presented as a writing surface. Our pages and our bodies have long converged in metaphor. Indeed, a page has a body, a header, and a footer; it might contain an appendix or index (from the Latin meaning 'indicator' or more specifically 'forefinger') or footnotes or frontispiece (from the late Latin *frontispicium*, from *frons*, 'forehead' and *spic-*, denoting 'see'); it might be part of a chapter (from the Latin *caput*, head); it may be part of a manuscript (from *manus*, hand), or it may be bound into a book with a spine (and the electronic page has access to memory). A material surface with boundaries, edges, and margins, for centuries the page has been made of skin, and bound in skin. And for centuries, the body has been metaphorized as book. Andreas Vesalius (1514–64), known for revolutionizing the pedagogy and study of anatomy, taught that not ancient books but the human body itself should be the primary text and ultimate authority in the study of human structure and today that tradition of the body as the Book of Nature is continued in digital anatomy projects.

While the body is a book to be read, the mind has been imagined as a page to be inscribed. As we have seen, the analogies we create to describe the human mind tend to reflect the technology of writing that we use to inscribe our thoughts. Recall Locke's characterization of the mind as white paper void of all characters in *Essay Concerning Human Understanding*. The page as trope for the inscription of both rational thought and morality – 'for white Paper receives any Characters'[13] – evokes and reinforces a system whereby education and integrity are embodied in the page itself: the page on the screen thus becomes emblematic of a moral shift. Recall, too, Vivian Sobchack's early concept of the hazards of the 'binary superficiality of electronic space,' in *Materialities of Communications*, a collection of essays concerned with media technologies, the body, and transitions within the academic field of the humanities. Arguing that the electronic text or virtual reality creates a 'quasi-disembodied state' and 'denies or prosthetically transforms' the body, she warned that the resulting 'lack of specific interest and grounded investment in the human body and enworlded action ... could well cost us all a future.' Furthermore, if we become 'merely ghosts in the machine,' she suggested, we 'can ignore AIDS, homelessness, hunger, torture, and all the other ills the flesh is heir to.'[14] Clearly, there is an element of hyperbole here, and my intent is not to point out the obvious embodiment of the computer user. These images encapsulate a common theme in such analyses of computer technology: electronic texts result in both disembodiment and immorality. In such works the texts of technology – texts unedited and unauthorized through any traditional system of determining literary excellence – are represented as replacing real humanity with unfeeling human-machines, and making redundant not only the body but also literature and all those creative endeavours that instil morals into the human psyche. K. Ludwig Pfeiffer's introduction to *Materialities of Communication* summarizes the complexity of the position:

> Certainly, one would like to know what kind of autodynamic wirings or analogues of them there are – if there are any – that test perception, guide behavior, evaluate experience, caring little or nothing for the pathetic semantic textures we weave around them. There are the brain, the hormones, and the other circuits that produce, in ways still fairly obscure, electric and chemical binarisms. But if it is one of the deadlocks of brain research that the steps from there to what still appears as meaningful cultural worlds *are extremely hard to take and have yet to be taken*, it behoves 'literary' people (like most of those in this volume) not to abandon prematurely some striving toward the 'nobleness of life' ... even if it consists only in 'literature.'[15]

The relationship between technology, human embodiment, and morals is complicated not only by a history of metaphors for body, text, and material page in an older tradition, but also by the growing prevalence of science and technology as more 'useful' disciplines than the study of literature.

While the analogies between page and body are ancient and associated with a tradition of education that values the spiritual side of humanity, the convergence of text, body, and mind as code in the electronic page or digital archive is more recent, and associated with the rise of the computational sciences. Vannevar Bush's article 'As We May Think,' which described in theory what would be realized in computer hypertext documents ('an enlarged intimate supplement to his memory') appeared in the July 1945 issue of the *Atlantic Monthly*. In 1948 Bush's colleague Norbert Wiener published a manifesto for the new technoscience of cybernetics, which explained both organic and machine processes as communications systems. Wiener and Bush had worked on a project together to solve partial differential equations with computers in 1940, which in turn influenced Wiener's conception of future computing machinery as an analogy between electronic computing and the mechanics of the nervous system.[16] Wiener's theories conflated both machine and text as symbolic structures of the human body: the chapter entitled 'The Individual as Word,' in his popularization of the theory of cybernetics, emphasiszed the metaphor of human body as the Book of Life. 'Earlier accounts of individuality were associated with some sort of identity of matter, whether of the material substance of the animal or the spiritual substance of the human soul,' he wrote. 'We are forced nowadays to recognize individuality as something which has to do with continuity of pattern, and consequently with something that shares the nature of communication.'[17] There was significant debate in the United States about the applicability of Wiener's information theory to genetics, but the metaphors of human life as text were nevertheless present and became, as E. Lily Kay demonstrates, dominant and potent metaphors 'in the general conceptualization of heredity as a genetic program, a scriptural technology.'[18] While representation of the body as information is a post-war phenomenon, however, the analogy of human bodies as textual, written documents was a very old one, and the image of the human body's creation as text was widespread. Indeed, some years earlier, in publishing their discovery in 1938 that the human genetic material was DNA in the form of long chains, Leeds researchers W.T. Astbury and F.O. Bell had borrowed the old metaphor to configure human life itself as text written on a scroll: 'Knowing what we know now from X-ray and related studies of the fibrous proteins ... how they can combine so readily with nucleic acid molecules and still maintain the fibrous configuration, it is but natural to assume, as a first working hypothesis at least, that they form the long scroll on which is written the pattern of life. No other molecules satisfy so many requirements.'[19] This discovery marks the beginning of a shift in the understanding of the human body as God's immutable text, which William Harvey had characterized as 'Nature's book ... so open, and legible' in his seventeenth-century text describing the mechanics of animal generation.[20] With the transcription of the invisible elements of the body made increasingly possible through technology and computers, we can not only 'know' but also re-write or program the body as well as characteristics of mind, identity, or consciousness.

The electronic page or digital archive is a technological condensation of text and body, human consciousness, and human reproduction. On the one hand, old metaphors for physical body and page suggest that the death of the book parallels the redundancy of humanity. As the page becomes immaterial so is the self depicted as immaterial, flickering in a state of virtuality, our humanity snagged on the edge of the screen separating world from data. On the other hand, the idea of the conscious mind freed from the body's limitations suggests the fulfilment of a long-standing desire for transcendence, of which the electronic archive is a secularized version.

The Electronic Page and Human Spirit

The relationship of the material page and the computer text to human identity is complicated by the various terms used by writers to connote vastly different notions of that unnameable, un-locatable, and un-measurable quotient that gives us our humanity: since writing began we have questioned where that aspect of our selves we call identity, consciousness, mind, rational thought, soul, spirit, or ghost resides. We have wondered whether it is immaterial or material, part and product of our bodies, or separate and distinct from gross matter. This very old question was a defining characteristic of early cyberpunk fiction which envisioned the consciousness as code living on without the body. William Gibson's *Mona Lisa Overdrive,* the third and last of the series that inaugurated the phrase 'data made flesh,' presents the ghost in the machine as a form of sentience akin to immortality. At the novel's end, Angela Mitchell, a cyborg being whose brain has been engineered and programmed to jack into cyberspace without hardware, is killed. Her consciousness is transferred to a powerful biochip called an aleph or 'soul-catcher.' Angie as sentient code inhabits a high resolution, three-dimensional virtual space along with the other inhabitants of cyberspace, 'ghosts,' artificial intelligences, and her lover Bobby (The Count) Newmark who has entered that world between life and death by programming his own consciousness from his decaying flesh into the aleph biochip. The re-presentations of ourselves and our environment, evolving from page to microchip to biochip, reiterate an old metaphor of capturing the soul or spirit in the text. In a remediation of the lines from Shakespeare's sonnet XVIII – 'So long as men can breathe, or eyes can see / So long lives this, and this give life to thee' – Angela and Bobby are immortalized in the electronic text as long as the aleph is provided with a source of power.

The dream of the human spirit as coded consciousness free from the constraints of the body is not only a literary construct but also appears in theoretical discourse in both the humanities and the sciences. In a 1996 lecture Toshiba Professor of Media Arts and Sciences at MIT Marvin Minsky claimed that in the near future it will be possible to transfer human memories, intact, to disk.[21] Also in 1996, head of British Telecom's research lab Peter Cochrane described a technology to capture human thoughts on a single silicon chip which he dubbed a Soul Catcher. 'Despite specialisation and an exponential growth in knowledge,' he wrote, 'we still see people of outstanding ability able [to] understand and contribute more than the average. Unfortunately, they die and their expertise is lost for all time. The question is, can we capture their expertise and presence for future generations? Do they have to die 100%?'[22] Somewhat tongue-in-cheek, *Business Week* later described the future technology as being able to make a rather straightforward decision by the year 2050 'to evacuate your biological body and take up residence in silicon circuits.'[23] The Soul Catcher would comprise 'wireless links to microsensors under your scalp and in the nerves that carry all five sensory signals' to record 'organized, online archives of everything that happens,' (so described by D. Raj Reddy, a professor of computer science at CMU). The article concludes, 'For people who chose not to inhabit silicon, virtual immortality could still ease the sense of futility that now haunts many people. Individuals would know their lives would not be forgotten, but would be preserved as a thread in a multimedia quilt that keeps a permanent record of the human race. And future generations would have a much fuller understanding of the past.' Much like the genetic 'map' that makes our cellular information live on generation after generation, this technology would purportedly transcribe consciousness by mapping the sequence of experience that is written upon our minds. The four-letter code of genetics, data storage miniaturization, microprocessor-controlled prosthetics, artificial intelligence, the global grid of communication/control systems – the fusion of our understanding of our own bodies with the virtual representation of life onscreen, all problematize what were once clear conceptual boundaries between categories of body-mind and text.

Jean Baudrillard has depicted our electronic media as stealing individual consciousness, which he characterized as a pilot disengaged from the grounding of body, community and reality, while the body 'appears simply superfluous, basically useless in its extension.' For Baudrillard, the world of freely available information is an obscene one: 'today there is a whole pornography of information and communication,' he suggested, 'that is to say of circuits and networks, a pornography of all functions and objects in their readability, their fluidity, their availability ...' Obscenity begins, he explains, when 'everything is exposed to the harsh and inexorable light of information and communication.'[24] Baudrillard's play on the word *obscene* as the movement away from the real body in real space also emphasizes a sense of indecency and moral offensiveness. And while claiming that 'this is not necessarily a negative value judgment,' he wrote of the disappearance of passion in the world of information – of hazard, chance, and vertigo as opposed to the passion, investment, desire, and expression of a previous era. Similarly, in his 1983 *Simulations*, Baudrillard criticized computer simulation as a spiritual crisis: the real, he explained, 'is produced from miniaturised units, from matrices, memory banks and command models.' Could 'the divinity,' he wondered, be 'volatilized into simulacra,' in the 'visible machinery of icons being substituted for the pure and intelligible Idea of God?' In Baudrillard's successive phases of the image, 'the reflection of a basic reality' is good: 'the representation is of the order of sacrament'; while the second order, which 'masks and perverts a basic reality,' inaugurates the age of simulacra and simulation and is 'an evil appearance – of the order of malefice.' Baudrillard's appraisal was from a position of despair over simultaneous changes to the signs for both our arts and our bodies. 'Digitality is [the] metaphysical principle (the God of Leibniz) and DNA its prophet' for the 'universe of structures and binary opposites' we have created, he suggested, finally concluding that simulation has damaged both our mental health and art itself, the very expression of our humanity: 'art is dead, not only because its critical transcendence is gone, but because reality itself ... has been confused with its own image ... simulation pushes us close to the sphere of psychosis ... The cool universe of digitality has absorbed the world of metaphor and metonymy. The principle of simulation wins out over the reality principle just as over the principle of pleasure.' Baudrillard's critique of 'the convergence of genetics and linguistics'[25] stands in direct opposition to the more ecstatic predictions of how the digital text will transform the human identity or spirit; but in many cases these oftentimes implausible claims may be reducible to the author's stance 'for' or 'against' the authority of science and technology.

The Archived Body

Floating through the skull and heart and lungs of Alexander Tsiaras's world we see every detail of our knowledge about the body and our ability to measure and define its function. We see here a body utterly and vividly intact. Only the soul has been removed ... whatever else is to be found in this collaboration between an executed criminal and an artist we are somehow deftly aided in the ancient search for the soul. It is the oddest of ironies. The atomized, digitized body of Joseph Paul Jernigan reconstituted and imbued with a mesmerizing beauty and realism is as good an argument for the tangibility of the soul as one can find in this cheerless age of cause and effect.

> – John Hockenberry, Introduction to Alexander Tsiaras,
> *Body Voyage™: A Three-Dimensional Tour of a Real Human Body*

A living virtual reality ... originates with the merger of genetics and simulation where blood turns into electricity ... [It] functions in the time of recombinant culture, whose sociology is based on splicing, cloning and sequencing ... The vanishing body has been resuscitated, just short of vacuity, as the circulating body. The body has become a circulating medium of exchange, coursing through the mediascape ... The perverted image (perverted as image exchange-value) and the ambivalent sign (fanatical and cynical) are the effects of the dependence of the mediascape on 'biological' bodies as image resources and image actualizers ... Welcome to the post-God era.

> Arthur Kroker and Michael Weinstein, *Data Trash*

As the body's most invisible elements and processes have become more legible and better understood, the page has become more complex and immaterial. Consider the difference, for example, between a page in which the ink under your fingertips is raised so you can actually feel the patterns of letters, to a page that exists only in random access memory until it is saved to a mysterious and invisible position on the hard drive as magnetized bits – a pattern of ones and zeros decipherable only by the machine. And consider the difference between William Harvey's startling announcement in 1616 of his discovery of blood circulation, to the discovery in 1989 by J.R. Riordan et al. that the majority of people suffering from cystic fibrosis have a small mutation in one microscopic DNA fragment, which causes three out of one gene's 250,000 base pairs (A-T or G-C) to be missing (that is, three out of some three billion base pairs in our DNA), a deletion which in turn results in the loss of *one* amino acid – one molecule – out of the 1,480 in the protein for which that gene is a 'blueprint.' Our bodies and our texts have become similarly coded bits (1,0 encodes machine texts, while A,C,G,T codes body texts) but conceptually our understanding of body and text has been reversed. The white page with its black ink that has always been visible and accessible in the codex book is now hidden, mysterious and invisible bits accessible to and understood by only the masters of technology. The components of living bodies, the creation of both life and thought that were once hidden and mysterious in the human body are now magnified, diagrammed, documented, transcribed, archived by the masters of technology. At the same time, the body becomes metaphorized as a text to be read, transcribed, and re-written. This fundamental change in our understanding of the body as page is illustrated most tellingly by current metaphors for study of the body no longer as Book of Nature authored by God but now as genetic blueprints co-authored by humans – the published, bound, and immutable work versus the page describing the plans for a work yet to be finished. Vesalius wrote of God the 'Author of the human fabric,' in his *Epitome* (1543) 'Concerning the Organs which Minister to the Propagation of the Species,' while Victor Spitzer and David Whitlock, directors of the Visible Human Project datasets, subtitled their *Atlas of the Visible Human Male* as 'Reverse Engineering of the Human Body.'[26] Similarly, the U.S. Department of Energy's overview of the Human Genome Project, 'To Know Ourselves,' introduces the human genome as 'The Recipe for Life.'[27] In the rhetoric of body as archive, the notion of media theorists that the natural human body has become redundant exists in direct contrast to the medical versions of its centrality.

The transcriptions of the human body through the Visible Human Project (VHP) and the Human Genome Project, both stored in the National Library of Medicine (NLM) at the National Institutes of Health (NIH) in Bethesda, Maryland, are the most comprehensive archival projects in medical history. The Visible Human Project is the outcome of the National Library of Medicine's long-range planning in 1986, which established the library's goal of 'building and disseminating medical image libraries much the same way it acquires, indexes, and provides access to the biomedical literature.'[28] The project effectively began in 1989, when the library's ad hoc Planning Panel on Electronic Image Libraries made the recommendation that the NLM build 'a digital image library of volumetric data representing a complete, normal adult male and female.'[29] Begun in 1990, the U.S. Human Genome Project involves the identification of all 60–80,000 genes in human DNA and the sequences of the three billion bases that make up human DNA, the storage of this information in databases, and the development of tools for the data. The datasets resulting from both projects are the human body paginated, represented as alphanumeric code, digitized, pixelated, and available online in the National Library of Medicine's electronic collections (www.nlm.nih.gov/databases/databases.html).

The Human Genome Project, like the Visible Human Project, is a process of making the interior workings of the body visualized, archived, and legible (that is, capable of being read or deciphered). The metaphor of transcription, the analogy between human life and text, is one that has dominated the rhetoric of human DNA since its discovery. In 1967 microbiologist Robert Sinsheimer, a key figure in the inception of the Human Genome Project, wrote *The Book of Life* in which he commented 'In this book are instructions, in a curious and wonderful code, for making a human being. In one sense – on a sub-conscious level – every human being is born knowing how to read this book in every cell of his body.'[30] The metaphor of body as book appears frequently in descriptions of the Human Genome Project. A National Human Genome Research Institute press release from 1998 equates books and life, for example, where 'changes in the spelling of the DNA letters can increase your chances of developing an illness, protect you from getting sick, or predict the way your body will handle medicines.'[31] Similarly, the statement published by the National Center for Biotechnology Information emphasized the Human Genome Project as a 'working draft' of the 'book of life.'[32]

The metaphor of the human as text informs not only promotional literature written for the public, but also actual practices. The library information for gene sequences as published documents indicates strikingly how much our bodies are actually and not only metaphorically perceived as pages of text. A search for 'chromosome 7' through GenBank at the National Center for Biotechnology website (www.ncbi.nlm.nih.gov) will result in a number of links to various sequences catalogued just as any other text in our library systems. The descriptor for accession number AC073349 describes the document as a 'working draft sequence' of a particular chromosome segment. This tiny portion of our bodies written in its four-letter alphabet is accessible online as a long scrolling document, approximately fifty single-spaced pages when printed. Writing the human body as page allows human technicians to position themselves as its authors. The catalogued segment has a title, 'The sequence of Homo sapiens clone,' and an author, R.H. Waterston. Another example is the National Institute of Health's patent on a cell line from a Hagahai person from Papua New Guinea, disclaimed five years after the initial application and only after much public criticism, in 1996. The claim of United States Patent No. 5,397,696 included the names of several 'Inventors'; the 'Title of Invention' was listed as 'Papua New Guinea human T-lymphotropic virus.' When the reproduction of our bodies, the most basic process of our creation, is described as a process of writing, and our 'code' as being 'transcribed' in our cellular DNA, genetic reproduction is seen as a process of writing, as program (from Latin *programma*, public notice; from Greek *prographein*, to set forth as a public notice: *pro-*, before + *graphein*, to write). The program – the code written to produce both the computer text and the body text – is the means to not only prolonging but also rewriting the fate of a human life.

This transcription of body as text is seen to be a threat to our humanity – not merely because we are taking ownership of human bodies in potentially exploitative ways, but also because the apparent literalizing of body as page is seen as displacing the position and value of human 'spirit.' This trepidation is perhaps best represented by the varying commentary upon the Visible Man archive for the National Library of Medicine. Creating this archive was a process of mapping by MRI and CT scans the fresh cadaver of executed murderer Joseph Paul Jernigan, freezing the body so the tissue offered the same resistance to the saw as did the bone, quartering the body and positioning it in blue gelatin, and finally milling away the surfaces of the frozen blocks from toes to head at 4 mm intervals and digitally photographing each newly exposed surface. At a resolution of 0.33 mm, the raw data totals fifteen gigabytes, or twenty-three CDs, as the literature frequently explains (the anonymous Visible Woman dataset, at higher resolution, is about forty gigabytes in size).

One of the most significant changes in the study of anatomy represented by the dissection of Jernigan is that the body has been cut not to reveal gross physical units such as a given muscle, organ, or tissue, but rather in cross-section as fine leaves of body. These digitized leaves of Jernigan's body now exist in the form of an enormous and readable book. Various 'fly-through' animations of the sliced body available for public viewing on the Internet are created in the same way that flip-page animations are: a sequence of consecutive images shown in rapid succession are interpreted by the eye as movement (see, for example, the link 'From head to toe: an animated trip through the Visible Human male cryosections' at the 'Visible Human Project® Gallery' www.nlm.nih.gov/research/visible/visible_gallery.html). The NPAC/OLDA Visible Human Viewer (www.dhpc.adelaide.edu.au/projects/vishuman/VisibleHuman.html) is a Java applet that allows the viewer to place something akin to a bookmark at any point on the body and then download that particular page as an axial, coronal, or saggital plane of the sliced body (see figures 36 and 37). This viewer, in turn, resulted in the publication of a physical 120-page, full-colour book, *Head 2 Heads: A Flipbook of Slices of Life* by Optical Toys (2000), which 'takes you through the human skull' (www.opticaltoys.com/head2heads.html). Not only the visual display but also the rhetoric of the body as a book appears frequently in the project literature: for example, project officer Michael Ackerman described one of the problems with the archival material as the absence of labelling for the various systems and organs of the digitized body: 'For a librarian, this is very unsettling. It's like having books lying all over the place not indexed or catalogued.'[33] The liner notes to *Body Voyage*, the CD-ROM published by Southpeak Interactive and featuring renderings of the data by photojournalist, Alexander Tsiaras, emphasize that the data of Jernigan's body comprises 'over a raw terabyte of data – the equivalent of five million typewritten pages.' *Life* magazine's feature article on 'The Visible Man' comments: 'Jernigan had no idea that his body would itself become a textbook.' The fold-out article 'A Technicolor Gatefold of the Digital Man' in this issue featuring Tsiaras's work is labelled 'The Whole Body Catalogue' and claims that Tsiaras used '15 gigabytes of computer data from a real body – equal to 20 million typewritten pages – to compose this picture.'[34]

Whether the electronically archived body is equivalent to five million or twenty million pages, what is most prized about the collection is its legibility. The systematized visibility of this body as text provoked comment by supporters on Jernigan's 'immortality'[35] through being transcribed as alphanumeric text, as if in imitation of the poets' conceit of achieving immortality through the lines inscribed upon a page. Simultaneously, the body digitized and transcribed as computer text has generated some critical commentary on the violation of transcribing human embodiment. Catherine Waldby, for example, writes that 'the violence of anatomy is the violence of a particular kind of writing practice, a set of techniques that destroy the fleshly body along particular analytic lines in order to inscribe its trace in various knowledge media.'[36] Neal Curtis suggests that, like Franz Kafka's 'In the Penal Colony,' the VHP 'reinforces the submission of the body to the law.' Curtis's complaint has to do with the VHP's 'confusion of a "complete" body with an *anatomically* complete body' – that the people involved do not recognize the dataset as a product of a 'techno-medical' or 'techno-scientific' discourse. Such procedures can return a 'form of life' to the body that is determined by those technical discourses, he continues: 'These techniques can reanimate the body and rebirth it, but can life be reduced to a simple reconstruction of motility? Clearly not.' In the Visible Man there is 'cruor,' the blood of wounds, but no 'sanguis,' the blood of life, Curtis explains. True enough: these are images of a dead body. However, the fact that these images present no 'exposure to sensible presence' or 'the irreducible vitality of sanguis' does not adequately demonstrate, for me, 'the universalism and determinism of such a discourse' – as if technology and science are realms populated by people who do not recognize difference, who are complicit in a project of 'silencing indeterminacy.'[37] These criticisms, as inflated as the rhetoric of praise mentioned above, rarely acknowledge the benefits of such technology to human lives.[38] Nor do they adequately address the fact that both executed criminals *and* people who donate their bodies for research have been dissected, studied, and reproduced as image for centuries prior to this particular technology. Rather, such arguments seek to demonstrate that inscription of the body through computer technology is a debasement of humanity.

Writing on the 'electronic abbatoir' though not specifically referring to the VHP, Kroker and Weinstein describe archivalists as 'vampiring organic flesh, and draining its fluids into cold streams of telemetry' where archived body parts are 'violently detached from the body organic ... disguised in the binary functionality of data and pooled into larger circulatory flows.' The aim of this bio-power, they suggest, is 'the transformation of human experience into the dull codes of binary functionality.' The 'violent metastasis that is cyber-culture,' is in direct opposition to a spiritual life: 'In the beginning was the Word,' they write, 'but in the end there is only the data byte.'[39] Waldby similarly explains in her book about the VHP that 'after all, when a body can be rendered into data and thus cross the interface into the digital afterlife, what prevents the process from effecting some form of reversal, the digital revenant who rematerialises in real space ... the kind of cyberspace summoned up by the VHP connotes the supernatural ... an afterlife of the abject, the corpse which cannot or will not relinquish vitality ... like those other animate corpses, vampires and zombies, to be vitalised by the will of another, actively prevented from a full death.' The suggestion, again, is that the inscription of technology interferes with human spirit: 'For the biomedical imagination,' she writes, 'this arresting and deferral of death might count as a gain on the side of life ... Another step in the gradual mastery of matter, bringing it closer to the negentropy of programmable matter, the assimilation of all materiality by the metaphysics of code' but the VHP also seems to create 'a new form of death-in-life, a new and horrifying destination for our own failing bodies, and a place from which they might return in uncanny form.'[40] However disturbing we might find such treatment of dead bodies as archive, however, we need to acknowledge that these texts are ultimately only images, numbers, and letters displayed on a dynamic electronic page – no matter how 'animated' they may seem. The rhetoric of a ghastly humanity floating through cyberspace and coming to life again as soulless vampires or zombies here again centres upon who we would prefer to metaphorize as 'author' of humanity.

Of Books and Spirit

As the self-proclaimed 'greatest contribution to anatomy since Vesalius's 1543 publication of *De Humani Corporis Fabrica*,'[41] the Visible Human Project has developed a logo incorporating an illustration from Vesalius's text (see figures 38 and 39). The image depicts a history of the various imaging techniques used to 'read' the human body and re-present it as text – from lettering discrete muscle groups to representing all structures as pixels and voxels. It also serves to emphasize the project's subject as defying mortality. Flayed, frozen, sliced, and digitized, the body in this image is that of a handsome-faced and well-muscled man in a classical pose, gesturing as if in exposition, vital and full of life despite lacking skin and a good portion of his brain. In contrast, the final panels of Vesalius's anatomized body gradually stripped of parts and life spirit show the same subject slumped in an attitude of dejection (see figure 40). In all likelihood a condemned criminal as well, this cadaver's mortality and sin are subtly emphasized by the rope supporting or hanging the body by the neck. Vesalius's sequence of images showing the slow loss of life spirit seem to function as an explicitly moral reminder to the early-modern reader of the inevitable decay of the flesh – or that the punishment for the sinner is eternal death.

The Visible Human Project logo, in contrast, depicts life with a morality conferred by the gift of the body to medical science, as expressed by numerous writers and perhaps most enthusiastically by John Hockenberry who claims that the convicted murderer Jernigan gave his body not only to science, 'but to humanity as well,' concluding, 'Whatever tragic legacy Joseph Paul Jernigan left in life, in death he has found grace.' The slogan on this page of the *Body Voyage* codex, 'Even Lazarus never looked this good,' attests to the desire in popular media – so much at odds with the theorists critical of the supposed redundancy of the body – to imbue the body archived in the cybertext with a soul, with immortality, with a high moral purpose. The slogan also attests that sophisticated technology has increasingly challenged the established authority of 'God's Book': the biblical story of Lazarus and religious faith ('he that believeth in me, though he were dead, yet shall he live,' John 11: 25) is here overshadowed by the suggestion that Jernigan has been raised from the dead through his gift to science and the subsequent technological rendering by human authors. While the eighteenth-century criminal was judged by God and sentenced to eternal punishment, the twentieth-century convict's salvation is conferred by science and technology. The spirit of the cyborg is at stake in the environment of the electronic page because life and death, previously the realm of God or nature, now are in the realm of the human-authored page.

In his introduction to the 1994 *Queen's Quarterly* issue entitled *The End of the Book?* Boris Castel writes of the 'intimate reader/writer exchange, now under assault from the increasingly noisy signals of our surrounding electronic web,' claiming that 'the printed page and circuit-driven technologies are not kindred, but powerful antagonists. Human intelligence and creativity will be the losers in our Faustian pact with an increasingly seductive electronic devil.'[42] This issue is a celebration of and lament for the containers for Western humanist knowledge, both the libraries and the bound pages of the book, once hand-made at great expense for the promotion of humanist values. The issue is illustrated throughout with evocative photographs of splendidly ornate European libraries and wooden bookshelves replete with ancient leather-bound texts. One caption reads: 'The magnificent Waldsassen monastery library in Bavaria, built in a century when knowledge was celebrating its triumphs.' Robert Fulford here nostalgically reflects upon 'The Ideology of the Book,' commenting that 'since the Enlightenment, Western civilization has made the book the shrine of modernity, the place where we store and locate our ideals.' Fulford expresses concern that costly new forms of information on computer networks and CDs will endanger the 'great historical movement, the gradual broadening of knowledge, outward from its original owners, the princes and priests, toward all of humanity.' What are 'our ideals' here? Fulford seems to be valorizing only a specific kind of knowledge that stops short of computer technology. He concludes with a call to arms:

> When I read about what Gates and his competitors are preparing for us, I sometimes think about those pioneers of information technology, the monks who preserved part of the wisdom of antiquity during the centuries when

hardly anyone else seemed to care about it ... 'A monastery without a library is like a castle without an armoury,' [a monk in Normandy in the year 1170] wrote. 'Our library is our armoury. Thence it is that we bring forth the sentences of the Divine Law like sharp arrows to attack the enemy. Thence we take the armour of righteousness, the helmet of salvation, the shield of faith, and the sword of the spirit ... ' In the environment created by onrushing technology, scholars, librarians, teachers, writers – all those who take responsibility for generating and spreading knowledge – may well find themselves called to a similar battle. They will need to be shielded by faith in the value of their endeavours, and by the sword of the spirit.[43]

What are we to make of this invocation of righteousness, salvation, faith, and spirit to attack the amorphous enemy technology? The religious references by both Castel and Fulford indicate a threat to human spirit but it is tempting to suggest that what is depicted is as much a threat to the humanist claim on the dissemination of knowledge and morality.

In his afterword to *The Future of the Book*, Umberto Eco invokes the words of Claude Frollo in Victor Hugo's *Hunchback of Notre Dame:* '"Ceci tuera cela" (The book will kill the cathedral, the alphabet will kill images).' Eco remarks that one significant issue raised by the participants in the symposium which instigated this collection of essays 'that *ceci* (the computer) *tuera cela* (the book).'[44] The phrase is a sign of the supercession or destruction of one medium of communication by another, but it is also worth noting that what was at stake for the clergy that Frollo represents was the material housing of spirituality – where knowledge of it would originate, where it would be taught and learned, how it would be disseminated to the people. Printed books challenged some of the authority the cathedral represented. In the same way, the dissemination of knowledge through the electronic page is threatening – or invigorating – because in our secular world the 'cathedral' of human consciousness or identity has for centuries been represented by the mundane and unthreatening codex or paper page. The page, material representation of mind and/or spirit, is the cathedral of the humanities. The electronic page, imagined as infiltrating human consciousness or displacing human spirit through its 'ectoplasmic arrival and departure,' also is imagined as invading and displacing the 'spiritual' values of the humanities.

Concluding Remarks

> Know then thyself, presume not God to scan;
> The proper study of mankind is Man.
> Placed on this isthmus of a middle state,
> A being darkly wise, and rudely great:
> With too much knowledge for the sceptic Side,
> With too much weakness for the stoic's pride,
> He hangs between; in doubt to act, or rest,
> In doubt to deem himself a god, or beast;
> In doubt his mind or body to prefer,
> Born but to die, and reasoning but to err;
> ...
> Sole judge of truth, in endless error hurled:
> The glory, jest, and riddle of the world!

Alexander Pope, *An Essay on Man*

On the website of the Center for Advanced Information Processing (CAIP) at Rutgers Univerisity, the Visible Man sprints, skips rope, does push-ups, and shadowboxes to the *Rocky* soundtrack. If ever there was reason to imagine a ghost in the machine, these videos of the flayed 'corpse' of Joseph Jernigan eerily reanimated are it. Billed as *Rocky 3000*, the video begins with superimposed titles reading 'They froze him ... and cut him into thin slices ... for the benefit of medical science ... we put the pieces back together ... And now he is ALIVE.'[45] A more recent demonstration of computer modelling and animation at CAIP shows the Visible Man in an arid landscape at night, running from a tank that follows, aims, fires, and blasts him apart into four neatly sliced segments.[46] A virtual Frankenstein's creature, created piecemeal from dead matter, this moving thing might seem to be haunted by the spectre of Cartesian dualism or even by all the ills of an Enlightenment ideal of progress; but surely these are merely macabre and ingenius jokes made from digital images. Aren't they? Are these movies innocent jests in good fun, though perhaps poor taste? Or do they evidence a lack of moral government inherent in a world where ungoverned technology reigns supreme? The reanimated Visible Man instantiates the moral crossroads which for many critics is represented by the materialistic techno-human. Is the representation of dead matter on the computer screen meaningless, or does it violate some intangible spirit that once occupied the body?

Moreover, the 'life' in this uncanny representation seems to focus the all too obvious continuities between 'artificial' and 'natural' life, especially at a time when investigations into creating artificial intelligence using complex networked computer systems are ongoing.

The Visible Man is a visual representation of one of the last frontiers for that little pilot that historically Western culture has imagined to be the conscious identity or soul within the human body. As 'Rocky,' this avatar of a man who was once Joseph Paul Jernigan is steered and animated by computer programmers. That is, if this animation can be said to have an animating spirit, it does not reside in the pineal gland: it resides in networked software. To draw further on this cyborg pilot image, a consistent trope since the Visible Man was first rendered as voxels has been the notion of the 'fly-through,' an animation that seems to take the viewer on a journey through the spaces of the virtual body. In this iteration, the viewer is cast as 'pilot' navigating the cyberspatial corpse. On the medical-technological side is R.M. Satava, who extols the Visible Human Project as one stage in a series of advances for virtual surgical simulation: 'For generations, pilots have 'pushed the envelope' of their aircraft simulators, harmlessly 'crashing' in virtual space, all the while learning the limits their machines' performance and the inadequacies of their own capabilities. Repeated practice under mentoring enabled them to hone their skills to their optimal performance, making the American fighter pilot the uncontested champion of the sky.' Within the proper education content, supervision, and evaluation, this is the promise of virtual reality simulators for surgeons.' Implicitly, the uncontested champion (and pilot) of the human body is to be trained surgeons.[47]

The same metaphor has been used by Catherine Waldby to critique the 'violence inherent in the practice of anatomy.' She has observed that the 'point of view is that of a pilot in a tiny space ship, looping and zooming through the enclosed spaces of the body's interior with a complete mobility.' The problem with virtual anatomy, Waldby proposes, is inherent in the communication of medical knowledge: 'anatomy is not a process of illustration but a way of spatialising flesh as communicable knowledge.' Further, 'writing of the body as intelligible, communicable terrain, and hence useful for medicine [is] to refer to how the fleshly body is worked over to produce intelligible and reproducible traces, anatomical drawings and diagrams ... the violence of anatomy is the violence of a particular kind of writing practice, a set of techniques that destroy the fleshly body along particular analytic lines in order to inscribe its trace in various knowledge media.' Not surprisingly, Waldby in a Foucauldian analysis of the Virtual Human Project's cultural import invokes the long history from the fifteenth to the nineteenth century 'in which medical knowledge was intimately bound up with penal and sovereign power,' and with material communications: 'the beginning of modern anatomical practice, conventionally dated from the anatomical texts produced by Vesalius in the sixteenth century,' Waldby writes, 'is inextricable from the coming of the mechanically printed book.' Vesalius's anatomical images here are used to support the notion that 'a similar sensibility and set of institutional power relationships seem to be at work here, a confluence of penal and scientific power that designates Jernigan's dead body as a legitimate object of spectacular and public violence.'[48] Waldby is not wrong of course, though the generalization ignores the sanitization inherent in these digital images: in twentieth-century North America there is no real public violence where dead bodies might be visible in the streets, no stench of incipient decay and ripening guts, no blood pooling and splashing in the public theatre of anatomical dissection, no heads hacked off before our eyes, no crunch of bone, or squishy invasion of the real decomposing bodies that were such powerful reminders of governing authority in earlier times. What is more of interest to me, however, is the polarity of these two views of the visual imaging that Satava deems 'progress.' Clearly, the new medium of computer-mediated communications represents a new contest for authority and government that conflates the imagery of body, communications, and social control.

To conclude, this study has questioned the stance we self-consciously have labelled postmodern in explicit distinction from the Enlightenment, the modern, and the early modern; it counters claims for the supposedly new 'cyborg identity' or 'technological consciousness' as an inevitable evolution from an early modern identity. The cyborg is not the oversimplified outcome of an Enlightenment ontology or a techno-iteration of a Cartesian dualism. The cyborg inscribes the contested ground of what it means to be human; it is the persistent question of who gets to pilot or govern the ship. Will the material knowledge of technology and science steer our morality, or will the more 'spiritual' knowledge of the humanities and the arts? Obviously, neither/and/both. The question, however, has been repeated for centuries, ever since humans began to see that creativity, knowledge, will, and understanding might be – probably were – the result of mechanical/material processes. There is no simple Enlightenment ontology except insofar as three hundred years have not made us any more confident about where our humanity resides in a material world. Pope's *Essay on Man* reminds us that to be human is to be uncertain of the very concept of humanity. His rhetoric of paradox and antithesis in the excerpt that begins this section exposes the impossibility of defining mankind in simple binary terms of mind/ matter, ignorance/reason, god/beast, truth/error. Rather than glorifying reason, Pope's poem problematizes the limits of human knowledge and reason: 'Go, wond'rous creature!' Pope exclaims, 'mount where Science guides, / Go, measure earth, weigh air, and state the tides,' noting that not even so gifted a scientist as Isaac Newton can know the workings of his own mind, or fully comprehend his own beginning and end (Epistle II, ll. 19–38). 'Trace Science then,' he cautions, 'with Modesty thy guide; / First strip off all her equipage of Pride' (Epistle II, ll. 43–4). That is to say, Enlightenment thinkers had sophisticated and contradictory reactions to the projects of science and the proposed purposes of human knowledge; Enlightenment thought – contrary to what so much of cyborg, postmodernist, or poststructuralist theory assumes – was characterized by diverse, intricate, complex, and persistently contested interpretations of the boundaries and binaries of being human that we are supposedly only now beginning to transgress.

Moreover, the ontological anxiety occasioned by our union with technology is heavily influenced by a textual tradition quite apart from medicine and science that suggests the Cartesian rational spirit or soul (*res cogito*) is separate from the rest of the natural world (*res extensa*), in opposition with a medical and scientific tradition that treats human consciousness and identity as material mechanism. The cyborg presents a problem of definition: when nature, machine, and human mind are defined by the same terms and essentially conceived as the same substance, we imagine that we ourselves have changed irrevocably and physically evolved to a posthuman form. The notions of the continuity of human mind with natural mechanical processes, however, were particularly obvious in materialist writings of the seventeenth and eighteenth centuries, which, as today, informed both optimistic and pessimistic characterizations of the human spirit.

Our current technology is undeniably a unique and powerful influence on our understanding of ourselves and our place in the world, but it seems unnecessarily limiting to exaggerate the novelty of the issues of identity and social values raised by the cyborg figure. While today's techno-culture and 'mechanical identities' can be understood by examining the unique aspects of contemporary conditions wrought by technological change, the cyborg is also a creature of history. If we are to interpret cyborgs politically and culturally, we need to resituate our studies to examine the conditions of their origin. It is of course tempting to suggest that the cyborg represents something new and radically different, but it is equally important to attempt to understand the extent to which our present constructs of identity are dependent upon, and indeed reconfigurations of, those of the past. There are two important strands to this history, which since the early modern period have converged and diverged, resisting, adopting, adapting, and reincorporating congruent images of humans and machines: the first is a medical-technological tradition of (metaphorical) mechanics; the second is a literary tradition of (mechanical) metaphors. Cyborg history is only tangentially a history of automata, only in part a history of Cartesian identities: what has been neglected is the history of mechanical physiology and the contemporaneous history of the literature of sensibility.

The living cyborg today is not a Cartesian automaton with a soul but more accurately a descendent of Willis's human engine, moved by the energy of atoms or particles, and steered by a circuit of communications; the metaphorical cyborg of fiction and film may be rather more 'Cartesian' in its dualism of mind and body, but it would be careless to limit its history to dualism alone. The most significant mechanistic assumption that delineates the man-machine, the sensible man of feeling, the monstrous female womb, and the cyborg is that in its most basic form, the matter of living beings and of inert materials is essentially the *same*: identity, consciousness, or soul, like the body, is a product of physical mechanism. Newtonian mechanism inspired a progression of interrelated conjectures in medical-philosophical texts about the material properties of organic matter and motion. That is, while many historians have imagined a point where mechanistic theories were replaced by animism or vitalism, it might be more appropriate to say in the context of cyborg history that the overriding limitation of early mechanistic philosophies – the problem of what animated organic matter – gradually came to be understood as mechanisms of electrical and chemical processes rather than hydraulic ones. We have never truly lost the image of the human body and identity as a mechanistic outcome of material organization. The man-machine became the sensible machine as soul became aether; the sensible machine became cyborg as aether became electrical charge.

The investigations into the physiology of consciousness, animate motion, and vitality produced the mechanical definitions for living bodies that we use today in science and medicine. They also instantiated nascent cyborg narratives of both optimism for future progress due to mechanization or technology, and resistance to the idea of human spirit reduced solely to physical mechanism. Through the seventeenth and eighteenth centuries, the evolving physiological trope of the 'governor' of the man-machine resulted in changing concepts of human individuality characterized specifically by communications not only from the top down, but also from the body members back up to the governor. The governor of the body thus became both material and autonomous: God and the immortal soul as the governor of the human machine – at least in physiological terms – would be replaced by mechanisms of communication. Simultaneously, as *homme machine* became *machine sensible,* the soul as imagination, creative genius, or sensitive individual spirit became the property of literature and poetry. The steersman remains a powerful metaphor in postmodern discourse: what is both reinforced and destabilized by contemporary communications systems is the language of government, identity, and communications that was composed for the man-machine in the seventeenth and eighteenth centuries. But if human identity was once imagined as 'stamped' onto the wax tablet of the mind or of the womb by traces of the imagination, the supposed instability and ethereality of our texts and our machines implies both a freedom from embodiment and a potential monstrosity wrought by mechanical reproduction of the human being. The predicted cultural effects of the steered organism, the cyborg, have been primarily metaphorical and political constructs based upon the early modern tropes for human spirit or mind extending through and beyond the body, to the body politic. Cyborg studies have been dominated by the anticipation of profound changes to human mind and society due to networked and unregulated communications systems – systems that are characterized by their very lack of governor or steersman. Despite the supposed shifts in human identity wrought by our evolution to the post-human cyborg, however, these versions of human mind/soul and society continue to use a metaphoric language anticipated and, indeed, prefigured by the early modern discourses of mechanist and materialist philosophy.

Notes

1 Introduction

1 D.S. Halacy, *Cyborg*, 7.
2 David Rorvik, *As Man Becomes Machine*, 17.
3 Donna Haraway, 'Manifesto,' 82.
4 David Tomas, 'Feedback and Cybernetics,' 21.
5 Donna Haraway, 'Manifesto,' 81.
6 Sherry Turkle, *Life on the Screen*, 268.
7 Sven Birkerts, 'The Fate of the Book,' in *Tolstoy's Dictaphone*, 190.
8 Bruce Mazlish, *The Fourth Discontinuity*, 6–7.
9 Claudia Springer, *Electronic Eros*, 33.
10 E.J. Dijksterhuis, *The Mechanization of the World Picture*.
11 Chris Hables Gray, Steven Mentor, and Heidi J. Figueroa-Sarriera, 'Cyborgology,' 13n2.
12 Fredric Jameson, 'Postmodernism and Consumer Society,' 125.
13 Jean Baudrillard, *Simulations*, 83, 142, 43.
14 Jean-François Lyotard, *The Postmodern Condition*, xxiii–xiv.
15 Donna Haraway, 'Manifesto,' 66.
16 Anne Balsamo, 'Reading Cyborgs Writing Feminism,' in Kirkup et al., 149.
17 Claudia Springer, 'The Pleasure of the Interface,' 306.
18 Claudia Springer, *Electronic Eros*, 19.
19 Sherry Turkle, *Life on the Screen*, 17.
20 Joseba Gabilondo, 'Postcolonial Cyborgs,' 424.
21 Chela Sandoval, 'New Sciences: Cyborg Feminism and the Methodology of the Oppressed,' 408.
22 Robbie Davis-Floyd and Joseph Dumit, eds., *Cyborg Babies*, 14.

23 Bruce Grenville, ed., *The Uncanny*, 9–10.
24 Chris Hables Gray, *Cyborg Citizen*, 13–14.
25 Steve Mann, 'The Post-Cyborg Path to Deconism.'
26 Steve Mann, James Fung, Mark Federman, and Gianluca Baccanico, 'Panopdecon.'
27 Donna Haraway, 'Manifesto,' 67.
28 Jean Baudrillard, 'The Ecstasy of Communication,' in *The Anti-Aesthetic*, 130.
29 Claudia Springer, 'Pleasure of the Interface,' 306.
30 Sadie Plant, 'Coming across the Future.'
31 Allison Fraiberg, 'Of Aids, Cyborgs, and Other Indiscretions,' n.p.
32 Dale Spender, *Nattering on the Net*, 246.
33 Michael Heim, *Metaphysics*, 101, 103.
34 Sven Birkerts, *The Gutenberg Elegies*, 180–1.
35 Arthur Kroker and Michael A. Weinstein, *Data Trash*, 2.
36 Vivian Sobchack, 'The Scene of the Screen,' 104–6.
37 Jay David Bolter, *Writing Space*, 221.
38 Jay David Bolter and Richard Grusin, *Remediation*, 254.
39 David Brande, 'The Business of Cyberpunk,' 510, 513.
40 William Macauley and Angel Gordo-López, 'From Cognitive Psychologies to Mythologies,' 437.
41 Michael Heim, *Metaphysics*, 101.
42 Catherine Waldby, *The Visible Human Project*, 153.
43 Claudia Springer, *Electronic Eros*, 19.
44 David Cox, 'Sight Unseen / Seeing, Mapping, Communicating, n.p.'
45 David Brande, 'The Business of Cyberpunk,' 512.
46 Sandy Stone, 'Split Subjects, Not Atoms,' 399, 398.
47 Chris Hables Gray, Steven Mentor, and Heidi J. Figueroa-Sarriera, 'Cyborgology,' 7.
48 Sadie Plant, 'Coming across the Future,' 31, 34.
49 Donna Haraway, *Modest_Witness*, 51.
50 As several commentators have noted, literary and visual tropes of the cyborg or human in cyberspace replicate and reinforce centuries-old narratives and ideologies about transcendence, patriarchy, gender, and humanity. See, e.g., Jennifer González, 'Envisioning Cyborg Bodies'; Rosi Braidotti, 'Cyberfeminism'; Margaret Wertheim, *The Pearly Gates of Cyberspace*.
51 Jonathan Sawday, 'Forms,' 173, 174.
52 Gill Kirkup, 'Introduction to Part One,' in *The Gendered Cyborg*, 8.
53 Donna Haraway, *Modest_Witness*, 23–39.

54 Jonathan Sawday, 'Forms,' 184.
55 Donna Haraway, *Modest_Witness*, 33.
56 Donna Haraway, 'Manifesto,' 65–6, 68, 69, 70.
57 Donna Haraway, *Primate Visions*, 139.
58 Donna Haraway, 'Foreword' to *The Cyborg Handbook*, xvi, xiv.
59 Donna Haraway, *Modest_Witness*, 51.
60 Julien Offray de la Mettrie, *Machine Man and Other Writings*, 33–4, 35, 8.
61 See, e.g., Gray et al., *The Cyborg Handbook*, 521.
62 Chris Hables Gray and Steven Mentor, 'The Cyborg Body Politic: Version 1.2,' 459.
63 Robert Hooke, *Micrographia*, n.p.
64 Thomas Willis, *The Anatomy of the Brain and Nerves*, 121.
65 Robert Markley, 'Boundaries: Mathematics, Alienation, and the Metaphysics of Cyberspace,' 56.

2 Matter, Mechanism, and the Soul

1 Richard Bentley, *The Folly and Unreasonableness of Atheism*, 'Epistle Dedicatory,' n.p. 13, 14–15, 22, 28.
2 For example, Helva in Anne McCaffrey's *The Ship Who Sang* is a human brain implanted in a titanium intergalactic scout ship; HAL in Stanley Kubrick and Arthur C. Clarke's *2001: A Space Odyssey* is an artificial in-telligence who rebels against the astronauts piloting the spaceship that he governs; the film *Virus* (1999) starring Jamie Lee Curtis and Donald Sutherland depicts a ship infected by an intelligent electrical life form that controls the ship and subsumes the captain, turned evil cyborg in exchange for wealth and power; Gene Roddenberry's television series *Andromeda* portrays a more benign sentient starship, Andromeda Ascendant, whose artificial intelligence system is represented by voice, hologram, video display, or avatar. Less dramatically, the artificial intelligence systems known as autopilots have been in use for decades to control the steering system of ships, aircraft, and spacecraft.
3 Robert Burton, *The Anatomy of Melancholy*, 148.
4 Galen's physiology was dominant in European medicine but there was also a tradition in Greek philosophy where the tissues and the flesh were the seat and agent of thought or the soul. See Friedrich Solmsen, 'Tissues and the Soul.'
5 For a summary of changing concepts of the centuries-old hollow nerve theory during the Enlightenment, see Edwin Clarke, 'The Doctrine of the

258 Notes to pages 38–52

Hollow Nerve in the Seventeenth and Eighteenth Centuries'; see also Eric
T. Carlson and Meribeth M. Simpson, 'Models of the Nervous System in
Eighteenth Century Psychiatry.'

6 John Harris, *Lexicon Technicum*, vol. I, 'Preface,' n.p.

7 Robert Boyle, 'Some Considerations Touching the Usefulness of Experimental Natural Philosophy, Essay 5,' in *The Works*, 537.

8 Thomas Willis, *Two Discourses*, 18.

9 René Descartes, *The Philosophical Writings*, 106.

10 Bentley, *The Folly and Unreasonableness of Atheism*, 46, 45.

11 John Locke, *Essay Concerning Human Understanding*, 6.

12 Andrew Baxter, *An Enquiry into the Nature of the Human Soul*, 1–2.

13 La Mettrie, *Machine Man*, 15, 26.

14 Allison Muri, 'Of Shit and the Soul.'

15 Manfred E. Clynes and Nathan S. Kline, 'Cyborgs and Space,' reprinted in
The Cyborg Handbook, 31, 32, 33.

16 Marshall McLuhan, *Understanding Media*, 4, 57, 80.

17 See, e.g., AScribe Newswire.

18 U.S. Congress, Office of Technology Assessment, 'The Human Genome
Project and Patenting DNA Sequences.'

19 See, e.g., Thomas A. Bass, 'Gene Genie.'

20 Walter Charleton, *Natural History of Nutrition, Life, and Voluntary Motion*, 4,
183–7.

21 John Maubray, *The Female Physician*, 1724), 25, 26–7.

22 Robert E. Schofield, *Mechanism and Materialism*.

23 John W. Yolton, *Thinking Matter*, 92.

24 Marie Boas, 'The Establishment of the Mechanical Philosophy,' 520.

25 GUL, Cullen MSS, No. 5, cited by Arthur Donovan, 'Pneumatic Chemistry,'
220.

26 RCPE MSS, C. 15, Vol. I, fol. 130^v, cited by Donovan, ibid.

27 Betty Jo Teeter Dobbs, *The Janus Faces of Genius*, 8, 12.

28 John Hunter, *A Treatise on the Blood, Inflammation, and Gun-shot Wounds*, 78,
86, 136, 89, 90.

29 See the *Oxford English Dictionary* online, 2nd ed. (1989). http://dictionary
.oed.com.

30 Robert Boyle, 'The Excellency of Theology Compared with Natural Philosophy,' in *The Works*, vol. 3, 451, 162; 'The Christian Virtuoso,' in *The Works*,
vol. 5, 41.

31 Burndy MS 16. For this discussion of Newton's concept of vegetation as
opposed to mechanism I am indebted to Betty Jo Teeter Dobbs, 'Newton's
Alchemy and His Theory of Matter.'

32 Cited by Dobbs, ibid., 517, 526.

33 Harris, 'Preface' to *Lexicon Technicum,* n.p.

34 Robert Boyle, 'A Free Enquiry into the Vulgarly Receiv'd Notion of Nature,' in *The Works,* vol. 4, 358.

35 Cited in René Descartes, *Treatise of Man,* 4.

36 René Descartes, *Treatise on Man,* in *The Philosophical Writings,* vol. 1, 100–1, 101.

37 René Descartes, *Principles of Philosophy,* ibid., 288.

38 Robert Boyle, 'The Excellency of Theology Compared with Natural Philosophy,' in *The Works,* vol. 3, 437.

39 Quoted by Sergio Moravia, 'From *Homme Machine* to *Homme Sensible:* Changing Eighteenth-Century Models of Man's Image,' 48.

40 Ibid., 46.

41 Robert Boyle, 'Some Considerations Touching the Usefulness of Experimental Natural Philosophy,' in *The Works,* vol. 1, 446.

42 Robert Boyle, 'A Free Inquiry Into the Vulgarly Received Notion of Nature,' in *The Works,* vol. 4, 363, 420, 408.

43 Ibid., 409, 412.

44 Ibid., 362.

45 Thomas Willis, 'Preface' in *Two Discourses,* 24.

46 Thomas Willis, *Dr Willis's Practice of Physick,* 1–6, 13.

47 Isaac Newton, *Opticks,* 345, 328, 350.

48 La Mettrie, *Machine Man,* 39, 33, 34.

49 Isaac Newton, *Opticks,* 326, 328.

50 Isaac Newton, *Sir Isaac Newton's Mathematical Principles of Natural Philosophy,* 547.

51 John Locke, *Essay Concerning Human Understanding,* 2.21.73.

52 Richard Bentley, 'Matter and Motion Cannot Think,' in *The Folly,* 25–6.

53 Richard Bentley, 'A Confutation of Atheism from the Structure and Origin of Origin of Humane Bodies,' in ibid, 6.

54 Isaac Newton, *Papers and Letters,* 302–3.

55 Joseph Addison, *The Spectator* (18 July, 1711), 427.

56 Samuel Clarke, *The Leibniz–Clarke Correspondence,* 14, 117.

57 Cotton Mather, *The Christian Philosopher,* 222, 87.

58 Tobias Smollett, *The History and Adventures of an Atom,* 3, 36, 162, 207–8, 209–10.

59 Sergio Moravia, 'From *Homme machine* to *Homme sensible,*' 55, 58.

60 Theodore M. Brown, 'From Mechanism to Vitalism in Eighteenth-Century Physiology,' 179, 184. Brown cites Robert Schofield's *Mechanism and Materialism,* 209.

61 See Christopher Lawrence, 'The Nervous System and Society in the Scottish Enlightenment,' 19–40.

62 Oliver Goldsmith, *An History of the Earth, and Animated Nature*, vol. 8, 170.

63 Thomas Beddoes, *A Lecture Introductory to a Course of Popular Instruction on the Constitution and Management of the Human Body*, 28.

64 Richard Lovett, *Philosophical Essays*, 208, 290.

65 George Rousseau, 'The Perpetual Crises of Modernism and the Traditions of Enlightenment Vitalism,' 31.

66 Jessica Riskin, *Science in the Age of Sensibility*, 69–71.

67 Peter Hanns Reill, 'The Legacy of the "Scientific Revolution,"' 23–43, 43.

68 See, e.g., Arthur R. Salomon, David W. Voehringer, Leonard A. Herzenberg, and Chaitan Khosla, 'Understanding and Exploiting the Mechanistic Basis for Selectivity of Polyketide Inhibitors of F_oF_1-ATPase.' This is only one arbitrarily chosen example among innumerable articles describing molecular 'mechanisms of action' applicable to the cellular mechanisms of living organisms.

69 See, e.g., Paola Bertucci and Giuliano Pancaldi, eds., *Electric Bodies*; and also Jessica Riskin, *Science in the Age of Sensibility*.

70 William Watson, 'A Sequel to the Experiments and Observations Tending to Illustrate the Nature and Properties of Electricity,' 745–6.

71 Browne Langrish 'The Crounean Lectures on Muscular Motion,' 31, 32.

72 Aram Vartanian, *La Mettrie's* L'homme machine, 119–20.

73 Elizabeth Haigh, 'Vitalism, the Soul, and Sensibility,' 40.

74 James Makittrick Adair, *A Philosophical and Medical Sketch of the Natural History of the Human Body and Mind*, 15–16.

75 Richard Lovett, *Philosophical Essays*, 240, 244–5, vii–viii.

76 Henry Cavendish, 'An Attempt to Explain Some of the Principal Phænomena of Electricity,' 585–6.

77 William Emerson, *The Principles of Mechanics*, iii–iv, iii, 192–8.

78 W. Bowe, 'Some Account of the Life of Mr. William Emerson,' xvii.

79 William Emerson, *Principles of Mechanics*, 204–5.

80 Adam Walker, *A System of Familiar Philosophy*, 1–2, 32, 322.

81 Ibid., 365, 367, 391.

82 James Parsons, 'The Crounian Lectures on Muscular Motion,' iv–v.

83 Anthony Carlisle, 'The Croonian Lecture: On Muscular Motion,' 1–2, 12.

84 A.P.W. Philip, 'On the Relation which Subsists between the Nervous and Muscular Systems,' 67–8, 68.

85 Daniel C. Dennett, 'The Practical Requirements for Making a Conscious Robot,' 133–5.

86 Ibid., 135–6.

87 See 'Cog Project Publications.'

88 Richard Bentley, 'Matter and Motion Cannot Think,' in *The Folly*, 20.

3 Some Contexts for Human Machines and the Body Politic

1 Norbert Wiener, *Cybernetics*, 15–16, 19.
2 Stelarc, 'From Psycho-Body to Cyber-Systems, 120. Stelarc's event for *Amplified Body/Laser Eyes and Third Hand*, performed in Houston and Tokyo in 1986, and Toronto and Roskilde in 1987, attempts to display the individual performer in an electronic circuitry extending from the body, whereby a machinery of motion, sound, and light are involuntarily or voluntarily stimulated by Stelarc's body: electrical discharges of his body, its blood flow, its movement and position, all tracked by various sensors and transducers, are converted to a mechanical 'choreography of controlled, constrained, and involuntary motions' (ibid., 118).
3 N. Katherine Hayles, 'Virtual Bodies and Flickering Signifiers,' 72.
4 N. Katherine Hayles, 'Embodied Virtuality or How to Put Bodies Back into the Picture,' 1.
5 Ross Farnell, 'In Dialogue with "Posthuman" Bodies: Interview with Stelarc,' 119–20.
6 Marshall McLuhan, cited in George Sanderson and Frank MacDonald, eds., *Marshall McLuhan*, 12.
7 William R. Clark and Michael Grunstein, *Are We Hardwired?* 138.
8 Steven Pinker, *How the Mind Works*, 24–5.
9 Norbert Wiener, 'Cybernetics,' 15.
10 Dean E. Wooldridge, *The Machinery of the Brain.*
11 Victor Weisskopf, *Knowledge and Wonder*, 220, 264.
12 Karl Pearson, *The Grammar of Science*, 44.
13 Cited in Laura Otis, *Networking*, 11, 24.
14 La Mettrie, *Machine Man*, 15.
15 John Locke, *Essay Concerning Human Understanding*, 6, 104, 81.
16 Immanuel Kant, *Anthropology from a Pragmatic Point of View*, 58.
17 See Cicero, *De oratore*, II, lxxxvi, 351–4, cited in Frances Yates, *Selected Works*, vol. 3, 2.
18 Manfred E. Clynes and Nathan S. Kline, 'Cyborgs and Space,' 30–1.
19 Marshall McLuhan to John I. Snyder, Jr., August 4, 1963, in Marshall McLuhan et al., *Letters of Marshall McLuhan*, 291.
20 Marshall McLuhan, *The Gutenberg Galaxy*, 30.
21 Chris Crittenden, 'Self-Deselection: Technopsychotic Annihilation via Cyborg,' 127.
22 Marshall McLuhan, *Gutenberg Galaxy*, 30.
23 Jean Baudrillard, 'The Ecstasy of Communication,' 126–34, 128, 129, 130.
24 Arthur Kroker and Michael A. Weinstein, *Data Trash*, 6, 103, 109.

25 Claudia Springer, *Electronic Eros*, 59, 79.

26 Vivian Sobchack, 'The Scene of the Screen,' 104.

27 Jean Baudrillard, *The Transparency of Evil*, 63, 60, 68, 126, 135, 137, 138.

28 Mark Poster, 'Cyberdemocracy,' 205, 203, 213–14, 214.

29 Recent examples include Chris Hables Gray, *Cyborg Citizen*, and Peter Ludlow, 'New Foundations: On the Emergence of Sovereign Cyberstates and Their Governance Structures.'

30 Thomas Willis, *The Anatomy of the Brain and Nerves*, 114.

31 See, e.g., Alexander Monro, *Experiments on the Nervous System*, 43.

32 For this summary of the body politic I am indebted to David G. Hale, 'Analogy of the Body Politic,' in *Dictionary of the History of Ideas*, 67–70.

33 Plato, *The Republic* Book IV 444b, 686, and 444d, 687.

34 Plato, *Timaeus* 44d, 1173, 42d–e, 1171.

35 Hale, 'Analogy of the Body Politic,' 67–70.

36 William Harvey, *De Motu Locali Animalium*, 97, 111, 149, 147, 111.

37 William Harvey, *The Circulation of the Blood and Other Writings*, 3.

38 Christopher Hill, 'William Harvey (No Parliamentarian, No Heretic) and the Idea of Monarchy,' 98.

39 Christopher Hill, 'William Harvey and the Idea of Monarchy,' 56.

40 Gweneth Whitteridge, 'Introduction' to William Harvey, *Disputations Touching the Generation of Animals*.

41 William Harvey, *Anatomical Exercitations, Concerning the Generation of Living Creatures* (London: printed by James Young, for Octavian Pulleyn, 1653), 459, cited in Hill, 'William Harvey and the Idea of Monarchy,' 56.

42 William Harvey, *The Circulation of the Blood*, 45.

43 William Harvey, *Disputations*, 8.

44 The images of the heart as king or royal fortress of the body are clearly derived from Aristotle; see e.g., J.G. Curtis, *Harvey's Views on the Use of the Circulation of the Blood*, 152.

45 Bodl. Oxf. MS Carte 45, fol. 212.

46 Thomas Willis, *Two Discourses*, n.p.

47 René Descartes, *The Philosophical Writings*, vol. 2, 56. This translation is from the second edition (in Latin) published in 1642 with a number of corrections to the text.

48 Ibid., vol. 1, 100–1.

49 Ibid., vol. 2, 58, 61.

50 Ibid., vol. 1, 345–6.

51 Ibid., vol. 1, 108.

52 Thomas Willis, *Two Discourses*, 23, 24,

53 Thomas Willis, *The Anatomy of the Brain and Nerves*, 115–16, 126–7.

54 Translated by Samuel Pordage as *The Anatomy of the Brain and Nerves* in 1681

and included in the volume *The Remaining Medical Works of … Dr Thomas Willis*. The edition used here is the facsimile of the 1681 translation.

55 Translated by Samuel Pordage in 1683 as *Two Discourses Concerning the Soul of Brutes, Which Is that of the Vital and Sensitive of Man*. The edition used here is a facsimile of the 1683 translation.

56 Thomas Willis, *The Anatomy of the Brain and Nerves*, 91, 111, 91.

57 Otto Mayr, *Authority, Liberty, and Automatic Machinery in Early Modern Europe*, 164, 165.

58 Thomas Willis, *The Anatomy of the Brain and Nerves*, 114–15.

59 Otto Mayr, *Authority, Liberty, and Automatic Machinery*, 190–1; see also Chapter 6 of Mayr's *The Origins of Feedback Control*.

60 Norbert Wiener, 'Cybernetics,' 15.

61 Thomas Willis, *Two Discourses*, 5, 22, 24.

4 The Man-Machine: Communications, Circulations, and Commerce

1 Steven Pinker, *How the Mind Works*, 4.

2 Thomas Willis, *The Anatomy of the Brain and Nerves*.

3 Ibid., n.p. 73, 126.

4 Ibid., 96–7, 98, 127–8, 106:

5 Thomas Willis, *Two Discourses*, Preface, n.p.

6 Ibid., 95, 22.

7 Thomas Willis, *The Anatomy of the Brain and Nerves*, 110, 91, 110, 90, 93, 110, 101.

8 Ibid., 114, 119, 120, 121, 111.

9 Ibid., 111, 128, 126.

10 Ibid., 93, 128.

11 Ibid., 129.

12 Ibid., 130.

13 Thomas Hobbes, *Leviathan*, 9.

14 Ibid., xvii–xviii.

15 Ibid., 167, 155, 165.

16 Ibid., 120, 124–5.

17 Thomas Willis, *The Anatomy of the Brain and Nerves*, 136.

18 Ibid., 130.

19 Elisha Coles's *English Dictionary* (London: printed for Peter Parker, 1677) defines *metathesis* in a literary sense, as 'a transposition or change of letters.' Indeed as far as I am able to determine, Willis's book is the only medical text in Early English Books Online (which claims to contain 'virtually every work printed in England, Ireland, Scotland, Wales and British North America and works in English printed elsewhere from 1473–

1700') that uses the term in a supposedly chemical sense. The year 1873 is the earliest mentioned in the *OED* where *metathesis* is used in this sense, as in the interchange of an atom or group of atoms between two different molecules. *Metathesis* was rarely used for *metastasis*, referring to the movement of pain or disease from one part of the body to another, but this is clearly not the sense in which it is used here. Willis also defined *metastasis* separately.

20 Thomas Willis, *The Anatomy of the Brain and Nerves*, 130.
21 See Robert G. Frank, 'Thomas Willis and His Circle.'
22 Thomas Willis, *Two Discourses*, n.p.
23 Ibid., Preface, n.p.
24 Ibid., 5.
25 For biographical information see Aram Vartanian, *La Mettrie's* l'Homme machine: *A Study in the Origins of an Idea*, and Ann Thomson, *Materialism and Society*.
26 La Mettrie, *Machine Man*, 33.
27 Thomas Willis, *Two Discourses*, 22, 38, 22.
28 Ibid., 24.
29 Ibid., 32, 23–4, 33.
30 Ibid., 56.
31 Ibid., 27, 32.
32 Ibid., 33, 34.
33 Ibid., 41, 38. Preface, n.p., 52, Preface, n.p., 42.
34 Ibid., 123.
35 R.J. Tarrant, 'Aspects of Virgil's Reception in Antiquity,' 65–6.
36 Tim Harris, *London Crowds in the Reign of Charles II*, 79–82.
37 Thomas Willis, *Two Discourses*, 42, 43.
38 Sheila Lambert, *Printing for Parliament*, i.
39 On politics and authority in literature during this period, see Steven N. Zwicker, *Lines of Authority*. See also David Norbrook, *Writing and the English Republic*; on periodical publishing, see Carolyn Nelson and Matthew Seccombe, *Periodical Publications, 1641–1700*; on pamphlet publishing, see Joad Raymond, *Pamphlets and Pamphleteering in Early Modern Britain*; on the book trade, see John Barnard, 'London Publishing, 1640–1660.'
40 Richard Atkyns, *The Original and Growth of Printing*, 7, 5–7.
41 Roy Porter, 'Print Culture,' Chapter 4 in *The Creation of the Modern World*, 72.
42 Important treatments of print culture and the marketplace of books in the eighteenth century include: Raymond Stephanson, 'The Sexual Traffic in Male Creativity,' Chapter 3 in *The Yard of Wit*; Thomas Keymer, *Sterne, the Moderns, and the Novel*; Linda Zionkowski, *Men's Work*, 'Gray, the Market-

place, and the Masculine Poet,' 'Aesthetics, Copyright, and "the Goods of the Mind,"' and 'Territorial Disputes in the Republic of Letters'; William Warner, *Licensing Entertainment*, and 'Licensing Pleasure'; J.S. Peters, 'The Bank, The Press, and the "Return of Nature"'; Martha Woodmansee, 'The Genius and the Copyright'; and finally, Pat Rogers, *Grub Street*, remains an authority on that chaotic literary landscape.

43 See Martin C. Battestin, 'The Critique of Freethinking from Swift to Sterne.'

44 William Warner, *Licensing Entertainment*, 181.

45 La Mettrie, *Machine Man*, 8, 13, 14–15, 26, 27.

46 Jonathan Swift, *A Tale of a Tub*, 125, 48, 49, 80, 129.

47 Alexander Pope, *Peri Bathous*, in *Alexander Pope*, 233–6.

48 Linda Zionkowski, 'Territorial Disputes,' 4, 5, 8.

49 Alexander Pope, *The Spectator* 408 (18 June 1712), in *The Spectator: A New Edition*, 704, 705, 706, 707.

50 Joseph Addison, *The Spectator* 409 (19 June 1712), ibid., 707, 710.

51 Jonathan Swift, *Gulliver's Travels*, 175.

52 John Arbuthnot and Alexander Pope, *Memoirs of the Extraordinary Life, Works, and Discoveries of Martinus Scriblerus*, 138, 138–9, 139–40, 138–41.

53 See also Raymond Stephanson, 'Richardson's "Nerves"'; John Mullan, *Sentiment and Sociability*; Robert Erickson, 'Written in the Heart'; Carol Houlihan Flynn, 'Running Out of Matter'; G.J. Barker-Benfield, *The Culture of Sensibility*; Ann Jessie Van Sant, *Eighteenth-Century Sensibility and the Novel*; Anne Vila, *Enlightenment and Pathology*; and Anita Guerrini, *Obesity and Depression in the Enlightenment*.

54 George Cheyne, *The English Malady*, 3–4, 34, 20, 14–15, 36, 14.

55 S.A. Tissot, *An Essay on Diseases*, 16–17, 27.

56 Mark Akenside, *The Pleasures of the Imagination*, ll. 96–101.

57 Laurence Sterne, *A Sentimental Journey*, 4, 20, 21.

58 Martin C. Battestin, 'The Critique of Freethinking from Swift to Sterne,' 406.

59 Laurence Sterne, *The Life and Opinions of Tristram Shandy, Gentleman*, 271, 55–6, 365, 209, 144–5.

60 Chris Hables Gray and Mark Driscoll, 'What's Real about Virtual Reality?' 39.

61 Jay David Bolter and Richard Grusin, *Remediation*, 231, 248.

62 Marshall McLuhan, *Understanding Media*, 80.

63 For examples of the dire proclamations for humanity's downfall due to computer-mediated communications, see works such as Boris Castel, ed., *The End of the Book?* special issue of *Queen's Quarterly*; Sven Birkerts, *The Gutenberg Elegies*; or Arthur Kroker and Michael Weinstein, 'Global Algorithm 1.4.'

64 Julien Offray de La Mettrie, *Histoire naturelle de l'âme* 151, 250 (my translation).

65 La Mettrie, *Machine Man*, 39.

66 Denis Diderot, *D'Alembert's Dream*, in *Rameau's Nephew and D'Alembert's Dream*, 184–5, 157.

67 See Sergio Moravia, 'From *Homme Machine* to *Homme Sensible*,' 45–60.

68 Richard Lovett, *Philosophical Essays*, 36, 240–5.

69 Erasmus Darwin, *Zoonomia*, 109, 1, 5, 10, 11.

70 Ibid., 109.

71 William Belcher, who was confined for seventeen years in a British madhouse, described himself as a 'Victim to the Trade of Lunacy' in *Belcher's Address to Humanity*, 10.

72 William Belcher, M.D., *Intellectual Electricity*, v–vi.

73 Ibid., 35–6, 41.

74 Ibid., n.p.

5 The Woman-Machine: Techno-Lust and Techno-reproduction

1 Anne McCaffrey, *The Ship Who Sang*, 7, 93.

2 Claudia Springer, *Electronic Eros*, 114.

3 For examinations of the mother-ship as monstrous techno-mother, see Barbara Creed, '*Alien* and the Monstrous-Feminine'; Mary Ann Doane, 'Technophilia'; and Kelly Hurley, 'Reading Like an Alien.'

4 Laura Briggs and Jodi I. Kelber-Kaye, 'There Is No Unauthorized Breeding in Jurassic Park,' 92.

5 Mary Ann Doane, 'Technophilia,' 120.

6 Allison Muri, 'Of Shit and the Soul,' 86–7.

7 Bruce Sterling, *Schismatrix Plus*, 34, 205.

8 Adele Clarke, 'Modernity, Postmodernity and Reproductive Processes,' 140.

9 Ibid., citing Rosalind Pollack Petchesky et al., 147.

10 Janice Raymond, *Women as Wombs*, xii.

11 Monica J. Casper, 'Fetal Cyborgs and Technomoms on the Reproductive Frontier,' 195–6.

12 Rosalind Pollack Petchesky, 'Foetal Images,' 63, 64.

13 Susan Merrill Squier, *Babies in Bottles*, 95, 96.

14 Susan Merrill Squier, 'Negotiating Boundaries,' 102.

15 Donna Haraway, *Modest_Witness*, 186.

16 Rayna Rapp, 'Real-Time Fetus / The Role of the Sonogram in the Age of Monitored Reproduction,' 47.

17 Rosalind Pollack Petchesky, 'Foetal Images,' 63.
18 Donna Haraway, *Modest_Witness*, 179, 178, 180.
19 Robbie Davis-Floyd and Joseph Dumit, *Cyborg Babies*, 3–4.
20 Thomas Laqueur, 'Orgasm, Generation, and the Politics of Reproductive Biology,' 35.
21 John Aitken, *Principles of Midwifery*, 61.
22 James Grantham Turner, 'Lovelace and the Paradoxes of Libertinism,' 73, 71.
23 Raymond Stephanson, 'Richardson's "Nerves,"' 268.
24 Samuel Richardson, *Clarissa*, vol. 4, 226.
25 Ibid., vol. 5, 223. I am grateful to Gordon Fulton (University of Victoria) for reminding me of these instances of the man-machine in *Clarissa*.
26 George Cheyne, *The Natural Method*, 281.
27 Walter Charleton, *Enquiries into Human Nature*, vol. VI. Preface, n.p.
28 William Emerson, *The Principles of Mechanics*, 205–6.
29 Oliver Goldsmith, *A Survey of Experimental Philosophy*, vol. 1, 252–3.
30 Samuel Richardson, *Clarissa*, vol. 2, 188. While 'passive' might be gendered as a female trait in the context of the argument I present here, the term *passive machine* is also used by Belton to describe his submission to his doctor's ministrations (ibid., vol. 7, 187).
31 La Mettrie, *Machine Man*, 8.
32 See, e.g., Raymond Stephanson, 'Richardson's "Nerves."'
33 John Dunton, *Bumography*, i.
34 Ibid., iii.
35 Ibid., vii.
36 Ibid. x.
37 Ibid., xi–xii.
38 Jonathan Swift, *Miscellanies*, 139.
39 *Poems on Several Occasions*, 88–9.
40 Frances Brooke, *History of Lady Julia Mandeville*, vol. 2, 111.
41 William Combe, *Devil Upon Two Sticks*, vol. 2, 88–90.
42 *Louisa Mathews*, vol. 3, 95–7.
43 Ludmilla Jordanova, 'Naturalising the Family,' in *Nature Displayed*, 167.
44 Joseph Addison, *The Spectator* 15 (17 March 1711), vol. II, 62, 62–3, 63.
45 Samuel Richardson, *Pamela*, vol. 3, 41–2.
46 Jane Brereton, *Poems on Several Occasions*, 275–6.
47 Anonymous, 'The School of Venus,' in *When Flesh Becomes Word*, Bradford K. Mudge, ed., 46–7.
48 John Cleland, *Memoirs of a Woman of Pleasure*, vol. 1, 67–8.
49 Physician of Bath, *A Safe-Conduct*, vol. 1, 40, 41, 42–3.

50 Ibid. 28–9.
51 Ibid., 29, 30.
52 Catharine Macaulay, *A Remarkable Moving Letter!* 8–9.
53 Two other images of the sexualised woman-machine amalgam are James
     Gillray's 'Nature display'd, shewing the Effect of the Change of Seasons on
     the Ladies Garden,' which depicts a woman with a smokestack in place of
     her head, described in Ludmilla Jordanova, *Nature Displayed*; and
     'L'Horlogère,' a woman's head combined with a clock as curvaceous body,
     described in Jennifer González, 'Envisioning Cyborg Bodies.'
54 John Cleland, *Memoirs of a Woman of Pleasure*, vol. 2, 128–9.
55 Ibid., vol. 1, 104, 166.
56 Ibid., vol. 2, 69, 123–4, 205.
57 Ibid., vol. 1, 118, 188–9, 190–1.
58 Ibid., vol. 2, 191, 190, 192, 194.
59 Ibid., 195, 183, 251.
60 See Peter Sabor's 'Explanatory Notes' in his *Memoirs of a Woman of Pleasure*,
     189–204.
61 John Cleland, *Memoirs of a Woman of Pleasure*, vol. 2, 251.
62 Robert Boyle, 'A Free Enquiry into the Vulgarly Receiv'd Notion of
     Nature,' in *The Works*, vol. 4, 373.
63 *Pleasure for a Minute: or, the Amorous Adventure*, 16.
64 George Wallis, *The Art of Preventing Diseases*, 280.
65 Peter Chamberlen, *Dr Chamberlain's Midwifes Practice*, 23, 31.
66 Charles Goodall, *The Royal College of Physicians of London*, 463–4, 465, n.p.
67 Peter Chamberlen, *Dr Chamberlain's Midwifes Practice*, n.p.
68 William Smellie, *A Treatise on the Theory and Practice of Midwifery* (London:
     Wilson, 1752), 115, quoted in Angus McLaren, *Reproductive Rituals*, 25.
69 Peter Chamberlen, *Dr Chamberlain's Midwifes Practice*, 32.
70 James Keill, *The Anatomy of the Humane Body*, 95.
71 James Handley, *Mechanical Essays*, 36.
72 Nicolas Venette, *Conjugal Love*, 34–6 (my emphasis).
73 Aristotle, 'Generation of Animals,' Book II, *The Complete Works*, 1148–9, 1149,
     1150.
74 William Harvey, *Disputations*, 443, 443–4, 445, 447.
75 Thomas Willis, *The Anatomy of the Brain and Nerves*, 77.
76 Willis, 'Of the Affections in Particular,' in *Two Discourses*, 52, 55.
77 Aristotle, pseud., *Aristotle's Masterpiece*, 12.
78 H.D. Jocelyn and B.P. Setchell, trans., *Regnier de Graaf*, 9.
79 Ibid., 79, 81.

80 Jane Sharp, *The Midwives Book*, 32, 63, 75.
81 Nicholas Venette, *Conjugal Love*, 80.
82 La Mettrie, *Machine Man*, 16, 29.
83 Raymond Stephanson has examined the extensive sexual tropes of male creativity and the brain-womb, in his *The Yard of Wit: Male Creativity and Sexuality 1650–1750*, to demonstrate the complex associations of conception and creativity as male 'labour' in the expanding marketplace of letters in eighteenth-century Britain.
84 Robert Erickson, '"The Books of Generation,"' 74–8.
85 Daniel Turner, *The Force of the Mother's Imagination*, 6.
86 John Maubray, *The Female Physician*, iv, xi, 178, 179, 28.
87 Dennis Todd, *Imagining Monsters*, 2–4, 21, 36. For more on Toft's hoax, see also Glennda Leslie, 'Cheat and Impostor.'
88 Lemuel Gulliver, pseud, *The Anatomist Dissected*, 5–6.
89 James Blondel, *The Strength of Imagination in Pregnant Women Examin'd*, 47.
90 Walter Charleton, *Natural History of Nutrition, Life, and Voluntary Motion*, 143–4, 144.
91 'An Account of What Hath Been of Late Observed by Dr Kerkringius Concerning Eggs to be Found in All Sorts of Females, by Dr Kerkringius,' *Philosophical Transactions* 7 (1672): 4022.
92 Daniel De Superville, 'Some Reflections on Generation, and on Monsters ...,' trans. Phil. Hen. Zollman, *Philosophical Transactions* 41 (1739–41): 301.
93 George Cheyne, *The Natural Method*, 278.
94 John Stedman, *Physiological Essays*, 90, 91.
95 James Graham, *Dr Graham's famous work!* 15, 15–16.
96 Ibid., 16–17, 16.
97 Ibid., 51–2, 71–2.
98 Henry Bracken, *The Midwife's Companion*, 44, 75–6, 129.
99 William Smellie, *A Treatise on the Theory and Practice of Midwifery*, 252.
100 John Burton, *A Letter to William Smellie*, 117, 147, 192.
101 Jean Astruc, *Elements of Midwifry*, xlviii.
102 Thomas Cooper, *A Compendium of Midwifery*, 3, 6–8.
103 John Leake, *A Course of Lectures*, n.p.
104 New Westminster Lying-in Hospital Minutes, London Metropolitan Archives. For more on Leake's hospital see Philip Rhodes, *Dr John Leake's Hospital*.
105 John Leake, *Introduction to the Theory and Practice of Midwifery*, 59–60.
106 John Leake, *A Lecture Introductory to the Theory and Practice of Midwifery*, 26.
107 John Aitken, *Principles of Midwifery*, n.p.

6 Cyborg Conceptions: Bodies, Texts, and the Future of the Human Spirit

1   Mary Shelley, *Frankenstein*, 66, 66–7, 96–7, 146.
2   Maureen N. McLane, *Romanticism and the Human Sciences*, 95.
3   Bruce Clarke, *Energy Forms*, 4, 175, 50.
4   Laura Otis, *Networking*, 221, 3.
5   Harold Innis, *The Bias of Communication*, 11.
6   See Walter J. Ong, *Orality and Literacy*.
7   Theodore Adorno, 'The Schema of Mass Culture,' 55, 83.
8   Sven Birkerts, *The Gutenberg Elegies*, 190.
9   Arthur Kroker and Michael A. Weinstein, 'Global Algorithm 1.4.' n.p.
10 Sadie Plant, 'Coming across the Future,' 35.
11 Christopher J. Keep, 'The Disturbing Liveliness of Machines,' 165, 179.
12 For historically sensitive treatments of how the notion of 'page' has been redefined by digital communications, see Peter Stoicheff and Andrew Taylor, eds., *The Future of the Page*; David Thorburn and Henry Jenkins, eds. *Rethinking Media Change*; or Mary E. Hocks and Michelle Kendrick, eds. *Eloquent Images*.
13 John Locke, *Essay Concerning Human Understanding*, 81.
14 Vivian Sobchack, 'The Scene of the Screen,' 104, 100, 104, 106.
15 K. Ludwig Pfeiffer, 'The Materiality of Communication,' in *Materialities of Communication*, 11.
16 See Norbert Wiener, *Cybernetics*.
17 Norbert Wiener, *The Human Use of Human Beings: Cybernetics and Society*, 103.
18 E. Lily Kay, 'Cybernetics, Information, Life,' 90.
19 W.T. Astbury and F.O. Bell, 'Some Recent Developments in the X-ray Study of Proteins and Related Structures,' 114.
20 William Harvey, *Disputations Touching the Generation of Animals*, 8.
21 Cited in N. Katherine Hayles, *How We Became Posthuman*, 13.
22 Peter Cochrane, 'Dead 100%: Capturing the Soul of Man,' 14.
23 'The Mind Is Immortal,' 100.
24 Jean Baudrillard, 'The Ecstasy of Communication,' 128–9, 130.
25 Jean Baudrillard, *Simulations*, 3, 8, 11–12, 103, 151–2, 107.
26 Victor Spitzer and David Whitlock, *Atlas of the Visible Human Male*.
27 Human Genome Program and the U.S. Department of Energy, 'To Know Ourselves' http://www.ornl.gov/hgmis/publicat/tko/index.html.
28 The National Library of Medicine (NLM), 'Fact Sheet.'
29 National Library of Medicine (U.S.), 'Electronic Imaging: Report of the Board of Regents.'
30 Robert Sinsheimer, *The Book of Life*, 5.

31 'The National Human Genome Research Institute (NHGRI),' press release http://www.nhgri.nih.gov:80/NEWS/Finish_sequencing_early/cracking _the_code.html.

32 The National Center for Biotechnology Information (NCBI), 'A New Gene Map of the Human Genome.' http://www.ncbi.nlm.nih.gov/genemap.

33 Cited in David Wheeler, 'Creating a Body of Knowledge,' A14.

34 *Life* (Feb. 1997): 44, 38.

35 *Life* magazine states that 'Jernigan is back. In an electronic afterlife, he haunts Hollywood studios and NASA labs, high schools and hospitals' (ibid.), 41; *The Economist* (U.S.) 341.7988 (19 Oct. 1996) uses the heading 'Virtual Immortality'; The *National Library of Medicine Newsletter* 50.6 (1995) reports that 'an anonymous 59-year-old Maryland woman who donated her body to science is now immortalized on the Internet'; the *Baltimore Sun* (29 Nov. 1995) described Jernigan as having 'won a measure of computerized immortality'; the *Denver Post* (6 June 1994) suggests the project 'promises eternal life for the participants.'

36 Catherine Waldby, 'Virtual Anatomy,' 89.

37 Neal Curtis, 'The Body as Outlaw,' 262, 263, 261, 263, 262–3, 263, 264.

38 One of the first uses of the VHP data was by SUNY researchers in developing their '3-D virtual colonoscopy,' a non-invasive imaging technology using a helical CT scanner and 3-D software to examine the colon. I would argue that simulating the human body electronically instead of using a 72-inch colonoscope is a demonstration of medical science's valuing rather violating or debasing human dignity and life.

39 Arthur Kroker and Michael A. Weinstein, *Data Trash*, 134, 154.

40 Catherine Waldby, *The Visible Human Project*, 155, 155–6.

41 Victor Spitzer and David Whitlock, *Atlas*, xi.

42 Boris Castel, Introduction to *The End of the Book?* 777.

43 Robert Fulford, 'The Idealogy of the Book,' 803, 809, 810–811.

44 Umberto Eco, Afterword to *The Future of the Book*, 295.

45 Nikhil Gagvani, et al., 'Volume Animation of the Visible Man.'

46 Vikas Singh and Nicu Daniel Cornea, 'Spatial Transfer Functions.'

47 R.M. Satava, 'Accomplishments and Challenges of Surgical Simulation,' 239.

48 Catherine Waldby, 'Virtual Anatomy,' 89, 100, 89, 86, 90, 88.

References

Adair, James Makittrick. *A Philosophical and Medical Sketch of the Natural History of the Human Body and Mind*. Bath: printed by R. Cruttwell, 1787.

Addison, Joseph, and Richard Steele. *The Spectator: A New Edition Reproducing the Original Text Both as First Issued and as Corrected by Its Authors*, 3 vols. Ed. Henry Morley. London: Routledge, 1891; reprinted New Brunswick, NJ: Rutgers Center for Electronic Texts in the Humanities, 2003. http://tabula.rutgers .edu/spectator.

Adorno, Theodor. 'The Schema of Mass Culture.' In *The Culture Industry*. Ed. J.M. Bernstein, trans. Nicholas Walker, 53–84. London: Routledge, 1991.

Adorno, Theodor, and Max Horkheimer. *Dialectic of Enlightenment*. Trans. John Cumming. New York: Herder and Herder, 1972 [1947].

Aitken, John. *Principles of Midwifery, or Puerperal Medicine*. Edinburgh: sold at the Edinburgh Lying-in Hospital, for the benefit of that charity, 1784.

Akenside, Mark. *The Pleasures of the Imagination. A Poem. In Three Books*. In *The Poems (1772)*. Printed by J. Chaney, for A. Dodd, 1738.

'An Account of What Hath Been of Late Observed by Dr Kerkringius Concerning Eggs to be Found in All Sorts of Females, by Dr Kerkringius.' *Philosophical Transactions* 7 (1672): 4018–26.

Arbuthnot, John. *An Essay Concerning the Nature of Ailments, and the Choice of Them, According to the Different Constitutions of Human Bodies*. London: printed for J. Tonson, 1731.

Arbuthnot, John, and Alexander Pope. *Memoirs of the Extraordinary Life, Works, and Discoveries of Martinus Scriblerus*. Ed. Charles Kerby-Miller. Oxford: Oxford University Press, 1988.

Aristotle. *The Complete Works of Aristotle*, 2 vols. Ed. Jonathan Barnes. Princeton: Princeton University Press, 1984.

Aristotle, pseud. *Aristotle's Masterpiece, or, The secrets of Generation Displayed in All the Parts Thereof ... Very Necessary for All Midwives, Nurses, and Young-Married Women.* London: printed for W.B., 1694.

AScribe Newswire. 'Researchers Connect Life's Blueprints with Its Energy Source.' *Smalltimes: News about MEMs, Nanotechnology and Microsystems.* http://www.smalltimes.com/document_display.cfm?document_id=5441.

Astbury, W.T., and F.O. Bell. 'Some Recent Developments in the X-ray Study of Proteins and Related Structures.' *Cold Springs Harbor Symposium for Quantitative Biology* 6 (1938): 109–12.

Astruc, Jean. *Elements of Midwifry. Containing the Most Modern and Successful Method of Practice in Every Different Kind of Labour.* Trans. with additions and explanatory notes by S. Ryley. London: printed for S. Crowder, 1766.

Atkyns, Richard. *The Original and Growth of Printing: Collected out of History, and the Records of this Kingdome. Wherein Is also Demonstrated, that Printing Appertaineth to the Prerogative Royale; and Is a Flower of the Crown of England.* London: printed by John Streater for the author, 1664.

Bailey, Nathan. *Dictionarium Britannicum: Or a More Compleat Universal Etymological English Dictionary Than Any Extant ... Collected by Several Hands.* London: printed for T. Cox, 1730.

– *An Universal Etymological English Dictionary ... The Second Edition, With Large Additions.* London: printed for E. Bell, J. Darby, A. Bettesworth, F. Fayman, J. Pemberton ..., 1724.

– *An Universal Etymological English Dictionary ... By N. Bailey, ... The Four-and-Twentieth Edition, Carefully Enlarged and Corrected by Edward Harwood, D.D.* London: printed for J. Buckland, W. Strahan, J.F. and C. Rivington, W. Owen, T. Caslon ..., 1782.

Balsamo, Anne. 'Reading Cyborgs Writing Feminism.' *Communication* 10 (1988): 331–44; reprinted in Kirkup et al., 148–58.

Barker-Benfield, G.J. *The Culture of Sensibility: Sex and Society in Eighteenth-Century Britain.* Chicago: University of Chicago Press, 1992.

Barnard, John. 'London Publishing, 1640–1660: Crisis, Continuity, and Innovation.' *Book History* 4.1 (2001): 1–16.

Bass, Thomas A. 'Gene Genie.' *Wired* 3.08 (1995). http://www.wired.com/wired/archive/3.08/molecular.html.

Battestin, Martin C. 'The Critique of Freethinking from Swift to Sterne.' *Eighteenth Century Fiction* 15.3–4 (2003): 341–420.

Baudrillard, Jean. 'The Ecstasy of Communication.' In *The Anti-Aesthetic: Essays on Postmodern Culture.* Ed. Hal Foster, trans. John Johnston, 126–34. Port Townsend, WA: Bay Press, 1983.

– *Simulacra and Simulation.* Trans. Sheila Faria Glaser. Ann Arbor: University of Michigan Press, 1994.

– *Simulations.* Trans. Paul Foss, Paul Patton, and Philip Beitchman. New York: Semiotext(e), 1983.

– *The Transparency of Evil: Essays On Extreme Phenomena.* Trans. James Benedict. London: Verso, 1993.

Baxter, Andrew. *An Enquiry into the Nature of the Human Soul; Wherein the Immateriality of the Soul Is Evinced from the Principles of Reason and Philosophy ... The Second Edition,* 2 vols. London: printed [by James Bettenham] for the author and sold by A. Millar, 1737.

Beddoes, Thomas. *A Lecture Introductory to a Course of Popular Instruction on the Constitution and Management of the Human Body.* Bristol: printed by N. Biggs for Joseph Cottle, 1797.

Belcher, William. *Belcher's Address to Humanity: Containing, A Letter to Dr Thomas Monro; A Receipt to Make a Lunatic, and Seize His Estate; and A Sketch of a True Smiling Hyena.* London: sold by Allen and West, no. 15, Paternoster-Row; and by the author, 1796.

– *Intellectual Electricity, Novum Organum of Vision, and Grand Mystic Secret: ... By a Rational Mystic.* London, 1798.

Bentley, Richard. *The Folly and Unreasonableness of Atheism Demonstrated from the Advantage and Pleasure of a Religious Life, the Faculties of Human Souls, the Structure of Animate Bodies, & the Origin and Frame of the World.* London: printed by J.H. for H. Mortlock, 1693.

Bertucci, Paola, and Giuliano Pancaldi, eds. *Electric Bodies: Episodes in the History of Medical Electricity.* Bologna: University of Bologna, Department of Philosophy, 2001.

Birkerts, Sven. *The Gutenberg Elegies: The Fate of Reading in an Electronic Age.* New York: Fawcett Columbine, 1994.

– ed. *Tolstoy's Dictaphone: Technology and the Muse.* St Paul, MN: Graywolf Press, 1996.

Blondel, James. *The Strength of Imagination in Pregnant Women Examin'd: And the Opinion That Marks and Deformities in Children Arise from Thence, Demonstrated to Be a Vulgar Error. By a Member of the College of Physicians, London.* London: printed and sold by J. Peele, 1727.

Blount, Thomas. *Glossographia: or A Dictionary, Interpreting All Such Hard Words, Whether Hebrew, Greek, Latin, Italian, Spanish, French, Teutonick, Belgick, British or Saxon; As Are Now Used in Our Refined English Tongue.* London: printed by Tho. Newcomb, 1656.

Boas, Marie. 'The Establishment of the Mechanical Philosophy.' *Osiris* 10 (1952): 412–541.

Bolter, Jay David. *Writing Space: The Computer, Hypertext, and the History of Writing.* Hillsdale, NJ: Erlbaum, 1991.

Bolter, Jay David, and Richard Grusin. *Remediation: Understanding New Media.* Cambridge, MA: MIT Press, 2000.

Bowe, W. 'Some Account of the Life of Mr William Emerson.' In William Emerson, *Tracts: Containing: I. Mechanics, or the Doctrine of Motion. II. The Projection of the Sphere. III. The Laws of Centripetal and Centrifugal Force,* i–xxii. London: printed for F. Wingrave, 1793.

Boyle, Robert. *The Works of the Honourable Robert Boyle,* 5 vols. Ed. Thomas Birch. London: printed for A. Millar, 1744.

Bracken, Henry. *The Midwife's Companion, or, a Treatise of Midwifery: Wherein the Whole Art Is Explained.* London: printed for J. Clarke, 1737.

Braidotti, Rosi. 'Cyberfeminism with a Difference.' 1996. http://www.let.uu.nl/ womens_studies/rosi/cyberfem.htm.

Brande, David. 'The Business of Cyberpunk: Symbolic Economy and Ideology in William Gibson.' *Configurations* 2.3 (1994): 509–36.

Brereton, Jane. *Poems on Several Occasions: By Mrs Jane Brereton. With Letters to Her Friends, and An Account of Her Life.* London: printed by Edw. Cave, 1744.

Briggs, Laura, and Jodi I. Kelber-Kaye. '"There Is No Unauthorized Breeding in Jurassic Park": Gender and the Uses of Genetics.' *NWSA Journal* 12.3 (2000): 92–113.

Brooke, Frances. *History of Lady Julia Mandeville.* Vol. 2. London: printed for Rd. Dodsley, 1763.

Brown, Theodore M. 'From Mechanism to Vitalism in Eighteenth-Century Physiology.' *Journal of the History of Biology* 7.2 (1974): 179–216.

Burton, John. *A Letter to William Smellie, MD. Containing Critical and Practical Remarks upon His Treatise on the Theory and Practice of Midwifery.* London: printed for W. Owen, 1753.

Burton, Robert. *The Anatomy of Melancholy.* Ed. Holbrook Jackson. London: J.M. Dent, 1978 [1621].

Bush, Vannevar. 'As We May Think.' *Atlantic Monthly* 176 (July 1945): 101–8.

Business Week. 'The Mind Is Immortal.' *Business Week* (30 August 1999): 100.

Carlisle, Anthony. 'The Croonian Lecture: On Muscular Motion.' *Philosophical Transactions* 95 (1805): 1–30.

Carlson, Eric T., and Meribeth M. Simpson. 'Models of the Nervous System in Eighteenth Century Psychiatry.' *Bulletin of the History of Medicine* 43.2 (1969): 101–15.

Casper, Monica J. 'Reframing and Grounding Nonhuman Agency: What Makes a Fetus an Agent?' *American Behavioral Scientist* 37.6 (1994): 839–56.

– 'Fetal Cyborgs and Technomoms on the Reproductive Frontier: Which Way to the Carnival?' In Gray et al., *The Cyborg Handbook,* 183–202.

Castel, Boris, ed. *The End of the Book?* Special issue of *Queen's Quarterly* 101.4 (1994).

Cavendish, Henry. 'An Attempt to Explain Some of the Principal Phænomena of Electricity, by means of an Elastic Fluid.' *Philosophical Transactions* 61 (1771): 584–677.

Chamberlen, Peter. *Dr Chamberlain's Midwifes Practice: Or, A Guide for Women in that High Concern of Conception, Breeding, and Nursing Children.* London: printed for Thomas Rooks, 1665.

Chambers, Ephraim. *Cyclopedia: Or, An Universal Dictionary of the Arts and Sciences.* London: printed for D. Midwinter, W. Innys, C. Rivington, and others, 1741–3.

Charleton, Walter. *Enquiries into Human Nature in VI. Anatomic Praelections in the New Theatre of the Royal Colledge of Physicians in London.* London: printed by M. White, for Robert Boulter, 1680.

– *Natural History of Nutrition, Life, and Voluntary Motion.* London: printed for Henry Herringman, 1659.

Cheyne, George. *The English Malady: Or, a Treatise of Nervous Diseases of all Kinds, as Spleen, Vapours, Lowness of Spirits, Hypochondriacal, and Hysterical Distempers, &c.* Delmar, NY: Scholars Facsimiles and Reprints, 1976 [1733].

– *The Natural Method of Cureing the Diseases of the Body, and the Disorders of the Mind Depending on the Body.* London: printed for Geo. Strahan, 1742.

Clark, William R., and Michael Grunstein. *Are We Hardwired? The Role of Genes in Human Behavior.* Oxford: Oxford University Press, 2000.

Clarke, Adele. 'Modernity, Postmodernity and Reproductive Processes *ca.* 1890–1990 or, "Mommy, where do cyborgs come from anyway?"' In Gray et al., *The Cyborg Handbook,* 139–56.

Clarke, Bruce. *Energy Forms: Allegory and Science in the Era of Classical Thermodynamics.* Ann Arbor: University of Michigan Press, 2001.

Clarke, Edwin. 'The Doctrine of the Hollow Nerve in the Seventeenth and Eighteenth Centuries.' In *Medicine, Science and Culture: Historical Essays in Honor of Owsei Temkim.* Ed. Lloyd G. Stevenson and Robert P. Multhauf, 123–41. Baltimore: Johns Hopkins Press, 1968.

Clarke, Samuel. *The Leibniz-Clarke Correspondence.* Ed. H.G. Alexander. New York: Manchester University Press, 1956.

Cleland, John. *Memoirs of a Woman of Pleasure.* 2 vols. London: printed [by Thomas Parker] for G. Fenton [Ralph Griffiths], 1749.

Clericus, Daniel, and Jacob Mangetus. *Bibliotheca Anatomica, Medica, Chirurgica,* 3 vols. London: printed by John Nutt, 1711–14.

Clynes, Manfred E., and Nathan S. Kline. 'Cyborgs and Space.' *Astronautics* (1960); reprinted in Gray et al., *The Cyborg Handbook,* 29–33.

Cochrane, Peter. 'Dead 100%: Capturing the Soul of Man.' *Daily Telegraph* (16 April 1996): 14.

'Cog Project Publications.' http://www.ai.mit.edu/projects/humanoid-robotics-group/cog/publications.html.

Coles, Elisha. *An English Dictionary*. London: printed for Peter Parker, 1677.

Combe, William. *The Devil upon Two Sticks in England: Being a Continuation of Le Diable Boiteux of Le Sage*, 4 vols. London: printed at the Logographic Press; and sold by J. Walter and W. Richardson, 1790.

Cook, John. *The New Theory of Generation, According to the Best and Latest Discoveries in Anatomy*. London: printed for J. Buckland and others, 1762.

Cooper, Thomas. *A Compendium of Midwifery, under the Three General Heads of Theory, Practice, and Diseases*. London, 1766.

Corea, Gena. *The Mother Machine: Reproductive Technologies from Artificial Insemination to Artificial Wombs*. New York: Harper and Row, 1985.

Cox, David. 'Sight Unseen / Seeing, Mapping, Communicating.' *Ctheory*. Ed. Arthur and Marilouise Kroker. EVENT-SCENES: E095. 2001. http://www.ctheory.net/text_file?pick=228.

Crane, R.S. *The Idea of the Humanities*. Chicago: University of Chicago Press, 1967.

Creed, Barbara. '*Alien* and the Monstrous-Feminine.' In Jacobus et al., 122–35.

Crittenden, Chris. 'Self-Deselection: Technopsychotic Annihilation via Cyborg.' *Ethics and the Environment* 7.2 (2002): 127–52.

Curtis, J.G. *Harvey's Views on the Use of the Circulation of the Blood*. New York: Columbia University Press, 1915.

Curtis, Neal. 'The Body as Outlaw: Lyotard, Kafka and the Visible Human Project.' *Body & Society* 5.2 (1999): 249–66.

Darwin, Erasmus. *Zoonomia; or, the Laws of Organic Life*. London: J. Johnson, in St Paul's Church-yard, 1796.

Dennett, Daniel C. 'The Practical Requirements for Making a Conscious Robot.' *Philosophical Transactions* 349 (15 Oct. 1994): 133–5.

Descartes, René. *The Philosophical Writings of Descartes*. 3 vols. Trans. John Cottingham, Robert Stoothoff, and Dugald Murdoch. Cambridge: Cambridge University Press, 1985.

– *Treatise of Man*. Trans. Thomas Steele Hall. Cambridge, MA: Harvard University Press, 1972.

De Superville, Daniel. 'Some Reflections on Generation, and on Monsters.' Trans. Phil. Hen. Zollman. *Philosophical Transactions* 41 (1739–41): 294–307.

Dijksterhuis, E.J. *The Mechanization of the World Picture*. Trans. C. Dikshoorn. Oxford: Clarendon Press, 1961.

Davis-Floyd, Robbie, and Joseph Dumit, eds. *Cyborg Babies: From Techno-Sex to Techno-Tots*. New York: Routledge, 1998.

Dery, Mark. *Escape Velocity: Cyberculture at the End of the Century*. New York: Grove Press, 1996.

Diderot, Denis. *D'Alembert's Dream.* In *Rameau's Nephew and D'Alembert's Dream.* Trans. Leonard Tancock. London: Penguin, 1966.

– *Lettre sur les aveugles.* Geneva: E. Droz, 1951 [1749].

Dixon, Joan Broadhurst, and Eric J. Cassidy, eds. *Virtual Futures: Cyberotics, Technology and Post-Human Pragmatism.* New York: Routledge, 1998.

Doane, Mary Ann. 'Technophilia: Technology, Representation, and the Feminine.' In Jacobus et al., 110–21.

Dobbs, Betty Jo Teeter. *The Janus Faces of Genius: The Role of Alchemy in Newton's Thought.* Cambridge: Cambridge University Press, 1991.

– 'Newton's Alchemy and His Theory of Matter.' *Isis* 73.4 (1982): 511–28.

Donovan, Arthur. 'Pneumatic Chemistry and Newtonian Natural Philosophy in the Eighteenth Century: William Cullen and Joseph Black.' *Isis* 67.2 (1976): 217–28.

Duden, Barbara. *Disembodying Women: Perspectives on Pregnancy and the Unborn.* Trans. Lee Hoinacki. Cambridge, MA: Harvard University Press, 1993.

Dunton, John. *Bumography: Or, a Touch at the Lady's Tails, Being a Lampoon (Privately) Dispers'd at Tunbridge-Wells.* London, 1707.

Eco, Umberto. 'Afterword.' In *The Future of the Book.* Ed. Geoffrey Nunberg, 295–306. Berkeley: University of California Press, 1996.

Emerson, William. *The Principles of Mechanics.* London: printed for J. Richardson. 1758.

Erickson, Robert. '"The Books of Generation": Some Observations on the Style of the British Midwife Books, 1671–1764.' In *Sexuality in Eighteenth-Century Britain.* Ed. Paul-Gabriel Boucé, 74–94. Manchester: Manchester University Press, 1982.

– '"Written in the Heart": *Clarissa* and Scripture.' *Eighteenth-Century Fiction* 2.1 (1989): 17–52.

Farnell, Ross. 'In Dialogue with "Posthuman" Bodies: Interview with Stelarc.' *Body & Society* 5.2–3 (1999): 129–47.

Flynn, Carol Houlihan. 'Running Out of Matter: The Body Exercised in Eighteenth-Century Fiction.' In Rousseau, *The Languages of Psyche,* 147–85.

Fraiberg, Allison. 'Of AIDS, Cyborgs, and Other Indiscretions: Resurfacing the Body in the Postmodern.' *Postmodern Culture* 1.3 (1991).

Frank, Robert G. 'Thomas Willis and His Circle: Brain and Mind in Seventeenth-Century Medicine.' In Rousseau, *The Languages of Psyche,* 107–46.

Fulford, Robert. 'The Ideology of the Book.' *Queens Quarterly* 101.3 (1994): 801–11.

Gabilondo, Joseba. 'Postcolonial Cyborgs: Subjectivity in the Age of Cybernetic Reproduction.' In Gray et al., *The Cyborg Handbook,* 423–32.

Gagvani, Nikhil, Kundan Sen, Arindam Bhattacharya, and Andres Martinez

(Research Assistant). 'Volume Animation of the Visible Man.' Animation of the Visible Human Data Set by Applying Motion Capture Data. 2000. On the CAIP 'Volume Manipulation' page. http://www.caip.rutgers.edu/~hsundar/rocky_1.avi.

Gervaise, Isaac. *The System or Theory of the Trade of the World.* London: printed by H. Woodfall, 1720.

Gibson, William. *Mona Lisa Overdrive.* New York: Bantam Books, 1988.

– *Neuromancer.* London: HarperCollins, 1984.

Goldsmith, Oliver. *An History of the Earth, and Animated Nature*, 8 volumes. London: printed for J. Nourse, 1774.

– *A Survey of Experimental Philosophy, Considered in Its Present State of Improvement*, 2 vols. London, 1776.

González, Jennifer. 'Envisioning Cyborg Bodies: Notes from Current Research.' In Gray et al., *The Cyborg Handbook*, 267–79.

Goodall, Charles. *The Royal College of Physicians of London, Founded and Established by Law as Appears by Letters Patents, Acts of Parliament, Adjudged Cases, &c.: and An Historical Account of the College's Proceedings against Empiricks and Unlicensed Practisers.* London: printed by M. Flesher for Walter Kettilby, 1684.

Graham, James. *Dr Graham's famous work! A Lecture on the Generation, Increase, and Improvement of the Human Species.* London, 1784.

Gray, Chris Hables. *Cyborg Citizen: Politics in the Posthuman Age.* New York: Routledge, 2000.

Gray, Chris Hables, and Mark Driscoll. 'What's Real about Virtual Reality? Anthropology of, and in, Cyberspace.' *Visual Anthropology Review* 8.2 (1992): 39–49.

Gray, Chris Hables, and Steven Mentor. 'The Cyborg Body Politic. Version 1.2.' In Gray et al., *The Cyborg Handbook*, 453–67.

Gray, Chris Hables, Steven Mentor, and Jennifer Figueroa-Sarriera, eds. *The Cyborg Handbook.* New York: Routledge, 1995.

– 'Cyborgology: Constructing the Knowledge of Cybernetic Organisms.' Introduction to Gray et al., *The Cyborg Handbook*, 1–14.

Grenville, Bruce, ed. *The Uncanny: Experiments in Cyborg Culture.* Vancouver: Vancouver Art Gallery and Arsenal Pulp Press, 2002.

Guerrini, Anita. *Obesity and Depression in the Enlightenment: The Life and Times of George Cheyne.* Norman: University of Oklahoma Press, 2000.

Gulliver, Lemuel, pseud. *The Anatomist Dissected: Or the Man-Midwife Finely Brought to Bed.* Westminster: printed by and for A. Campbell, 1727.

Gumbrecht, Hans Ulrich, and K. Ludwig Pfeiffer, eds. *Materialities of Communication.* Stanford: Stanford University Press, 1994.

Haigh, Elizabeth. 'Vitalism, the Soul, and Sensibility: The Physiology of Théophile Bordeu.' *Journal of the History of Medicine* 31 (1976): 30–41.

Halacy, D.S. *Cyborg: Evolution of the Superman.* New York: Harper and Row, 1965.

Hale, David G. 'Analogy of the Body Politic.' In *Dictionary of the History of Ideas: Studies of Selected Pivotal Ideas.* Ed. Philip P. Wiener, 67–70. New York: Charles Scribner's Sons, 1973.

Hamilton, Edith, and Huntington Cairns, eds. *The Collected Dialogues of Plato, Including the Letters.* Princeton: Princeton University Press, 1963.

Handley, James. *Mechanical Essays on the Animal Oeconomy.* London: printed for A. Bettesworth, and C. Rivington, 1721.

Haraway, Donna. 'Foreword' to Gray et al., *The Cyborg Handbook.*

– 'A Manifesto for Cyborgs: Science, Technology, and Socialist Feminism in the 1980s.' *Socialist Review* 80 (1985): 65–107.

– *Modest_Witness@Second_Millennium. FemaleMan©-Meets_OncoMouse™: Feminism and Technoscience.* New York: Routledge, 1996.

– *Primate Visions: Gender, Race, and Nature in the World of Modern Science.* New York: Routledge, 1989.

Harris, John. *Lexicon Technicum: or, An Universal English Dictionary of Arts and Sciences.* London: printed for Dan. Brown, Tim. Goodwin, John Walthoe, Tho. Newborough, John Nicholson, Dan. Midwinter, and Francis Coggan, 1708.

Harris, Tim. *London Crowds in the Reign of Charles II: Propaganda and Politics from the Restoration until the Exclusion Crisis.* Cambridge: Cambridge University Press, 1987.

Hartley, David. *Observations on Man, His Frame, His Duties, and His Expectations.* 2 vols. London: printed by S. Richardson, 1749.

Harvey, William. *Anatomical Exercitations, Concerning the Generation of Living Creatures.* Trans. M. Llewellyn. London: printed by James Young, for Octavian Pulleyn, 1653.

– *The Circulation of the Blood and Other Writings.* Intr. Andrew Wear; trans. Kenneth Franklin. London: Everyman, 1993.

– *De Motu Locali Animalium.* Ed., trans., and intr. Gweneth Whitteridge. Cambridge: Cambridge University Press, 1959.

– *Disputations Touching the Generation of Animals.* Trans. and intr. Gweneth Whitteridge. Oxford: Blackwell Scientific, 1981.

Hayles, N. Katherine. 'Embodied Virtuality or How to Put Bodies Back into the Picture.' In *Immersed in Technology, Art and Virtual Environments.* Ed. Mary Anne Moser and Douglas McLeod, 1–28. Cambridge, MA: MIT Press, 1996.

– *How We Became Posthuman: Virtual Bodies in Cybernetics, Literature, and Informatics.* Chicago: University of Chicago Press, 1999.

– 'Virtual Bodies and Flickering Signifiers.' *October* 66 (1993): 69–92.

Heim, Michael. *The Metaphysics of Virtual Reality.* New York and Oxford: Oxford University Press, 1993.

Hill, Christopher. 'William Harvey and the Idea of Monarchy.' *Past and Present* 27 (1964): 54–72.

– 'William Harvey (No Parliamentarian, No Heretic) and the Idea of Monarchy.' *Past and Present* 31 (1965): 97–103.

Hobbes, Thomas. *Leviathan.* Ed. Richard Tuck. Cambridge: Cambridge University Press, 1991.

Hockenberry, John. Introduction to Alexander Tsiaras, *Body Voyage™: A Three-Dimensional Tour of a Real Human Body.* New York: Warner Books, 2000.

Hocks, Mary E., and Michelle Kendrick, eds. *Eloquent Images: Word and Image in the Age of New Media.* Cambridge, MA: MIT Press, 2003.

Hooke, Robert. *Micrographia, or some Physiological Descriptions of Minute Bodies, Made by Magnifying Glasses; with Observations and Inquiries Thereupon.* London: printed for J. Martyn and J. Allestry, 1665.

Human Genome Program and the U.S. Department of Energy. 'To Know Ourselves.' 1996. http://www.ornl.gov/hgmis/publicat/tko/index.html.

Hunter, John. *A Treatise on the Blood, Inflammation, and Gun-shot Wounds, by the Late John Hunter.* London: printed by John Richardson for George Nicol, 1794.

Hurley, Kelly. 'Reading Like an Alien: Posthuman Identity in Ridley Scott's *Alien* and David Cronenberg's *Rabid.*' In *Posthuman Bodies.* Ed. Judith Halberstam and Ira Livingston, 203–24. Bloomington: Indiana University Press, 1995.

Innis, Harold. *The Bias of Communication.* Toronto: University of Toronto Press, 1951.

Jacobus, Mary, Evelyn Fox Keller, and Sally Shuttleworth, eds. *Body Politics: Women and the Discourse of Science.* New York: Routledge, 1990.

Jameson, Fredric. 'Postmodernism and Consumer Society.' In *The Anti-Aesthetic: Essays on Postmodern Culture.* Ed. Hal Foster, 111–25. Port Townsend, WA: Bay Press, 1983.

Jocelyn, H.D., and B.P. Setchell, trans. *Regnier de Graaf On the Human Reproductive Organs. Journal of Reproduction and Fertility Supplement* 17. Oxford: Blackwell Scientific, 1972.

Johnson, Samuel. *Dictionary of the English Language, In Which the Words Are Deduced from their Originals.* London: printed for J. Knapton and others, 1756.

Jordanova, Ludmilla. *Nature Displayed: Gender, Science and Medicine, 1760–1820.* London: Addison-Wesley Longman, 1999.

Kant, Immanuel. *Anthropology from a Pragmatic Point of View.* Trans. M.J. Gregor. The Hague: Martinus Nijhoff, 1974.

Kay, E. Lily. 'Cybernetics, Information, Life: The Emergence of Scriptural Representations of Heredity.' *Configurations* 5.1 (1997): 23–91.

Keep, Christopher J. 'The Disturbing Liveliness of Machines: Rethinking the Body in Hypertext Theory and Fiction.' In *Cyberspace Textuality: Computer*

Technology and Literary Theory. Ed. Marie-Laure Ryan, 164–81. Bloomington: Indiana University Press, 1999.

Keill, James. *The Anatomy of the Humane Body Abridg'd: Or, a Short and Full View of All the Parts of the Body*. London: printed for Ralph Smith; and William Davis, 1703.

Keymer, Thomas. *Sterne, the Moderns, and the Novel*. Oxford: Oxford University Press, 2002.

Kirkup, Gill, Linda Janes, Kath Woodward, and Fiona Hovenden, eds. *The Gendered Cyborg: A Reader*. New York: Routledge, 2000.

Kroker, Arthur, and Michael A. Weinstein. 'Global Algorithm 1.4: The Theory of the Virtual Class.' *Ctheory: Theory, Technology and Culture* (1996). http://www.ctheory.net/text_file.asp?pick=35.

– *Data Trash: The Theory of the Virtual Class*. New York: St Martin's Press, 1994.

La Mettrie, Julien Offray de. *Machine Man and Other Writings*. Trans. and ed. Ann Thomson. Cambridge: Cambridge University Press, 1996.

– *Histoire naturelle de l'âme*. The Hague, 1745.

Lambert, Sheila. *Printing for Parliament, 1641–1700*. London: Swift Printers, 1984.

Langrish, Browne. 'The Crounean Lectures on Muscular Motion, Read before the Royal Society in the Year 1747.' *Philosophical Transactions* 44 (1746–7): i–66.

Laqueur, Thomas. 'Orgasm, Generation, and the Politics of Reproductive Biology.' *Representations* 14, *The Making of the Modern Body: Sexuality and Society in the Nineteenth Century* (Spring 1986): 1–41.

Lawrence, Christopher. 'The Nervous System and Society in the Scottish Enlightenment.' In *Natural Order: Historical Studies of Scientific Culture*. Ed. Barry Barnes and Steven Shapin, 19–40. Beverly Hills, CA: Sage, 1979.

Leake, John. *A Course of Lectures on the Theory and Practice of Midwifery: In which, Every Thing Essentially Necessary to the True Knowledge of that Art Will be Fully Explain'd, and Clearly Demonstrated; Particularly, those Operations Which are Conformable to the Principles of Mechanical Motion*. London, 1767.

– *Introduction to the Theory and Practice of Midwifery: ... to which Are Added, A Description of the Author's New Forceps*. London: printed for R. Baldwin, 1787.

– *A Lecture Introductory to the Theory and Practice of Midwifery: Including the History, Nature, and Tendency of that Science*. London: printed for R. Baldwin, 1776.

Leslie, Glennda. 'Cheat and Impostor: Debate Following the Case of the Rabbit Breeder.' *The Eighteenth Century: Theory and Interpretation* 27.3 (1986): 269–86.

Locke, John. *Essay Concerning Human Understanding*. Ed. Peter H. Nidditch. Oxford: Oxford University Press, 1975.

Louisa Mathews. By an Eminent Lady. 3 vols. London: printed for J. Lackington, 1793.

Lovett, Richard. *Philosophical Essays, in Three Parts. Containing I. An Enquiry Into the Nature and Properties of the Electrical Fluid.* Worcester: printed for the author, by R. Lewis, 1766.

– *The Subtil Medium Prov'd, or, That Wonderful Power of Nature … Call'd Sometimes Aether, but Oftener Elementary Fire, Verify'd.* London: printed for J. Hinton and others, 1756.

Ludlow, Peter. 'New Foundations: On the Emergence of Sovereign Cyberstates and Their Governance Structures.' In *Crypto Anarchy, Cyberstates, and Pirate Utopias.* Ed. Peter Ludlow, 1–24. Cambridge: MIT Press, 2001.

Lyotard, Jean-François. *The Postmodern Condition: A Report on Knowledge.* Trans. Geoff Bennington and Brian Massumi. Minneapolis: University of Minnesota, 1984.

Macaulay, Catharine. *A Remarkable Moving Letter!* London: at the Shakespeare Press by the Etheringtons; for Robert Faulder, 1779.

Macauley, William, and Angel Gordo-López. 'From Cognitive Psychologies to Mythologies: Advancing Cyborg Textualities for a Narrative of Resistance.' In Gray et al., *The Cyborg Handbook,* 433–44.

Mann, Steve. 'The Post-Cyborg Path to Deconism.' *Ctheory.* Ed. Arthur and Marilouise Kroker (2003). http://www.ctheory.net/text_file.asp?pick=368.

Mann, Steve, James Fung, Mark Federman, and Gianluca Baccanico. 'Panopdecon: Deconstructing, Decontaminating, and Decontextualizing Panopticism in the Postcyborg Era.' http://wearcam.org/panopdecon.htm.

Mann, Steve, and Hal Niedzviecki. *Cyborg: Digital Destiny and Human Possibility in the Age of the Wearable Computer.* Toronto: Doubleday Canada, 2001.

Markley, Robert. 'Boundaries: Mathematics, Alienation, and the Metaphysics of Cyberspace.' In *Virtual Realities and Their Discontents.* Ed. Robert Markley, 55–77. Baltimore: Johns Hopkins University Press, 1996.

Mather, Cotton. *The Christian Philosopher: A Collection of the Best Discoveries in Nature, With Religious Improvements.* London: printed for Eman. Matthews, 1721.

Maubray, John. *The Female Physician, Containing All the Diseases Incident to That Sex, in Virgins, Wives, and Widows; Together with Their Causes and Symptoms … To Which Is Added, the Whole Art of New Improv'd Midwifery.* London: printed for Stephen Austen, 1730.

Mayr, Otto. *Authority, Liberty, and Automatic Machinery in Early Modern Europe.* Baltimore: Johns Hopkins University Press, 1986.

– *The Origins of Feedback Control.* Cambridge, MA: MIT Press, 1970.

Mazlish, Bruce. *The Fourth Discontinuity: The Co-evolution of Humans and Machines.* New Haven: Yale University Press, 1993.

McCaffery, Larry, ed. *Storming the Reality Studio: A Casebook of Cyberpunk and Postmodern Fiction.* Durham: Duke University Press, 1991.

McCaffrey, Anne. *The Ship Who Sang.* New York: Ballantine Books, 1970
[1969].

McHale, Brian. 'POSTcyberMODERNpunkISM.' In Larry McCaffery, 308–23.

McLane, Maureen N. *Romanticism and the Human Sciences: Poetry, Population, and the Discourse of the Species.* Cambridge: Cambridge University Press, 2000.

McLaren, Angus. *Reproductive Rituals: The Perception of Fertility in England from the Sixteenth Century to the Nineteenth Century.* London: Methuen, 1984.

McLuhan, Marshall. *The Gutenberg Galaxy: The Making of Typographic Man.* Toronto: University of Toronto Press, 1962; reprint, 1997.

– *Understanding Media: The Extensions of Man.* New York: McGraw Hill, 1964; reprint, Cambridge: MIT Press, 1994.

McLuhan, Marshall, Matie Molinaro, Corinne McLuhan, and William Toye, eds. *Letters of Marshall McLuhan.* Toronto: Oxford University Press, 1987.

Monro, Alexander. *Experiments on the Nervous System, with Opium and Metalline Substances; Made Chiefly with the View of Determining the Nature and Effects of Animal Electricity.* Edinburgh: printed by Adam Neill and Company, London, 1793.

Moravec, Hans. *Mind Children: The Future of Robot and Human Intelligence.* Cambridge: Harvard University Press, 1988.

Moravia, Sergio. 'From *Homme Machine* to *Homme Sensible*: Changing Eighteenth-Century Models of Man's Image.' *Journal of the History of Ideas* 39.1 (1978): 45–60.

Mudge, Bradford K., ed. *When Flesh Becomes Word.* Oxford: Oxford University Press, 2004.

Mullan, John. *Sentiment and Sociability: The Language of Feeling in the Eighteenth Century.* Oxford: Clarendon Press, 1988.

Muri, Allison. 'Of Shit and the Soul: Tropes of Cybernetic Disembodiment.' *Body & Society* 9.3 (2003): 73–92.

National Center for Biotechnology Information. 'A New Gene Map of the Human Genome.' http://www.ncbi.nlm.nih.gov/genemap.

National Human Genome Research Institute. 'The National Human Genome Research Institute (NHGRI)' press release (Sept. 1998). http://www.nhgri.nih.gov:80/NEWS/Finish_sequencing_early/cracking_the_code.html.

National Library of Medicine. 'Fact Sheet: The Visible Human Project®.' Last updated 16 Feb. 2001. http://www.nlm.nih.gov/pubs/factsheets/visible_human.html.

– Board of Regents. 'Electronic Imaging: Report of the Board of Regents.' U.S. Department of Health and Human Services, Public Health Service, National Institutes of Health, 1990. NIH Publication 90-2197.

Nelson, Carolyn, and Matthew Seccombe. *Periodical Publications, 1641–1700:*

British Newspapers and Periodicals, 1641–1700. London: Bibliographical Society, 1986.

Newton, Isaac. *Opticks: Or, a Treatise of the Reflections, Refractions, Inflections and Colours of Light.* London: printed by W. Bowyer for W. Innys, 1717.

– *Papers and Letters on Natural Philosophy and Related Documents.* Ed. I. Bernard Cohen and Robert E. Schofield. Cambridge: Cambridge University Press, 1958.

– *Sir Isaac Newton's Mathematical Principles of Natural Philosophy.* Trans. Andrew Motte; ed. Florian Cajori. Berkeley: University of California Press, 1946.

Norbrook, David. *Writing and the English Republic: Poetry, Rhetoric, and Politics, 1627–1660.* Cambridge: Cambridge University Press, 1999.

Ong, Walter J. *Orality and Literacy: The Technologizing of the Word.* London: Routledge, 1982.

Otis, Laura. *Networking: Communicating with Bodies and Machines in the Nineteenth Century.* Ann Arbor: University of Michigan Press, 2001.

Parsons, James. 'The Crounian Lectures on Muscular Motion for the Years 1744 and 1745. Read before the Royal Society.' *Philosophical Transactions* 43 (1744–5): iii–86.

Pearson, Karl. *The Grammar of Science.* New York: Meridian Books, 1957 [1892].

Petchesky, Rosalind Pollack. 'Abortion in the 1980s: Feminist Morality and Women's Health.' In Ellen Lewen and Virginia Olesen, eds., 139–73, *Women, Health, and Healing: Toward a New Perspective.* New York: Tavistock, 1985.

– 'Fetal Images: The Power of Visual Culture in the Politics of Reproduction.' *Feminist Studies* 13 (1987): 263–92.

– 'Foetal Images: The Power of Visual Culture in the Politics of Reproduction.' In *Reproductive Technologies: Gender, Motherhood and Medicine.* Ed. Michele Stanworth, 57–80. (Minneapolis: University of Minnesota Press, 1987.

Peters, J.S. 'The Bank, the Press, and the "Return of Nature": On Currency, Credit, and Literary Property in the 1690s.' In *Early Modern Conceptions of Property.* Ed. John Brewer and Stusan Staves, 365–88. London: Routledge, 1996.

Pfeiffer, K. Ludwig. 'The Materiality of Communication.' In Gumbrecht and Pfeiffer, 1–12.

Philip, A.P.W. 'On the Relation which Subsists between the Nervous and Muscular Systems in the More Perfect Animals, and the Nature of the Influence by which It Is Maintained.' *Philosophical Transactions* 123 (1833): 55–72.

Physician of Bath. *A Safe-Conduct through the Territories of the Republic of Venus. Containing, a Practicable Proposal for the Prevention and Final Eradication of a Certain Disease. Also, Occasional Anecdotes (Never Before Published).* London: printed for the author, 1794.

Pinker, Steven. *How the Mind Works.* New York: W.W. Norton, 1997.

Plant, Sadie. 'Coming Across the Future.' In Dixon and Cassidy, 30–6.

Plato. *The Collected Dialogues of Plato, Including the Letters.* Ed. Edith Hamilton and Huntington Cairns. Princeton: Princeton University Press, 1963.

Pleasure for a Minute: or, the Amorous Adventure: A Tale. London: printed for A. Dodd, 1730.

Poems on Several Occasions: Together with Some Odes in Imitation of Mr Cowley's Stile and Manner. London: printed for Luke Stokoe; and George Harris, 1703.

Pope, Alexander. *Alexander Pope.* Ed. Pat Rogers. Oxford: Oxford University Press, 1993.

– *The Spectator* 408 (18 June 1712). In *The Spectator: A New Edition* ...; reprinted New Brunswick, NJ: Rutgers Center for Electronic Texts in the Humanities, 2003. http://tabula.rutgers.edu/spectator.

Porter, Roy. *The Creation of the Modern World: The Untold Story of the British Enlightenment.* New York: W.W. Norton, 2000.

Poster, Mark. 'Cyberdemocracy: Internet and the Public Sphere.' In *Internet Culture,* ed. David Porter, 201–17. New York: Routledge, 1996.

Quincy, John. *Lexicon Physico-Medicum: Or, a New Medical Dictionary ... The Eleventh Edition.* London: printed for T. Longman, 1994.

Rapp, Rayna. 'Real-Time Fetus / The Role of the Sonogram in the Age of Monitored Reproduction.' In *Cyborgs and Citadels: Anthropological Interventions in Emerging Sciences and Technologies.* Ed. Gary Lee Downey and Joseph Dumit, 31–48. Santa Fe: School of American Research Press, 2000.

Raymond, Janice. *Women as Wombs: Reproductive Technologies and the Battle over Women's Freedom.* San Francisco: HarperCollins, 1993.

Raymond, Joad. *Pamphlets and Pamphleteering in Early Modern Britain.* Cambridge: Cambridge University Press, 2003.

Reill, Peter Hanns. 'The Legacy of the "Scientific Revolution": Science and the Enlightenment.' In *Eighteenth-Century Science.* Ed. Roy Porter, 23–43. Cambridge: Cambridge University Press, 2003.

Rhodes, Philip. *Dr John Leake's Hospital: A History of the General Lying-in Hospital, York Road, Lambeth, 1765–1971: The Birth, Life, and Death of a Maternity Hospital.* London: Davis Poynter, 1977.

Richardson, Samuel. *Clarissa. Or, the History of a Young Lady: Comprehending the Most Important Concerns of Private Life. In Eight Volumes ... The Third Edition,* 8 vols. London: printed for S. Richardson, 1750–1 [1747–8].

– *Pamela: or, Virtue Rewarded,* 4 vols. London: printed for C. Rivington, in St Paul's Church-Yard; and J. Osborn 1741 [1740].

Riordan, J.R., et al. 'Identification of the Cystic Fibrosis Gene: Cloning and Characterization of Complementary DNA.' *Science* 245 (Sept. 1989): 1066–73.

Riskin, Jessica. *Science in the Age of Sensibility: The Sentimental Empiricists of the French Enlightenment.* Chicago: University of Chicago Press, 2002.

Robinson, Bryan. *A Dissertation on the Aether of Sir Isaac Newton.* Dublin: printed by S. Powell, 1743 (possibly 1746?).

Rogers, Pat. *Grub Street: Studies in a Subculture.* London: Methuen, 1972.

Rorvik, David. *As Man Becomes Machine: The Evolution of the Cyborg.* New York: Pocket Books, 1978 [1970].

Rousseau, George S. 'Nerves, Spirits and Fibres: Towards Defining the Origins of Sensibility.' In *Studies in the Eighteenth Century III: Papers Presented at the Third David Nichol Smith Memorial Seminar, Canberra 1973.* Ed. R.F. Brissenden and J.C. Eade, 137–57. Toronto: University of Toronto Press, 1976; reprinted in *Enlightenment Crossings: Pre- and Post-modern Discourses, Anthropological.* Manchester: Manchester University Press, 1991, 122–41.

– 'The Perpetual Crises of Modernism and the Traditions of Enlightenment Vitalism: With a Note on Mikhail Bakhtin.' In *The Crisis in Modernism: Bergson and the Vitalist Controversy.* Ed. Frederick Burwick and Paul Douglass, 15–75. Cambridge: Cambridge University Press, 1992.

– 'Towards a Semiotics of the Nerve: The Social History of Language in a New Key.' In *Language, Self and Society: A Social History of Language.* Ed. Peter Burke and Roy Porter, 213–75. London: Polity Press, 1991.

– ed. *The Languages of Psyche: Mind and Body in Enlightenment Thought.* Berkeley: University of California Press, 1990.

Rowland, Robyn. *Living Laboratories: Women and Reproductive Technologies.* Bloomington: Indiana University Press, 1992.

Sabor, Peter, ed., *Memoirs of a Woman of Pleasure.* Oxford: Oxford University Press, 1985.

Salomon, Arthur R., David W. Voehringer, Leonard A. Herzenberg, and Chaitan Khosla. 'Understanding and Exploiting the Mechanistic Basis for Selectivity of Polyketide Inhibitors of F_0F_1-ATPase.' *Proceedings of the National Academy of Sciences of the United States of America* (2000): 14766–71.

Sanderson, George, and Frank MacDonald, eds. *Marshall McLuhan: The Man and His Message.* Golden, CO: Fulcrum, 1989.

Sandoval, Chela. 'New Sciences: Cyborg Feminism and the Methodology of the Oppressed.' In Gray et al., *The Cyborg Handbook*, 407–21.

Satava, R.M. 'Accomplishments and Challenges of Surgical Simulation: Dawning of the Next-generation Surgical Education.' *Surgical Endoscopy: Ultrasound and Interventional Techniques* 15 (2001): 232–41.

Sawday, Jonathan. '"Forms Such as Never Were in Nature": The Renaissance Cyborg.' In *At the Borders of the Human: Beasts, Bodies and Natural Philosophy in the Early Modern Period.* Ed. Erica Fudge, Ruth Gilbert, and Susan Wiseman, 171–95. New York: Macmillan, 1999.

Schofield, Robert E. *Mechanism and Materialism: British Natural Philosophy in an Age of Reason*. Princeton: Princeton University Press, 1970.

Scutt, Jocelynne A. *The Baby Machine: Reproductive Technology and the Commercialisation of Motherhood*. London: Green Print, 1990.

Sharp, Jane. *The Midwives Book or the Whole Art of Midwifry Discovered*. Ed. Elaine Hobby. New York and Oxford: Oxford University Press, 1999.

Shelley, Mary. *Frankenstein; Or, the Modern Prometheus*. Ed. D.L. Macdonald and Kathleen Scherf. Peterborough, ON: Broadview Press, 2001.

Singh, Vikas, and Nicu Daniel Cornea. 'Spatial Transfer Functions: A Unified Approach of Volume Deformation to Modeling and Animation.' http://www.caip.rutgers.edu/vizlab_group_files/CURRENT/animSpatial.avi.

Sinsheimer, Robert. *The Book of Life*. Reading, MA: Addison-Wesley, 1967.

Smellie, William. *A Treatise on the Theory and Practice of Midwifery*. London: printed for D. Wilson, 1762 [1752].

Smollett, Tobias. *Expedition of Humphry Clinker*. London, 1771.

Smollett, Tobias (attributed to). *The History and Adventures of an Atom*. Ed. by O.M. Brack. Athens: University of Georgia Press, 1989 [1769].

Sobchack, Vivian. 'The Scene of the Screen: Envisioning Cinematic and Electronic "Presence."' In Gumbrecht and Pfeiffer, 83–106. Stanford: Stanford University Press, 1994.

Solmsen, Friedrich. 'Tissues and the Soul: Philosophical Contributions to Physiology.' *Philosophical Review* 59.4 (1950): 435–68.

Spender, Dale. *Nattering on the Net: Women, Power and Cyberspace*. Toronto: Garamond Press, 1995.

Spitzer, Victor, and David Whitlock. *Atlas of the Visible Human Male: Reverse Engineering of the Human Body*. Sudbury, MA: Jones and Bartlett, 1998.

Springer, Claudia. *Electronic Eros: Bodies and Desire in the Postindustrial Age*. Austin: University of Texas Press, 1996.

– 'The Pleasure of the Interface.' *Screen* 32.3 (1991): 303–23.

Squier, Susan Merrill. *Babies in Bottles: Twentieth-Century Visions of Reproductive Technology*. New Brunswick, NJ: Rutgers University Press, 1995.

– 'Negotiating Boundaries: From Assisted Reproduction to Assisted Replication. In *Playing Dolly: Technocultural Formations, Fantasies, and Fictions of Assisted Reproduction*. Ed. E. Ann Kaplan and Susan Squier, 101–15. New Brunswick, NJ, and London: Rutgers University Press, 1999.

Stedman, John. *Physiological Essays and Observations*. Edinburgh: printed for A. Kincaid & J. Bell, 1769.

Stelarc. 'From Psycho-Body to Cyber-Systems: Images as Post-human Entities.' In Dixon and Cassidy, 116–23.

Stephanson, Raymond. 'Richardson's "Nerves": The Physiology of Sensibility in *Clarissa*.' *Journal of the History of Ideas* 49.2 (1988): 267–85.

- *The Yard of Wit: Male Creativity and Sexuality, 1650–1750*. Philadelphia: University of Pennsylvania Press, 2004.
Sterling, Bruce. *Schismatrix Plus*. New York: Ace Books, 1996.
Sterne, Laurence. *The Life and Opinions of Tristram Shandy, Gentleman*. Ed. and intr. Ian Watt. Boston: Houghton Mifflin, 1965 [1759–67].
- *A Sentimental Journey through France and Italy*. Ed. and intr. Ian Jack. Oxford: Oxford University Press, 1984 [1768].
Stoicheff, Peter, and Andrew Taylor, eds. *The Future of the Page*. Toronto: University of Toronto Press, 2004.
Stone, Sandy. 'Split Subjects, Not Atoms; or, How I Fell in Love with My Prosthesis.' In Gray et al., *The Cyborg Handbook*, 393–406. First published in *Configurations* 2.1 (1994): 173–90.
Swift, Jonathan. *Gulliver's Travels*. Ed. and intr. Paul Turner. Oxford: Oxford University Press, 1998 [1726].
- *A Tale of a Tub, to which is Added the Battle of the Books and the Mechanical Operation of the Spirit*, 2nd ed. Ed. and intr. A.C. Guthkelch and D. Nichol Smith. Oxford: Oxford University Press, 1968 [1704].
- *Miscellanies in Prose and Verse. In Two Volumes*. Dublin, 1728.
Tarrant, R.J. 'Aspects of Virgil's Reception in Antiquity.' In *The Cambridge Companion to Virgil*. Ed. Charles Martindale, 56–72. Cambridge: Cambridge University Press, 1997.
Thomson, Ann. *Materialism and Society in the Mid-Eighteenth Century: La Mettrie's Discours Préliminaire*. Geneva and Paris: Librairie Droz, 1981.
Thorburn, David, and Henry Jenkins, eds. *Rethinking Media Change: The Aesthetics of Transition*. Cambridge, MA: MIT Press, 2003.
Tissot, S.A. *An Essay on Diseases Incident to Literary and Sedentary Persons. With Proper Rules for Preventing Their Fatal Consequences*. London: printed for J. Nourse, 1769.
Todd, Dennis. *Imagining Monsters: Miscreations of the Self in Eighteenth-Century England*. Chicago: University of Chicago Press, 1995.
Tomas, David. 'Feedback and Cybernetics: Reimaging the Body in the Age of the Cyborg.' In *Cyberspace/Cyberbodies/Cyberpunk: Cultures of Technological Embodiment*. Ed. Mike Featherstone and Roger Burrows, 21–43. London: Sage, 1995.
Toulmin, Stephen. *Cosmopolis: The Hidden Agenda of Modernity*. New York: Free Press, 1990.
Turner, Daniel. *The Force of the Mother's Imagination upon Her Foetus in Utero, Still Farther Considered: In the Way of a Reply to Dr Blondel's Last Book, Entitled, The Power of the Mother's Imagination over the Foetus Examined*. London: printed for

J. Walthoe, R. Wilkin, J. and J. Bonwicke, S. Birt, J. Clarke, T. Ward and E. Wicksteed, 1730.

Turner, James Grantham. 'Lovelace and the Paradoxes of Libertinism.' In *Samuel Richardson: Tercentenary Essays*. Ed. Margaret Anne Doody and Peter Sabor, 70–88. Cambridge: Cambridge University Press, 1989.

Turkle, Sherry. *Life on the Screen: Identity in the Age of the Internet*. New York: Touchstone, 1995.

U.S. Congress, Office of Technology Assessment. 'The Human Genome Project and Patenting DNA Sequences.' Unpublished draft document. April 1994. 370 pp. http://www.georgetown.edu/research/nrcbl/bioethics/nirehg/nirehgdigcol.htm.

Van Sant, Ann Jessie. *Eighteenth-Century Sensibility and the Novel: The Senses in Social Context*. Cambridge: Cambridge University Press, 1993.

Vartanian, Aram. *La Mettrie's L'homme machine: A Study in the Origins of an Idea*. Princeton: Princeton University Press, 1960.

Vaucanson, Jaques de. *An Account of the Mechanism of an Automaton, or Image playing on the German-Flute*. Trans. J.T. Desaguliers. London: printed by T. Parker, 1742.

Venette, Nicolas. *Conjugal Love, or, The Pleasures of the Marriage Bed*. Facsimile. New York: Garland, 1984 [1687].

Vesalius, Andreas. *De Humani Corporis Fabrica Libri Septem*. Basilae: ex officina Ioannis Oporini, 1543.

Vila, Anne. *Enlightenment and Pathology: Sensibility in the Literature and Medicine of Eighteenth-Century France*. Baltimore: Johns Hopkins University Press, 1998.

Waldby, Catherine. 'Virtual Anatomy: From the Body in the Text to the Body on the Screen.' *Journal of Medical Humanities* 21.2 (2000): 85–107.

– *The Visible Human Project: Posthuman Medicine and Informatic Bodies*. London: Routledge, 2000.

Walker, Adam. *A System of Familiar Philosophy in Twelve Lectures*. London: printed for the author, 1799.

Wallis, George. *The Art of Preventing Diseases, and Restoring Health, Founded on Rational Principles, and Adapted to Persons of Every Capacity*. London: printed for G.G.J. and J. Robinson, 1793.

Warner, William. *Licensing Entertainment: The Elevation of Novel Reading in Britain, 1684–1750*. Berkeley: University of California Press, 1998.

– 'Licensing Pleasure: Literary History and the Novel in Early Modern Britain.' In *The Columbia History of the British Novel*. Ed. John Richetti, John Bender, Deirdre David, and Michael Seidel, 1–22. New York: Columbia University Press, 1994.

Watson, William. 'A Sequel to the Experiments and Observations Tending to Illustrate the Nature and Properties of Electricity.' *Philosophical Transactions* 44 (1746–7): 704–49.

Weisskopf, Victor. *Knowledge and Wonder: The Natural World as Man Knows It.* Cambridge, MA: MIT Press, 1979.

Wertheim, Margaret. *The Pearly Gates of Cyberspace: A History of Space from Dante to the Internet.* New York: W.W. Norton, 1999.

Wheeler, David. 'Creating a Body of Knowledge.' *Chronicle of Higher Education* (2 February 1996): A14.

Wiener, Norbert. 'Cybernetics.' *Scientific American* 179 (1948): 14–19.

– *Cybernetics: Or Control and Communication in the Animal and the Machine.* Cambridge: Technology Press, 1948.

– *The Human Use of Human Beings: Cybernetics and Society.* Boston: Houghton Mifflin, 1950.

Williams, Sarah. '"Perhaps Images at One with the World are Already Lost Forever": Visions of Cybrog Anthropology in Post-Cultural Worlds.' In Chris Hables Gray et al., *The Cyborg Handbook*, 379–90.

Willis, Thomas. *The Anatomy of the Brain and Nerves.* Trans. Samuel Pordage; ed. William Feindel. Montreal: McGill University Press, 1965.

– *Dr Willis's Practice of Physick Being the Whole Works of That Renowned and Famous Physician Wherein Most of the Diseases Belonging to the Body of Man are Treated of.* London: printed by H. Clark, for T. Dring, C. Harper, and J. Leigh, 1684.

– *Two Discourses Concerning the Soul of Brutes, which Is that of the Vital and Sensitive of Man.* Trans. Samuel Pordage. Gainesville: Scholars Facsimiles and Reprints, 1971.

Wolveridge, James. *Speculum Matricis; Or, the Expert Midwives Handmaid.* London: printed by E[dward] Okes, 1671.

Woodmansee. Martha. 'The Genius and the Copyright: Economic and Legal Conditions of the Emergence of the "Author."' *Eighteenth-Century Studies* 17 (1984): 425–8.

Wooldridge, Dean E. *The Machinery of the Brain.* New York: McGraw-Hill, 1963.

Yates, Frances. *Selected Works.* Volume 3. *The Art of Memory.* London: Routledge, 1999.

Zionkowski, Linda. 'Aesthetics, Copyright, and "the Goods of the Mind."' *British Journal for Eighteenth-Century Studies* 15 (1992): 163–74.

– 'Gray, the Marketplace, and the Masculine Poet.' *Criticism* 35.4 (1993): 589–608.

– *Men's Work: Gender, Class, and the Professionalization of Poetry, 1660–1784.* New York: Palgrave, 2001.

– 'Territorial Disputes in the Republic of Letters: Canon Formation and the
   Literary Profession.' *The Eighteenth Century: Theory and Interpretation* 31.1
   (1990): 3–22.
Zwicker, Steven N. *Lines of Authority: Politics and English Literary Culture, 1649–
   1689.* Ithaca: Cornell University Press, 1993.

Illustration Credits

Index

nerve theory, 15, 37, 104, 111, 121–2

nervous system: animated by aether/ subtle spirit, 64; as communications system, 31, 46, 87–8, 89–91, 98, 101, 112, 115, 125, 135–6, 252 (circuit providing feedback, 114–15, 123–6, 136; circuit of commerce and communications, 123–5; computer circuit, 92–3, 95, 234–5; telegraph wires, 94, 228; telephone wires, 93–4); as computer, 93; as distillation apparatus, 59–1, 78, 104, 203; as electric system, 90; as metropolis, 105, 123–6; as steam engine governed by feedback mechanism, 78

Newton, Isaac, 15, 27, 31, 35–6, 38, 45, 49, 51–2, 60, 61, 62, 64–6, 71–2, 73, 75, 114, 119, 153, 160, 163, 197, 251; *Opticks*, 26, 38, 52, 61, 64

Newtonian philosophy, 26, 27, 48, 49, 50, 71, 75, 78, 252

Niccol, Andrew, 172

obsolescence, human, 6, 10, 15

occult qualities, 46, 219

Ong, Walter J., 229

Onslow, Thomas, 213

Otis, Laura, 227

Parsons, James, 79–80

Pepys, Samuel, 140

Petchesky, Rosalind Pollack, 173, 175

Pfeiffer, K. Ludwig, 234

Phillips, Edward: *The New World of Words*, 35

Plant, Sadie, 12, 16–17, 230

Plato, 103; *Republic*, 103; *Timaeus*, 103

Pleasure for a Minute: or, the Amorous Adventure, 200

Pope, Alexander, 25, 27, 149–52, 157, 210; *An Essay on Man*, 248, 251; *Memoirs of Martinus Scriblerus*, 151–2; *Peri Bathous*, 147–8

post-cyborg, 11

post-Enlightenment, 6, 10, 21

Poster, Mark, 101

post-human, 5, 8, 10, 15, 21, 44, 89, 90, 95, 97, 158, 230–1

postmodern anthropology, 10

postmodern condition: metanarratives, 9; schizophrenia, 9, 84; simulacra, 7, 9, 114, 231, 238; transformation of reality, 9

postmodernism, as stereotype, 10–11

print culture, 8, 18, 24, 141–6, 147–9, 156–7, 211

prostheses, 12: artificial organs, 22, 24, 85; artificial valves or pacemaker and the cyborg, 22, 85; prosthetic extensions to female anatomy, 183–7

Pyun, Albert, 170

Quincy, John: *Medical Dictionary*, 202

Reddy, Raj, 237

Reill, Peter Hanns, 72

Richardson, Samuel: *Clarissa*, 178–9, 181, 193; *Pamela: or, Virtue Rewarded*, 190

Rig Veda, 103

Riordan, J.R., 239

Robinson, Bryan, 26; *A Dissertation on the Aether of Sir Isaac Newton*, 26

RoboCop, 4

Rorvik, David, 4

Rowland, Robyn, 166